CHAOS AND HARMONY

Other books by Trinh Xuan Thuan:

The Secret Melody
The Birth of the Universe: The Big Bang and After

CHAOS AND HARMONY

PERSPECTIVES ON SCIENTIFIC REVOLUTIONS OF THE TWENTIETH CENTURY

TRINH XUAN THUAN

TRANSLATION BY AXEL REISINGER

TEMPLETON FOUNDATION PRESS

PHILADELPHIA & LONDON

Templeton Foundation Press
300 Conshohocken State Road, Suite 670
West Conshohocken, Pennsylvania 19428
www.templetonpress.org

Original French edition: Chaos et l'harmonie © 2000 Librairie Arthème Fayard
English edition: © 2001 Oxford University Press, Inc.
2006 Templeton Foundation Press paperback edition

*Templeton Foundation Press helps intellectual leaders and others learn
about science research on aspects of realities, invisible and intangible.
Spiritual realities include unlimited love, accelerating creativity, worship,
and the benefits of purpose in persons and in the cosmos.*

Library of Congress Cataloging-in-Publication Data

Thuan, Trinh Xuan.
    [Chaos et l'harmonie. English]
    Chaos and harmony: perspectives on scientific revolutions of the twentieth century /
Trinh Xuan Thuan ; translated by Axel Reisinger.
        p. cm.
    Originally published: Oxford ; New York : Oxford University Press, 2001.
    Includes bibliographical references and index.
    ISBN-13: 978-1-932031-97-3 (alk. paper)
    ISBN-10: 1-932031-97-9
  1. Astrophysics—Philosophy. 2. Chaotic behavior in systems. I. Title.

    QB461.T5613 2006
    523.01--dc22

                                                          2005051185

Printed in the United States of America

06 07 08 09 10   10 9 8 7 6 5 4 3 2 1

To the memory of my father,
and to all in search of beauty and harmony.

« CONTENTS »

« FOREWORD »

As we enter the dawn of the twenty-first century, we are grappling with profound changes in the way we perceive the world. After dominating Western thought for three hundred years, the Newtonian view of a fragmented and mechanistic universe is receding in favor of a world that is holistic, indeterministic, and teeming with creativity.

For Newton, the universe boiled down to a gigantic machine composed of inert particles subject to blind forces. If a system could be characterized at any particular instant, its entire past history could be re-created and its future predicted with just a few physical laws. The future was embedded in the present and the past, and time was effectively abolished. This gave rise to a curious dichotomy: On the one hand, the laws of nature were immutable and timeless; on the other, the world was evolving and contingent. The laws of physics were oblivious to the arrow of time, yet thermodynamics and psychology forced time to move inexorably forward. An abandoned castle falls into ruins, a flower wilts, and our hair turns gray as time moves on, never the other way around. The universe was shackled in a straitjacket that precluded any creativity and innovation. Everything was preordained, and no surprise was allowed. That prompted Friedrich Hegel to utter his famous outcry: "Nothing is ever new in nature." The world was a place where reductionism reigned supreme. All it took to understand it in its entirety was to decompose it into simpler components and study their behavior. Indeed, the whole was precisely the sum of its parts–nothing more, nothing less. There was a direct relation between cause and effect. The magnitude of the effect was invariably proportional to the intensity of the cause and could be predicted accurately.

Constraining and sterile determinism, with its rigid and dehumanizing reductionism, prevailed until the end of the nineteenth century. They came to be challenged, transformed, and ultimately swept aside by a far more exhilarating and liberating view in the twentieth century. A new dimension began to creep into numerous scientific disciplines. Contingency took on a prominent place in fields as diverse as cosmology, astrophysics, geology, biology, and genetics. Reality was no longer to be determined solely by natural laws applied to particular initial conditions. From then on, it would also depend critically on a series of historical and contingent events. Some of these events influenced reality on the most profound levels–indeed were the origin of our very own existence. Such was the case of the chunk of rock that came crashing into Earth some 65 million years ago, causing the extinction of the dinosaurs and allowing new breeds of our ancestral mammals to flourish. That chance collision was directly responsible for man's emergence. In the eighteenth century, Laplace had articulated the dream that human intelligence could "encompass in a single formula all motions, from the largest bodies in the universe to the lightest atoms," and that "nothing would be left unexplained . . . and the future as well as the past would be clear before its eyes." That dream was left in tatters.

It was not just historical events that were responsible for the liberation of nature. Physical laws themselves lost much of their rigidity. With the advent of quantum mechanics in the early part of the twentieth century, happenstance and whim became full-fledged components of the subatomic world. In place of the staidness of deterministic predictability came the thrill of quantum uncertainty. Narrow and simplistic reductionism was thrown overboard; from localized and compartmentalized, reality became holistic and global. Even the macroscopic world would not be spared. Chaos theory brought chance and uncertainty not only into our everyday life but even to planets, stars, and galaxies. Randomness invaded a world that had long been regulated in its minutest details. A simple cause-and-effect relationship could not capture the whole story. The amplitude of the effect could no longer be assumed proportional to the magnitude of the cause. Indeed, some phenomena proved so sensitive to initial conditions that an infinitesimally small change in initial conditions could lead to such large variations in the subsequent behavior of a system that any prediction became hopeless. Henri Poincaré stated as early as 1908: "A cause so tiny as to escape us results in a considerable effect which we are unable to predict; we then attribute this effect to chance." The world had come a long way from Laplace's dream.

Freed from its deterministic yoke, nature can now give free rein to her creativity. The immutable laws of physics are only general guidelines around which she can experiment and improvise. They define the realm of what is allowed; they suggest possibilities. It is entirely up to nature to actualize them. She decides her own destiny. To manufacture complexity, she plays on deviations from equilibrium, for only then can the unexpected come into existence. Symmetry is interesting only to the extent that it is broken. Matter produces novelty when it is out of equilibrium. Perfect order is barren, while controlled disorder is fertile and chaos is pregnant with innovations. Nature keeps inventing; she creates beautiful and varied forms that can no longer be characterized by straight lines and simple geometrical shapes, but rather by far more complex forms. Benoit Mandelbrot coined the word *fractal* to describe them. Matter organizes itself according to laws and principles of complexity. It acquires "emergent" properties that cannot be explained in terms of its constituents. Reductionism is all but dead. Nature's newfangled freedom casts a new light on the old dichotomy between intemporal, immutable, and eternal physical laws and the temporal, changing, and contingent world. Nature is enfolded *in* time because it can innovate and create on the basis of laws that are rooted *outside* time.

I wanted to retrace the development of ideas that led to this new vision of the world. The book is organized along the following lines:

Chapter 1 deals with truth and beauty. Scientific work is often thought of as cold and impersonal, devoid of any aesthetic feeling. Nothing is further from the truth.

Chapter 2 tells the story of the solar system. It illustrates how every level of reality is determined by the combined effects of determinism and serendipity, of chance and necessity. As we witness the birth of the sun and planets, we are made to appreciate how comet and asteroid impacts with Earth are responsible not just for the beauty of spring flowers and the soothing glow of the moon, but even for our very existence.

Chapter 3 talks about chaos theory. Through examples drawn from astrophysics, meteorology, biology, and medicine, we will realize how chaos helps nature bring about its potentialities as it fashions reality.

Chapter 4 shows us how nature uses subtle symmetry principles to impart to the physical world a profound unity. These principles have enabled us to meld electricity with magnetism, and time with space. We will become familiar with "black holes," in which the coupling of space and time manifests itself in the most bizarre effects.

Chapter 5 ushers us into the world of atoms, where quantum indeterminacy predominates. Reality is no longer objective but actually depends on the observer. We will see how symmetry principles can help us restore some order in the dizzying proliferation of particles by guiding us toward a unified theory of nature's fundamental forces. We will review the latest theory, which asserts that elementary particles are simply vibrations of infinitesimally small bits of "strings" in a ten-dimensional space!

Chapter 6 tells how nature, capitalizing on the freedom conferred by chaos and quantum indeterminacy, exercises her creativity. We will examine how she actualizes the potentialities embedded in the laws of physics to create life. A living organism is more than the sum of its constituent atoms and molecules. As such, it can never be explained on the basis of pure reductionism. We will have to invoke "emergent" principles of self-organization and complexity that operate holistically and globally on the organism as a whole.

Chapter 7 deals with man's "unreasonable effectiveness" in grasping the universe. Not only have the laws of physics been tuned with exquisite precision to allow the emergence of life, but they have also made it possible for conscience to arise. Why are these laws mathematical in nature? Could it be that man tries to understand the universe simply to give it meaning?

This book is aimed at the layman. It does not assume any technical expertise. It will serve those who are curious about the extraordinary advances of science in the twentieth century, as well as their philosophical and theological implications. I have tried to be as precise and rigorous as possible without getting bogged down in technical jargon. I often resort to analogies to explain difficult scientific concepts. For convenience, a glossary at the end of the book lists unfamiliar terms together with brief definitions. While some concepts are admittedly rather esoteric, I made every effort to describe them in as friendly and accessible a manner as possible. I also included a number of figures and photographs, which will hopefully reinforce my points and brighten the reader's experience.

Trinh Xuan Thuan

« ACKNOWLEDGMENTS »

Françoise Grandgirard typed the text with help from Jeanne Gruson. Both read and critiqued the text, helping me improve its clarity. I owe them both a profound debt of gratitude. I thank Claude Durand for reading the manuscript from end to end and for his invaluable advice. Jean Staune read the original version of the text and provided useful comments, while Hélène Guillaume helped to shape it into its final form. I am very much in their debt. I am also grateful to Chantal Balkowski and Georges Alecian for inviting me to the Department of Extragalactic Astronomy and Cosmology at the Paris-Meudon Observatory, to Laurent Vigroux for his hospitality in the Department of Astrophysics at the Saclay Research Center, and to the University of Paris for awarding me a visiting professorship while I was putting the finishing touches to this book.

CHAOS AND HARMONY

« I »

Truth and Beauty

A MAN AND A WOMAN

It is a beautiful spring day in Paris. At a sidewalk café, a man is enjoying a glass of beer while reading a newspaper. At the next table, a woman is sipping coffee while watching passersby. They haven't noticed each other yet. Suddenly, the man turns his head and his gaze meets the woman's. At that instant, a remarkable series of events is set in motion. The golden light of the sun reflects off the woman's slender body and penetrates the man's eyes. Traveling at a speed of 300,000 kilometers per second, 10,000 billion particles of light (called photons) rush in through his pupils. First they traverse an oval-shaped body called the lens, then a transparent and gelatinous substance, before they strike the retina.

THE DANCE IN THE RETINA

In the retina, more than 100 million rod- and cone-shaped cells go to work. Covering the retina like darts bristling out of a dartboard, some of these rods receive large amounts of light from the bright areas of the woman's body, such as her moist lips highlighted by vivid red lipstick. Others receive less light, because it comes from more subdued parts of the woman, such as her discreetly made-up cheeks. While rods are sensitive to very dim light, cones require brighter light. Both rods and cones contain light-sensitive pigments that respond differently to different levels of light coming from the woman's body. All rods have the same type of pigment. Cones, however, come in three types, each containing a different visual pigment. One type absorbs best in the blue, another in the green-yellow, and the third in the orange-red. Visual pigment molecules are each composed of 20 carbon atoms, 28 hydrogen atoms, and 1 oxygen atom. They respond to light by engaging in a kind of strange ballet. When at rest, in the absence of light, each such molecule is attached to a protein and is all crumpled up. But as soon as light strikes it (the light reflected by the woman hits 30 million billion molecules in the man's eye every second), the molecule in the retina separates from the protein and straightens out. After a while, it crumples back up until the next photon arrives.

NEURONS SPRING INTO ACTION

All these events took less than $1/1,000$ of a second since the moment the man's gaze met the woman's. But the man is still not "conscious" of the woman's presence, because the information carried by the particles of light has yet to reach his brain. The frenetic dance of the molecules in his retina must fire up neurons, first in his eyes and then in his brain. Molecules on the surface of neurons also change shape, blocking the flow of sodium ions (particles with a positive electrical charge) in the surrounding liquid, which triggers an electrical current propagating from neuron to neuron, from the eye all the way to the brain. In the cerebral cortex, each neuron processes the information transmitted by thousands of neurons before relaying it in turn to thousands of other neurons farther up the chain. A great many of the hundreds of billions of neurons in the man's brain, interconnected in an incredibly complex network, participate in relaying the information. The flow of potassium and sodium stops depending on whether or not it is blocked by neurons. Electrical currents race furiously through neural networks, exciting swarms of neurons relaying signals that go on to excite yet more neurons. Current crackles everywhere. After a few thousandths of a second, an image is reconstructed in the man's brain: He finally sees the woman. He notices her short blond hair, her deep blue eyes, her dark brown dress molding her body, her slightly tilted head with a pensive look.

The woman turns her head, meets the man's eyes, gives him a faint smile, and offers a cheerful "Hello." Instantly, a multitude of air molecules start jiggling about. The vibrations transmit the sound of the woman's vocal cords to the man's ears. Only two meters separate them, and the sound arrives in $1/150$ of a second. The drum (a 1-millimeter-thick membrane) in each of the man's ears begins to vibrate. The vibrations are transmitted to the liquid in the cochlea, a structure shaped like a snail's shell. That is where sounds are decoded. A thin membrane starts oscillating in synchronism with the vibrating liquid. This membrane contains an array of fibers of various thicknesses, much like the strings of a harp. The harp resonates in unison with the woman's sensuous voice and reconstructs the relatively high pitch of the syllable "hel-" and the deeper one of the syllable "-lo." Eventually, the sounds are passed on to the auditory nerve, which conveys the information to the cerebral cortex. And the man finally hears the word "hello."

All theses processes are quite well understood. Neurobiology unveils more and more secrets of the brain each passing day. What remains a complete mystery is what causes the lightning-quick thought that crosses the man's mind: "She is so beautiful!"

NATURE IS BEAUTIFUL

What is beauty? Not only do we have no clue about how our brain apprehends beauty, but we are even less able to describe it in precise terms. It is even more of a challenge to speak of beauty in the context of science, which is exactly what I am going to attempt to do. The popular wisdom is that scientific work is a purely rational pursuit from which any emotion is banned. Physics is widely perceived as a precise and exact science in which there is no place for aesthetic contemplations. Aesthetic judgments are supposedly irrelevant in science; all that is left are cold and impersonal facts. The truth is that scientists are no less sensitive to Nature's beauty than artists. My frequent visits to various observatories have never dulled the intense and always renewed pleasure I experience when I find myself in sites of exceptional beauty, far removed from the lights of civilization. I am in absolute awe every time I see the majestic and arid splendor of the Arizona desert, where the Kitt Peak Observatory is located, or the desolation of the moonscape, stripped of any vegetation, on the summit of Mauna Kea, an extinct volcano in Hawaii where huge telescopes have popped up like mushrooms. My heart always starts racing when the spiral arms of a galaxy billions of light-years away take shape on a monitor screen hooked up to a telescope.

If Nature is so beautiful, why should the theories that describe it not be so too? Why should scientists be less prone than poets to letting themselves be guided by aesthetic considerations in addition to rational arguments? Some of the greatest scientists have answered the question unequivocally. The French mathematician Henri Poincaré (1854–1912), for one, stated: "Scientists do not study Nature for utilitarian reasons. They do it because they find it pleasurable; and they find it pleasurable because Nature is beautiful. If Nature were not beautiful, it would not be worth studying, and life would not be worth living." Poincaré even offered a definition of beauty to which I will return later on: "I speak of an inner beauty that stems from the harmonious order of the parts, which pure intelligence has the ability to grasp." A cri de coeur expanded upon later on by Werner Heisenberg (1901–1976), one of the fathers of quantum mechanics: "If Nature leads us to mathematical forms of great beauty and simplicity—by 'forms' I mean coherent systems of hypotheses, axioms, and the like—which nobody had foreseen before, we cannot help but think that they must be real, that they reveal a true side of Nature. . . . You must have experienced it too: The almost frightening simplicity and totality of the interconnections which

Nature displays before us and for which we were not at all prepared." Albert Einstein (1879–1955) himself wrote at the end of his first paper on general relativity: "Anyone who understands the present theory could not miss its magic." "Harmonious order," "simplicity," "coherence," "magic": These are all words defining "beauty" in science, a concept which I will now try to further articulate.

THE THINGS OF LIFE AND THE RELATIVITY OF BEAUTY

The beauty a physicist speaks of is quite different from what a musician experiences when listening to a sonata by Mozart or a fugue by Bach. Nor is it the same as what an art lover reacts to when admiring the dancers of Degas, the apples of Cézanne, or the water lilies of Monet. It is not even the same as what our earlier character was feeling upon noticing the beautiful woman sitting at the next table. Feminine beauty obeys criteria that are notoriously dependent on cultural, psychological, or even biological contexts. The plump bodies of the women painted by Rubens or Renoir are no longer considered the paradigm of beauty. In the 1960s, Twiggy's skinny figure was considered attractive. The beauty of oriental women is different from that of their Western counterparts, even though massive advertising campaigns for cosmetic products have spread Western standards throughout the world, which has led to such absurd practices as some Asian women having the shape of their eyelids rounded. There are fashionable trends even in the world of art. Van Gogh died in poverty, unable to sell his canvases. Half a century later, people trip over one another buying his paintings at astronomical prices. Aesthetic perceptions change from one culture to another. The style of a painting of Mount Fuji by Hokusai has little in common with Cézanne's rendition of the Sainte-Victoire mountain. The timeless magnificence of the Taj Mahal in India is quite different from the splendor of the cathedral of Chartres. It would be downright presumptuous for anyone to define what constitutes beauty. Like love and hatred, we recognize it when it takes over our soul, no matter how difficult it may be to describe the experience in words.

Beauty is in the eye of the beholder. It is a cliché, perhaps, but so true. It can spring up around any street corner and find its way into ordinary objects in our everyday lives, provided we are receptive. A simple flower, a tree that only yesterday was completely unremarkable because we were preoccupied with other issues, suddenly evokes an overwhelming aesthetic sense. As the German philosopher Arthur Schopenhauer (1788–1860) put it so eloquently, at such times we consider "neither the place, nor the time, nor the why, nor the purpose of things, but quite simply and purely their essence"; we do so because we then allow "neither abstract thought nor any principle of reasoning to clutter our conscience; instead, we turn all the power of our mind toward intuition." Schopenhauer went on to argue: "When we are completely engrossed by it and our conscience is filled by a natural object, be it a landscape, a tree, a rock, a building, or anything else; . . . the moment we forget our own individuality, our own will, and we remain as pure subject, as a clear mirror of the object, in such a way that everything happens as though the object existed in and of itself without anyone being able to perceive it, when it is impossible to distinguish between the object and intuition itself, when they both merge into a single entity, a single conscience completely filled and dominated by a unique and intuitive vision; in short, when we sever all ties with will: that is when what we grasp is no longer any particular thing in its individuality but, rather, the idea, the eternal form."

If there is no objective criterion for beauty in human creation, should we expect to discover one in scientific work? Is there a way to forge an aesthetic system in science to judge Nature's beauty and her organization? Perhaps the answer is yes, for unlike the relative beauty of women and things, the appeal of a physical theory is universal. It can be appreciated by any scientist, regardless of ethnic origin or cultural heritage. A Vietnamese physicist can extol the virtues of general relativity with as much passion as any of his French or American colleagues.

In spite of Schopenhauer's exhortations to disregard reason and let intuition be our guide in grasping beauty, I will in fact attempt the hazardous feat of trying to define the concept of beauty in a physical theory. I will refrain from offering a precise definition, which would be doomed to failure. I shall, instead, simply list and illustrate the characteristics a scientific theory must exhibit in order to be beautiful.

BEAUTY IN SCIENCE

To begin with, the word *beautiful* does not refer to the purely plastic beauty of equations carefully laid down on a piece of paper, even though I confess that even that sight elicits in me a certain sense of abstract beauty, similar to what I feel when I look at a page filled with characters lovingly drawn by a Chinese calligrapher. The poet and painter Henri Michaux made expert use of the plastic beauty of Chinese characters in his ink drawings. Nor is beauty the same thing as what physicists and mathematicians talk about when they use the word *elegance*. A mathematical proof or a result in physics are "elegant" when they are derived with a minimum number of steps. A theory can be beautiful without the benefit of elegant solutions. By any measure, the theory of general relativity is one the most beautiful intellectual constructs ever produced by the human mind. Yet, in most cases, its solutions hardly qualify as elegant. The mathematical derivations are extremely complicated. That does not prevent it from being perhaps the most beautiful theory ever devised. A theory is beautiful when it has an air of inevitability. It is the same feeling some people experience when listening to a fugue by Bach. Not a single note could be changed without upsetting the overall harmony. The same can be said of the *Mona Lisa* by Leonardo da Vinci. Not a single stroke of the brush could be altered without destroying the perfection of the painting. So it is for a theory. The moment Einstein accepted the physical principles at the basis of his theory of gravitation, he no longer had any choice:

General relativity was inevitable. As he himself wrote: "The main appeal of the theory lies in the fact that it is self-contained. Should any of its conclusions be invalidated, the entire theory would have to be rejected. It is impossible to modify it without jeopardizing the entire structure." The inevitability of a beautiful theory is so overwhelming that when it bursts onto the scene, physicists often wonder how they could have missed it before.

The second quality of a beautiful theory is its simplicity. We are not talking here about the simplicity of the equations themselves, as measured for instance by the number of symbols they contain, but rather of the simplicity of the underlying ideas. As an example, Isaac Newton (1642-1727) needed only three equations—one for each dimension of space—while general relativity requires a total of fourteen. Yet the latter is more beautiful because it rests on simpler fundamental concepts, which we will discuss later on. The heliocentric universe of Copernicus (1473-1543), in which planets move in an orderly fashion along elliptical orbits around the Sun, is simpler than the geocentric model of Ptolemy (ca. 90—ca. 168), where Earth occupies a central place and the planets describe circles whose centers themselves describe other circles. A theory that is simple uses a minimum number of hypotheses. It does not carry excess baggage. It satisfies the postulate of simplicity stated by William of Occam (ca. 1285-1349): "What is not necessary is useless."

CONFORMITY WITH THE WHOLE

The final quality—the most important one, in my opinion—is to conform with Nature's intricacies. It must allow beauty and truth to merge into one. Indeed, a physical theory has no reason for existing unless it reveals new connections in Nature that can be verified by observations or laboratory experiments, unless it lays bare before our eyes "the almost frightening simplicity and totality of the interconnections of Nature," as Heisenberg put it so well. A theory that cannot be verified experimentally belongs not in the realm of science but of metaphysics. Intellectual speculations remain sterile as long as they are not rooted in the forms of Nature. Heisenberg defined beauty as it was perceived in antiquity as "conformity of the parts between themselves and with the whole." Relativity theory is beautiful because it managed to connect and unite fundamental physical concepts that until then had remained completely distinct—time, space, matter, and motion. Matter curves space, and the curvature of space dictates how motion proceeds. The Moon follows an elliptical trajectory around the Earth because Earth's mass causes the space around it to curve. In turn, motion determines the behavior of space-time. An elementary particle traveling at nearly the speed of light sees its time stretch out and its space shrink. The slowing down of time is not a utopian fountain of youth: Particles hurtling around accelerators, such as the one at CERN in Geneva, Switzerland, have been shown to indeed live longer than when they are at rest. And detailed observations have demonstrated that the path of starlight bends when it grazes the sun exactly as if space were curved in its vicinity.

The beauty of a theory is all the more compelling when it reveals a host of new and unexpected connections as researchers explore its deeper implications. General relativity meets this criterion to the utmost degree. Its richness never ceases to amaze us. Einstein himself was stunned when he realized that his equations implied an expanding universe. Just as a stone tossed in the air cannot remain frozen in place, the universe cannot be static: It must either expand or contract. Back in 1915, every astronomical observation suggested that the universe was static. That prompted Einstein to modify his equations so as to conform with the then-prevailing view. He would later regret this action and call it "the greatest blunder of my life" when the American astronomer Edwin Hubble (1889–1953) discovered in 1929 that the universe is in fact expanding. Einstein had failed to place enough trust in the beauty and truth of his own equations. General relativity has kept on delivering wonderful treasures ever since. It is the pillar on which the big bang theory rests. It has enabled cosmologists to go back in time and describe how the universe evolved out of a huge primordial explosion that also gave birth to space-time. It has allowed us to conceive of regions of space where gravity is so powerful and space so strongly curved that not even light can escape—these regions have been dubbed "black holes." It also tells us that massive galaxies can curve space so as to bend the light emitted by distant objects, creating cosmic mirages. Astronomers refer to these galaxies as "gravitational lenses," because they bend and focus light much as the lens in our eye does.

Inevitable, simple, congruent with the whole: Those are the hallmarks of a beautiful theory. It is, in fact, this aesthetic yearning for congruity with the whole that has spurred on physicists of the last two centuries to search for a Theory of Everything that could encompass all physical phenomena in the universe and unify the four fundamental forces of nature.

Before embarking on a search for the holy grail in physics—the Theory of Everything—with beauty as our guide, we must first become acquainted with the power of contingency. Beauty can lead to truth only if we learn to distinguish what is fundamental from what is fortuitous. Beauty alone cannot be a reliable guide in constructing a theory if we fail to take into account the intervention of chance. Nature is governed both by fundamental laws and by accidental events without deep significance. The history of the formation of the solar system is a perfect example to illustrate how random events can shape Nature. Accordingly, we are about to travel back 4.6 billion years to witness live the birth of the solar system. What happened then is of crucial importance not only because it eventually culminated in our own existence but also because it can teach us to distinguish necessity from contingency.

« 2 »

Contingency and Necessity

The Formation of
the Solar System

THE SUN IS BORN OF THE DEATH OF A STAR

The solar system (Figure 1) was formed about 4.6 billion years ago in a system named the Milky Way, one galaxy among hundreds of billions of galaxies that populate the observable universe. Near the periphery of this galaxy, roughly two-thirds of the way out toward the periphery, some 30,000 light-years from the galactic center, a massive star is reaching the end of its life. It is running out of the fuel that generates its energy and makes it shine. Gravity takes over and the core of the dying star collapses inward, while at the same time a tremendous explosion sends the outer layers hurtling out into space at thousands of kilometers per second. In its death throes, this massive star turns into a supernova shining as brightly as billions of suns. Near the supernova is a gas cloud composed almost entirely of the two chemical elements synthesized in the early moments after the big bang, during the first three minutes of the universe. Three-quarters of its mass (which amounts to two billion billion billion tons) are made of hydrogen, while helium makes up 23 percent. The remaining 2 percent include elements heavier than helium, the most abundant of which are carbon, nitrogen, and oxygen. They were created by nuclear reactions in the interior of earlier-generation stars. These stars seeded interstellar space during their explosive deaths. Hit broadside by the shock wave created by the nearby supernova, the cloud collapses under the effect of gravity. Its center becomes increasingly dense. From an almost perfect vacuum (10,000 billion billion times less dense than water), the density becomes twice, then ten times, even one hundred times

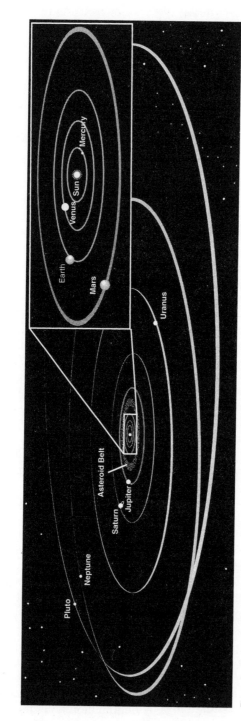

Figure 1. *The solar system.* The planets in the solar system exhibit regularities that any theory of the formation of the solar system must account for. With the exception of Pluto, all planets rotate around the Sun in the same direction as the Sun's rotation (from west to east), and lie in the same plane (called the ecliptic plane). Except for Venus and Uranus, all planets spin from west to east and, with only a few exceptions, all planetary moons also orbit in that same direction. The general direction of rotation from west to east is a consequence of the initial spin of the solar nebula that gave birth to the Sun and the planets.

Planets fall into two categories. Near the Sun are the terrestrial planets (Mercury, Venus, Earth, and Mars); with a small mass and a rocky surface made of heavy elements, they have little or no atmosphere. Much farther out are the giant planets (Jupiter, Saturn, Uranus, and Neptune); they are very massive, without a solid surface, and their very thick atmosphere is made primarily of light elements such as hydrogen and helium.

that of water. Atoms are herded into a restricted region of space, and their mutual collisions begin to heat up the gas. What was once a frigid cold (about 263°C below freezing) turns into increasingly intense heat. One hundred million years have elapsed since the cloud started collapsing. The density in its core is now 160 times greater than that of water. The 10-million-degree-temperature mark has just been passed. Hydrogen and helium atoms within the cloud collide furiously, liberating electrons, hydrogen nuclei (or protons), and helium nuclei. The extreme temperatures and densities initiate nuclear reactions. Protons unite four at a time to form helium nuclei. This fusion reaction produces energy that manifests itself in the form of radiation. The gas ball begins to shine. The Sun is born. The death of a massive star has given birth to a new celestial body that will ultimately be responsible for our own existence.

What is the source of energy that keeps its fire burning? It turns out that the mass of a helium nucleus is slightly less than that of four free protons before they fuse. The missing mass, which amounts to about 0.7 percent, has been converted to energy. As Einstein taught us, multiplying the missing mass by the square of the speed of light gives us the energy. In the case of just four protons fusing into one helium nucleus, the energy released is quite small—not even enough to run a small electric light. Yet the Sun shines bright and keeps life going here on our blue planet because hundreds of billion billion billion billion (10^{38}) fusion reactions take place every second. Some 400 million tons of hydrogen are being converted into helium every second in the heart of the Sun. The inward collapse of the gas ball is stopped dead in its track by the outflux of energy. An equilibrium is reached between the outgoing pressure of the radiation, which wants to cause the star to explode, and the effect of gravity, which makes it fall in on itself. The gas cloud—also called solar nebula—is endowed with motion. It rotates about itself, as do all interstellar clouds, galaxies, and stars. As it collapses, the speed of its rotation increases, much as an ice skater spins faster when he pulls his arms in close against his body. The centrifugal force due to the rotation flattens the nebula into a disk centered on the massive Sun, which gravity has shaped like a sphere (Figure 2).

THE DUST FROM RED GIANTS

Up to now, we have focused our attention on the gas that gave birth to the Sun. But there are other actors that also occupy center stage in this cosmic drama. Scattered in the gaseous nebula are innumerable grains of dust. Even though they are only $\frac{1}{10,000}$ of a millimeter in size, they are giants on an atomic scale. They result from the accretion of billions of atoms of silicon, oxygen, magnesium, and iron created in stellar furnaces and supernovae. They owe their cohesiveness to electromagnetic forces. These dust grains have a solid nucleus coated with a thin layer of ice. They were born long before and somewhere else. To be precise, they came from the envelope of stars called "red giants." These are enormous stars, roughly 100 times the

Figure 2. *The formation of the solar system.* *(a)* The initial phase of the formation of the solar neb-ula. Gases and planetesimals rotate around a central gas mass that collapses under its own gravity to give birth to the Sun. *(b)* The early solar system at 50 million years after its birth. In the inner regions of the solar system, planetesimals accrete to give rise to the terrestrial planets (Mercury, Venus, Earth, and Mars). The gas, composed of 98 percent of hydrogen and helium, gets heated by the intense energy from the Sun and escapes toward the outer regions of the solar system, where it forms the giant planets. As a result, Jupiter, Saturn, Uranus, and Neptune are enormous gaseous balls made almost entirely of hydrogen and helium (the planet Pluto is a special case—it is believed to be a very large asteroid orbiting near the edge of the solar system). In the inner regions, numer-ous chunks of rocky debris are left over (they are called "asteroids"); they collide violently against the newly formed terrestrial planets, carving huge craters on their surfaces. The crater-pocked sur-face of Mercury provides a record of this period of intense bombardment. *(c)* The formation of the planets reaches completion after about 100 million years. In the interior regions near the Sun, all the planetesimals have accreted into planets, and the bombardment of the planetary surfaces comes to a stop.

size of our sun, or 70 million kilometers in radius. Because their envelopes are cooler (3000°C, instead of the more than 6000°C at the surface of the Sun), red giant stars enabled chemical elements, freshly synthesized in their interior and expelled by stellar winds blowing at hundreds of kilometers per second, to condense in the form of dust. The tiny particles, swept and herded together by the supernova's shock wave, coalesced into solid grains, which in turn dispersed throughout the frigid environment of the gas cloud. They are still there by the time the solar nebula flattens into a disk a few light-hours in radius.

THE ACCRETION GAME

Cathedrals are constructed of stone. Planets, on the other hand, are made of aggregated dust whose cement is the electromagnetic force combined with gravity. Dust grains scattered in the gas disk coalesce under the action of gravity to create larger grains still (Figure 3). In turn, these exert an even greater gravitational pull, which snares other grains wandering by. The size and mass of the grains keep increasing. After a few tens of years, the tiny grains have grown to fine gravel. It is now the turn of the gravel pieces to continue the process. They grow to the size of candy, eggs, and eventually potatoes after about one hundred years. Nature relentlessly pursues her march toward complexity by playing the accretion game. The chunks of mat-ter—also called planetesimals or asteroids—become successively larger than a football, a stadium, a city block, a town, a county, an entire country, the moon, and the process keeps on going. As the size of these objects changes, so does their shape. The small planetesimals are lightweight, and their grav-ity is too weak to offset the electromagnetic forces that give them rigidity and hardness. Electromagnetism favors roughness and unevenness. That is why objects from the size of gravel to that of a country have irregular,

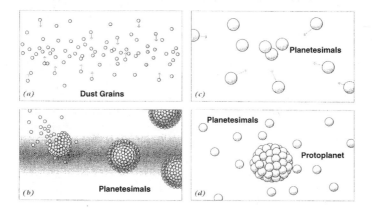

Figure 3. *The formation of the planets. (a)* Dust grains come together and coalesce into small rocky bodies distributed in a thin disk. *(b)* Gravity collects these small bodies into larger asteroid-size chunks called planetesimals. *(c)* Planetesimals, in turn, gather under the effect of their own gravity. *(d)* Planets are born.

potato-like shapes. Indeed, Mars's two satellites, Phobos and Deimos, have such a shape. When planetesimals reach the size of the moon, though, gravitational forces take over. Unlike electromagnetic forces, they abhor jagged lines and prefer the perfect shape of a sphere. As a result, planets and their larger satellites are invariably spherical.[1]

The growth process slows down markedly toward the end, as planetesimals become larger and larger. Nature requires a lot more time to build asteroids as large as a stadium than she does for football-size chunks, and even longer for more massive objects. The reason is quite simple. As planetesimals grow larger and larger, their numbers fall dramatically. While pieces of gravel are reckoned in the billions, moon-size objects can be counted on one hand. Fewer objects means fewer collisions and reduced accretion, hence slower growth.

HAZARDOUS ENCOUNTERS

The growth of planetesimals does not necessarily occur smoothly and uneventfully. It does not always proceed from small to large. From time to time, the accretion nurtured by gravity results not in union but in destruction. Hurled toward one another at great speed, planetesimals occasionally end up being pulverized in devastating collisions. The painstaking work of accretion accomplished up to that point is suddenly reduced to naught in an impact of untold violence.

Two football stadium–size asteroids come together. Instead of smoothly merging into a larger body, they disintegrate into millions of small pebbles. The impact dissipates an energy of tens of megatons of TNT. Its power exceeds several hundred Hiroshima-type bombs. All the effort expended to build dust grains into a planet is destroyed in a fraction of a second. It all must be started from scratch all over again. The larger and more massive the planetesimals, the greater the risks of destruction. A crash between two eighteen-wheelers on the highway causes far more damage than two bicycles bumping into each other. Thus, near the end of the process, encounters between planetesimals may be rare, but they turn extremely destructive. Growth slows to a crawl. While it took only a few hundred years to grow from dust particles to a football, 100 million years was necessary for planets to finally enter the stage. But Nature has all the time she needs, whether she waits for rare encounters to happen or rebuilds planetesimals when collisions do cause destruction. After 100 million years, a total of eight planets—the case of Pluto, the most distant one, is special—make their appearance and begin their endless merry-go-round around the young Sun (Figure 2).

THE SUN DOES SOME MOPPING UP

Near the Sun, the accretion process gave rise to four small planets. They are often referred to as "inner" planets because of their proximity to the Sun, or "terrestrial" planets because their properties resemble those of Earth. They are: Mercury, Venus, Earth, and Mars. Earth is the largest, with a diameter of some 12,700 kilometers. Venus is virtually identical in size, while Mars is about half as big, and Mercury a little more than a third. The terrestrial planets are rocky worlds, with little or no atmosphere. On the surface of Earth, the air that fills our lungs is made of 77 percent nitrogen and 22 percent oxygen. The atmosphere on Venus is one hundred times as dense as that on Earth. If you were to travel to Venus, you would experience an enormous pressure on your shoulders, as if you were a kilometer deep in the ocean. The atmosphere there would not be very good for your health, made as it is of 96 percent carbon dioxide and 4 percent nitrogen. The carbon dioxide traps solar heat, creating an overwhelming greenhouse effect that makes the surface of Venus about five times hotter than the temperature of boiling water (480°C). A similar fate is in store for our planet if we carelessly continue to spew out into the atmosphere vast amounts of carbon dioxide from car exhausts and plant smokestacks, and if we go on cutting trees that absorb carbon dioxide and enrich the air with oxygen through the process of photosynthesis.

The atmosphere on Mars, by contrast, is a hundred times less dense than that on Earth. It is so tenuous that satellites dispatched from Earth to explore the planet never have any trouble peering at the rust-colored, desertlike landscape. The Mariner space probes took pictures of dried-out riverbeds suggesting that water once flowed on the surface of Mars. Direct evidence for the presence of water in Mars's past comes from meteorites ejected from the planet's surface. Some of these eventually landed on Earth and were found to contain water-soaked clay. Wherever there is water is a good place to look for possible life. We now know that Martians exist only in the fertile imagination of science-fiction writers, and that the network of canals they supposedly built is only an optical illusion created by sandstorms. That does not preclude the possible existence of single-cell microorganisms buried deep in Mars's soil or under the Martian polar caps believed to contain frozen water. Breathing wouldn't be very pleasant on Mars, either—first because there is hardly any atmosphere to speak of, and second because what little there is has the same composition as that on Venus. As for Mercury, it has simply lost any trace of atmosphere at all.

One question immediately comes to mind: Why is the composition of the atmosphere and the soil (made primarily of silicates) on the terrestrial planets so different from the composition of the original solar nebula out of which they condensed? As we have seen, the composition of the nebula is the same as that of all the other stars and galaxies—namely, three quarters hydrogen, and most of the balance taken up by helium—except for a small pinch of heavier elements such as carbon, oxygen, and nitrogen, which account for no more than 2 percent of the total. As it turns out, the culprit is the forming Sun: Its burning fires chased hydrogen and helium out toward the edges of the solar system. The Sun acts as a huge circulating fan. The near regions are cleansed of light and volatile elements, including hydrogen and helium. There are two reasons for this purging action. First, the inner planets take the full brunt of the Sun's torrid heat, which causes the atoms in their atmosphere to move about feverishly. Second, their mass is relatively small and their gravity is not strong enough to keep in their grip the restless atoms, which end up escaping into space. Only heavier materials are left behind, materials that are not vaporized by the intense heat of the Sun— they are called "refractory"—such as silicates forming the crusts of Venus, Earth, and Mars. On Mercury, the planet closest to the Sun, not even silicates can withstand the fierce solar heat; they, too, are vaporized and blown away. Mercury ends up being made primarily of ferrous materials that can better resist the extreme temperatures imparted by the Sun; all its primordial atmosphere was driven out by the seething solar heat.

Venus, Earth, and Mars are not much better off. They, too, lost all their early vaporous blankets. The atmosphere they possess today evolved only later. It is made of gases (including water vapor) originally trapped within the planetesimals that accreted to build up the rocky core. These gases were subsequently released. On Earth, the condensation of large amounts of water vapor provoked massive deluges, creating the oceans that now cover two-thirds of the surface and give our planet its distinctive blue color. Life would eventually emerge from these oceans.

While the terrestrial planets lost 98 percent of their primordial atmosphere, exactly the opposite happened to the giant planets—Jupiter, Saturn, Uranus, and Neptune—gravitating in the more distant regions of the solar system (they are often referred to as "outer" planets). They captured the light and volatile gases expelled from the inner region by the action of the solar furnace. By far the biggest winner in this process was Jupiter: It gained so much gas that its mass now eclipses that of any other planet. Compared to Earth, it is 318 times more massive and its diameter is 11 times larger. Add the mass of all the other planets, their satellites, and all the asteroids and comets in the solar system, and you would still come up with only 40 percent of Jupiter's mass. Around a rocky core 10,000 kilometers (km) in radius, Jupiter has an atmosphere composed of hydrogen and helium 60,000 km thick.

Jupiter's atmosphere is believed to be gaseous through the first 20,000 km only. At greater depths, hydrogen gas should liquefy under the enormous pressure of the regions above. The upper reaches of the atmosphere were analyzed in detail by NASA's Galileo probe in December 1995. The spacecraft released a package of various scientific instruments that parachuted down to measure temperature, pressure, and chemical composition. The instruments did not last very long. Crushed by the enormous pressure exerted by the upper layers, they became silent after just a few hours. But that was long enough for them to transmit valuable data about the Jovian atmosphere. The other three giant planets—Saturn, Uranus, and Neptune— were not left out. They, too, have a rocky core surrounded by a thick atmosphere of hydrogen and helium.

Solar heat loses much of its intensity at such great distances from the

Sun. While the temperature at high noon, when the Sun is directly over-head, is a torrid 327°C on Mercury and a very pleasant 27°C on Mars, it reaches only a chilly -123°C in the upper layers of Jupiter's atmosphere, and a downright frigid -210°C on Neptune. At such low temperatures, the motions of hydrogen and helium atoms become very sluggish, and gravity has no difficulty trapping huge quantities of these gases around the giant planets.

CHANCE AND NECESSITY

We have just traveled 4.6 billion years into the past to witness the birth of the solar system. We were blinded by the tremendous explosion of a supernova. We cheered when the Sun first turned on, dispensing its heat and energy. We were mesmerized by the accretion game of dust grains in the solar nebula flattened into a disk. We saw their size grow quite rapidly from gravel to potatoes, and then much more slowly from football to moon. Things became worrisome and brutal toward the end. The early gentle union of dust grains gave way to cataclysmic collisions of gigantic chunks of rock hurtling toward each other at speeds of tens of kilometers per second, causing them to shatter into a shower of gravel. We finally heaved a sigh of relief when, 100 million years later, Nature, after repeatedly repairing the damage, produced eight planets that began their majestic journey around the Sun.

We immersed ourselves in the solar system for an opportunity to reflect on what is fortuitous and what is necessary, to ponder on the unpredictable and the predictable, the particular and the general, to learn about contingency and necessity. Aesthetics and beauty are not much help in searching for the laws of Nature if we fail to distinguish the contingent from the universal. A fact qualifies as universal if it depends neither on time nor on location. As an example, certain events that led to the birth of the Sun and its retinue of planets are taken as universal because they occurred countless times during the 15 billion years since the big bang, not just in our own galaxy, but in others as well. The collapse of the interstellar gas cloud, the initiation of nuclear reactions at its core, which ignited the Sun, the formation of a protoplanetary disk, the advent of planetesimals, the accretion game that gave birth to the planets—all these processes fall in the realm of the necessary and universal. This whole sequence of events is believed to have occurred and will occur again many times in other corners of our vast universe; our own solar system is only one among many others in the cosmos. This explains the current active search for extrasolar planets around other stars in the Milky Way.

Some astronomers believe they have already detected what appears to be a solar system in the process of forming around a star known as Beta Pictoris, located some 450,000 billion kilometers from the Sun (Figure 4). The star is surrounded by a disk of dust and gas emitting copious amounts of infrared radiation. The disk is relatively young—it formed perhaps only a few

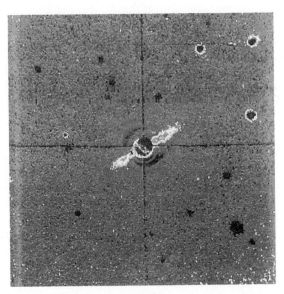

Figure 4. *A solar system in the process of forming.* Astronomers believe that they are witnessing the formation of a solar system around the star Beta Pictoris, photographed here in the infrared. The star (circular image at the center of the photograph) is surrounded by a disk of gas and dust radiating copious amounts of infrared energy. The disk is similar to the one that gave birth to our own solar system. It is seen edgewise, which gives it a rectilinear appearance. It is quite young, having formed perhaps only a few hundred million years ago. Among the planets that will some day orbit around Beta Pictoris, will there be one capable of harboring life? (Photo courtesy of NASA)

hundred million years ago, less than one-tenth the age of the solar system—and constitutes a marvelous proof that our understanding of how solar systems are born is correct. Planets are bound to form in the disk, and they will gravitate in a plane around Beta Pictoris, much as the planets in our own solar system pursue their endless course around the Sun in the zodiacal plane. Perhaps life and conscience might even emerge on one of these planets as it did on Earth.

Telescopes have so far failed to obtain direct photographs of planets around other stars outside our own solar system. The feat is extremely challenging, because planets are obscured by the blinding glare of the star around which they orbit. In principle, the Hubble telescope should be able to "see" planets within a radius of a few tens of light-years. Perhaps it will succeed someday. In the meantime, we are continuing to gather indirect evidence of planets as massive as Jupiter near certain stars in the Milky Way. Because of their gravitational action, planets cause very slight but measurable wobbles on the motion of the central star. By accurately measuring how the speed of the star varies with time, it is possible to infer the presence of one or several planets in its vicinity.

KEPLER'S SOLIDS

We have seen that a great many physical processes are universal. Yet contingency also plays a very important role in fashioning reality. If we were to ignore this element of chance and fail to take into account contingent events, we would run the risk of going down blind alleys in our quest for the laws of Nature. The young Johannes Kepler (1571–1630) learned that lesson the hard way when he tried in 1604 to understand how planets are arranged in the solar system.

Kepler was convinced that God is a geometer and that the beauty of mathematics must be reflected in the heavens in the way the planets are organized. Euclid had demonstrated many centuries earlier that in three-dimensional space there exist only five solid figures—the so-called Pythagorean solids—whose faces are identical. They are the tetrahedron, made of four triangles, the cube, made of six squares, the octahedron, made of eight triangles, the dodecahedron, made of twelve pentagons, and the isocahedron, made of twenty triangles. Back in Kepler's day, Uranus, Neptune, and Pluto had yet to be discovered. Only six planets were known at the time, separated by five intervals. Five solids, five intervals between planets: For Kepler, this had to be more than mere coincidence. Here was an opportunity to explain in one fell swoop both the number of planets and their distances from the Sun. Kepler went on to construct a solar system by nesting the five Pythagorean solids between the six spheres containing the planetary orbits. The complete sequence went as follows: Mercury, octahedron, Venus, isocahedron, Earth, dodecahedron, Mars, tetrahedron, Jupiter, cube, and, finally, Saturn (Figure 5). We may find it hard today to accept that one of the great pioneers of modern science, the man who discovered the laws governing the motion of planets, could have been so mistaken. We now know that there are nine planets, hence eight intervals, and that the layout of their orbits has nothing at all to do with the five Pythagorean solids. Kepler was wrong because he forgot to make a distinction between chance and necessity, between what is contingent and what is fundamental. He applied his sense of beauty and harmony to phenomena that were contingent. The number of planets and the way they are arranged around the Sun are not dictated by fundamental laws, but depend on a series of accidental events that determined how dust grains accreted in the solar nebula. In some other solar system around some other star, the number of planets and their orbits may turn out to be completely different. Beauty is a reliable guide only when it comes to universal phenomena.

THE ROUND OF THE SEASONS

Does this mean that we must ignore chance events in our attempts to explain the world? The answer is not at all, because Nature resorts as much to contingency as to necessity in fashioning reality. In order to explain natural phenomena, we must both discover the laws governing what is neces-

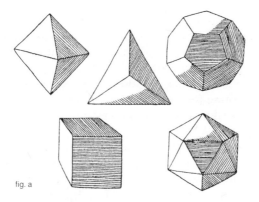

fig. a

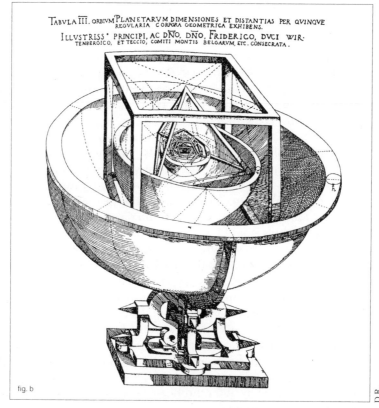

fig. b

Figure 5. *Kepler and the five perfect Pythagorean solids*. Kepler believed that the number of planets and their arrangement with respect to the Sun were not arbitrary, but reflected God's will. Because God was a geometer, the five perfect solids of Pythagoras *(a)* had to fit between the spheres of the six planets known at the time. Kepler proposed such a geometrical model of the solar system in 1596 in his work *The Cosmic Mystery (b)*. We know today that this model cannot be correct since there are now nine known planets in the solar system. Kepler went down the wrong path because he failed to realize that the number of planets and their respective distances to the Sun belonged not in the realm of necessity but of contingency.

sary and reconstruct fortuitous events. The round of the seasons illustrates well the importance of happenstance in our everyday lives. It constitutes vivid proof that contingent events can have a profound influence on reality.

Who, in our climes, has not marveled at the splendor of spring, with its brilliant colors, its festival of flowers, its trees festooned with budding leaves where birds set down to sing their loves? As the year progresses, the days become longer and the stormy heat of summer soon replaces the mildness of spring nights. Later, fall pushes summer aside and paints trees with gold, dazzling us with gushes of pastels and browns. The frosty winds of winter soon strip orchards and forests bare, and the birds take their songs to warmer climes. The appearance of lilacs in time announces the return of spring, and a new round of seasons starts all over again. One cycle lasts a year, the time it takes Earth to complete one revolution around the Sun.

EARTH IS NOT STANDING UP STRAIGHT

Why do seasons exist on Earth? Why does the beauty of spring flowers always give way to the oppressive heat of summer? Why must the mildness of autumn be replaced by the biting cold of winter? The answer is quite simple. It is because Earth doesn't stand up straight. It tilts to one side. Like any other planet, Earth is subject to two distinct motions: While it orbits around the Sun, it also spins on itself once a day, which explains why the darkness of night is followed by the brightness of daytime. The Earth rotates about an axis that passes through the North and South Poles and points toward the North Star. It also travels around the Sun in a plane called the ecliptic, or zodiacal plane. All planets, with the exception of Pluto, also move endlessly around the Sun in the same plane, which was formed as the solar nebula collapsed into a flattened disk.

As it happens, the spin axis of Earth is not perpendicular to the ecliptic, but is tilted at an angle of 23.5 degrees (Figure 6). It is this inclination that is responsible for the explosion of colors in the spring and the migration of swallows in the fall. In June, the tilt is such that the Northern Hemisphere has the Sun more directly overhead and is exposed to more of its heat, while the Southern Hemisphere faces away and receives less of its golden rays. While the French and Americans are busy getting a suntan on beaches and their children frolic in the water to cool off, their counterparts in Brazil don coats and gloves to fend off the assault of winter. Six months later, in January, the roles are reversed. Because the tilt of Earth relative to the ecliptic is constant, it is the Northern Hemisphere's turn to be starved of solar heat. The Swedes see their days getting shorter and shorter as snow blankets their towns, while the Chileans flock to the ocean shores for vacations. Thus does the round of seasons go on. The next time you complain about the stifling heat of an August afternoon, or about frost encrusting your windshield on a cold January day, whenever you delight in the beauty of fall foliage or marvel at a bed of spring flowers, just remember that it is all due to the fact that Earth doesn't stand up straight.

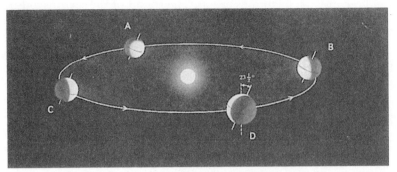

Figure 6. *The round of the seasons.* The inclination of Earth's spin axis (by 23.5 degrees relative to the axis perpendicular to its orbital plane) is responsible for the seasons. Earth maintains the same orientation (with the North Pole pointing toward the star Polaris) throughout its annual orbit around the Sun, so that the amount of solar illumination on a given spot on the globe varies continuously during the year. When Earth is at point C in its orbit, it is summer in the Northern Hemisphere, while winter grips the Southern Hemisphere. Six months later, when it is at point B, the situation is reversed.

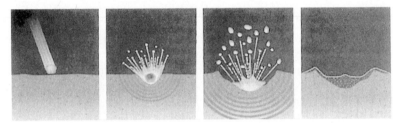

Figure 7. *The formation of craters.* (a) Most craters on the surface of the Moon (or Mercury) were created by the impact of rocky bolides crashing into the surface at a velocity of 10 km/s or more. (b) The bolide is vaporized by the impact and a shock wave propagates toward the interior of the celestial body. (c) The central explosion creates a huge cavity and ejects numerous pieces of debris, whose total mass can reach 100 times the mass of the crashing bolide. (d) The vast majority of the ejecta falls back into the crater; some gets scattered around, while the rest escapes the gravity of the celestial body and is lost in space. Some of that ejecta may eventually reach Earth.

UPSETTING COLLISIONS

We still have to figure out why Earth is tilted. To find the answer to that question, we need once again to travel back in time to the late stages of the formation of the solar system, some four billion years ago. Most of the rocky chunks have by then collected under the influence of gravity to form the planets. Not many are left over, and collisions between asteroids and planets (Figure 7), which were once commonplace as evidenced by the pockmarked surfaces of the Moon (Figure 8) and Mercury (Figure 9), have by now become increasingly rare. But all danger is not over yet. There are still a few large asteroids roaming around here and there, speeding at a few tens of kilometers per second, whose orbits cross those of the planets. One of them happened to hit Earth. Back when asteroids were plentiful and collisions far

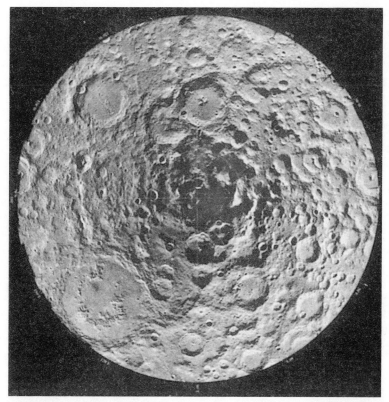

Figure 8. *The wounds of the Moon*. This mosaic of about 1,500 photographs returned by the space probe Clementine shows the region near the Moon's south pole. The numerous craters covering this region give it a ravaged and tortured relief that attests to the violence of the many asteroid impacts on the lunar surface during the first 100 million years in the history of the solar system. (Photo courtesy of NASA)

more frequent, impacts were just as likely to occur north and south of the equator. If a collision were to happen in the Northern Hemisphere, causing Earth to tilt one way, its effect would likely be offset by another impact in the Southern Hemisphere, which would straighten Earth right back up. But toward the tail end of the process, which lasted some 440 million years in total, the pool of asteroids dwindled so much that collisions became quite rare and subsequent corrections failed to materialize. That is how Earth was left tilted, giving mankind the round of the seasons and driving the migratory birds every year toward more propitious skies.

The last collision our planet sustained with a large asteroid was purely accidental. It falls squarely in the realm of contingency. It is absolutely not part of the fundamental laws of Nature. No physical law predestined Earth to be tilted at an angle of precisely 23.5 degrees. Chance could just as well have left it standing almost perfectly straight up. That is in fact what happened to Jupiter, which is tilted by only 3 degrees and experiences virtually

Figure 9. *The pockmarked surface of Mercury*. Mercury, the planet closest to the Sun, was photographed by the space probe Mariner 10 in 1974. All of Mercury's primitive atmosphere, heated by the intense solar radiation, was swept away toward the outer parts of the solar system. The average temperature on Mercury's surface is 350°C during the day, and falls to -170°C at night. Without any atmosphere to protect it, Mercury was exposed to incessant and violent bombardments by asteroids during the formation of the solar system. The surface of Mercury, which resembles that of the Moon, is covered with thousands of craters and basins, some as large as 1,300 km in diameter. Most of these craters and basins were named after artists and writers, such as Bach, Shakespeare, Goethe, and Mozart (in contrast to the lunar topography, which commemorates scientists, such as Copernicus, Tycho, Kepler, and Descartes). (Photo courtesy of NASA)

no seasons. Temperatures there vary very little during the 12 years it takes the giant planet to complete one orbit around the Sun. As it turns out, Jupiter would not have seasons even if it were more strongly tilted. Being essentially a large gas sphere without a solid surface, Jupiter is hardly in a position to support either fauna or flora, and it will never enjoy the luxuriance of a spring or the golden colors of an autumn.

THE SUN RISES IN THE WEST ON VENUS

Every planet in the solar system has its own inclination. Mercury, closest to the Sun, resembles Jupiter in that it is almost straight. Listing by less than 2 degrees, it looks like a soldier standing at attention. Venus has a tilt of 13 degrees. But what makes it very unusual is the direction of its spin: It is the only planet where the Sun rises in the west and sets in the east. It is as if it is turning upside down. The fact that most planets spin *in the same direction* is no accident. That was simply dictated by the overall spin of the primordial solar nebula that gave birth to the planets. But the particular spin direction—from east to west—is entirely contingent. The original nebula could just as well have rotated from west to east, in which case France, rather than Japan, would be the country of the Rising Sun; the Vietnamese and Chinese would be thought of as westerners, while the French and Germans would be Orientals.

Happenstance also caused Venus to spin in the opposite direction. It began its life spinning in the same direction as all the other planets. At some point, it must have sustained a particularly violent collision with a large asteroid under just the right conditions. The asteroid probably impacted Venus in a direction opposite to the planet's original spin, causing it to in effect switch into reverse. And the Sun has been rising there in the west ever since.

Mars is the next planet after Earth out from the Sun, and the one closest to us. At 25.2 degrees, its tilt is very similar to Earth's, and it, too, experiences seasons and cyclical temperatures. But Martian seasons are a far cry from their splendor here on Earth. Winter there causes a massive condensation of carbon dioxide, which constitutes 98 percent of the atmosphere. This results in an expansion of the polar cap, and in the coating of the Martian surface by a thin layer of dry ice. Come spring, the Sun's heat evaporates the dry ice, which returns to the atmosphere, and the polar cap shrinks back to its original size. By summer, the Martian soil, baked by the intense heat from the Sun, becomes much warmer than the atmosphere. The temperature difference between ground and air creates fierce sandstorms, similar to those occurring in deserts here on Earth, which can obscure the surface of Mars for weeks at a time and protect it from the curious gaze of earthlings. And so go the Martian seasons, marked by the rhythm of the expanding and receding polar caps and by the swirls of sandstorms. The explosion of flowers in the spring and the festival of colors in the fall are conspicuously absent. Even though a meteorite from Mars picked up in Antarctica seems to suggest that a primitive unicellular life may have existed on Mars some 3 billion years ago, the miracle of plant life as it evolved on Earth never took hold on the "red planet."

A PLANET LYING ON ITS SIDE

Let us turn our attention to the giant planets on the outskirts of the solar system. There is Saturn, inclined at 26.7 degrees, and Neptune, with a tilt of 29.6 degrees. Nothing much out of the ordinary so far, until we come up to Uranus: Its inclination is a whopping 98 degrees! In other words, it is practically lying on its side (Figure 10). Instead of being close to the zodiacal plane, as is true for most other planets, its equator is nearly perpendicular to it! This unusual situation gives rise to seasons exaggerated to the extreme. In the summertime, the South Pole faces the Sun for a period of 42 years (half the time for Uranus to complete one revolution around the Sun), while the North Pole is plunged into glacial darkness. Things get switched around half a Uranus year later for the next 42 earth-years. On Uranus, we would be treated to the magnificent spectacle of sunrise and sunset only twice during our lifetime. Here again, contingency is at the origin of this bizarre seasonal cycle. And here again too, a rocky bolide is believed to be the culprit. Experts suspect that a collision with a massive asteroid knocked Uranus on its side.

In short, purely random events, which are not predetermined by the laws of Nature, can have repercussions on all levels of reality. Here on Earth, contingent happenings have had a direct influence on our everyday lives. Our planet could have remained upright like Jupiter, and day- and nighttime temperatures would remain practically invariant all through the year. Or it might have ended up knocked to its side like Uranus, in which case we would have a long winter night lasting six months, followed by an equally long

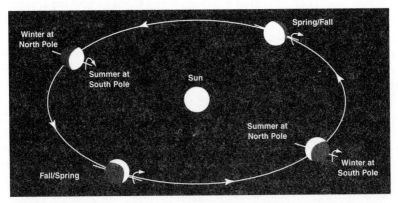

Figure 10. *The exaggerated seasons on Uranus.* Uranus practically lies on its side. Its spin axis deviates by only 8 degrees from the planet's orbital plane around the Sun. As a result, during summertime at the South Pole, the Sun does not set there for a full 42 years, while the North Pole is plunged into winter's glacial darkness for the same period. The situation is reversed 42 years later, when Uranus is on the diametrically opposite side of its orbit around the Sun.

bright and hot summer day. But fate was such that Earth is tilted by 23.5 degrees, and that is what gives us a chance to admire the pristine whiteness of snow-covered landscapes, the beauty of spring flowers, and the orgy of browns and golds in the fall. The word "astrology" takes on a brand-new meaning in this context: Our daily lives are ordained not by the positions of the planets at the time we were born, but by fortuitous celestial events that occurred in the distant past of the solar system.

Earth's seasons depend not just on the inclination of the planet. The shape of its orbit around the Sun also plays an important role. As Kepler taught us in 1609, the orbit is not a circle but an ellipse, with the Sun occupying one of the two foci. As a consequence, the Earth–Sun distance varies during the course of a year, by an amount slightly less than 1 percent. Small as this variation may be, it is enough to create differences in the seasons north and south of the equator. Contingency manifests itself once again. By a twist of fate, Earth is farther from the Sun just as summer arrives in the Northern Hemisphere, and it receives slightly less heat than it would otherwise. As a result, people who live north of the equator do not sweat nearly as much, while their counterparts in the Southern Hemisphere shiver more. Six months later, it is winter in the northern half and summer below. Once again, people above the equator are the winners. Since Earth is now slightly closer to the Sun, it receives a tad more than its normal share of solar rays. The upshot is that people who inhabit northern latitudes are in less of a deep freeze, while those in the southern ones see the mercury reach record highs.

And so does chance influence directly our well-being. In fact, it affects every aspect of reality.

THE POLE STAR IS NOT WHAT IT USED TO BE

Is this state of affairs permanent? Will people in the Northern Hemisphere forever enjoy a less extreme climate? That would indeed be the case if the spin axis of Earth remained firmly pointed toward the North Star, also known as Polaris or the celestial pole star. Among all the luminous dots twinkling in the night sky, the North Star, located in the Little Dipper (the constellation Ursa Minor), is the only one that seems to remain fixed in the sky as Earth rotates. As it turns out, the direction of Earth's spin axis is not constant. The Greek astronomer Hipparchus of Rhodes (180–125 B.C.) already knew that as early as 120 B.C. As children, we used to be fascinated by the wobble of a spinning top. Its rotation axis is not fixed. Instead, it describes a cone in space. The same is true of the spin axis of Earth, although a complete period is not counted in a few seconds but, rather, lasts some 26,000 years. Astronomers call this phenomenon "precession" of the equinoxes (Figure 11). Four thousand years ago, the axis connecting the North and South Poles pointed not toward Polaris but toward a star named Alpha-Draconis in the constellation Draco (the Dragon). In fourteen thousand years from now, our great-great-great . . . grandchildren will see Earth's rota-

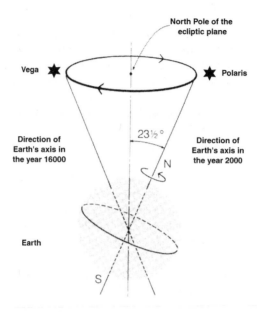

Figure 11. *The North Star will not be what it is today.* The gravitational forces exerted by the Moon and the Sun on Earth's equatorial bulge (the radius of Earth is about 20 km larger at the equator than at the poles because its spin causes centrifugal forces that are strongest near the equator and pull on the equatorial regions) try to straighten up Earth, which is inclined by 23.5 degrees with respect to the direction perpendicular to its orbital plane around the Sun. Instead of straightening up, Earth's spin axis describes a circular motion, called "precession," with a 26,000-year cycle. Polaris, which is currently within a degree of the spin axis, now plays the role of North Star. But it has not always had that distinction in the past and will lose it in the future. Fourteen thousand years from now, in the year 16,000, Vega will become the North Star, a role that Polaris will reclaim once again 12,000 years later, in the year 28,000.

tion axis pointed toward another star, named Vega, in the constellation Lyra.[2] At that time, Earth will be closer to the Sun during summertime in the Northern Hemisphere. It will then be the turn of the people living above the equator to suffer more extreme temperatures—more sweltering heat in the summer and deeper freezes in the winter.

Why does Earth's spin axis precess? In the case of an ordinary top, the phenomenon is caused by the gravity exerted by Earth. In the case of Earth itself, it is due to its gravitational interaction with the Sun and, more important, with the Moon.

We have witnessed earlier the process that led to the birth of the Sun out of the solar nebula. Where the Moon comes from was not nearly as clear, and debates on this topic continued to rage until the mid-1980s.

We are about to see how chance played once again a crucial role in endowing Earth with its satellite. Here is another case where contingency is going to play as determining a role as necessity in fashioning reality in all its aspects, from the ebb and flow of the ocean tides to the emergence of life itself.

A BIT OF DETECTIVE WORK ABOUT THE MOON

To figure out the origins of the Moon, astronomers have to be like Sherlock Holmes. They must gather the most significant pieces of evidence and reconstruct a plausible scenario that integrates them all in a coherent fashion. Let us go out, then, in search of clues that might point us in the right direction.

The first clue is that the Earth-Moon duo is abnormal. The Moon is far too large relative to Earth. To be sure, Jupiter, Saturn, Uranus, and Neptune all possess satellites whose size is comparable to that of the Moon, but these are giant planets that are respectively 318, 95, 15, and 17 times more massive than Earth. The diameter of the Moon (3,400 km) is roughly one quarter that of Earth (12,700 km). By contrast, Jupiter (143,000 km) and Uranus (51,000 km) are both 30 times larger than their largest moons, while the diameters of Saturn (120,500 km) and Neptune (49,500 km) dwarf those of their own satellites by a factor of 23 and 18, respectively.

The second clue is quite troubling: Among the terrestrial planets, only Earth has such a large moon (Figure 12). Mercury and Venus have no satellite at all. Mars has two of them, named Phobos and Deimos, but they are of such insignificant size—their diameters are only 28 and 16 km, respectively, which make them no more than large asteroids—that in the 1950s the Soviet astrophysicist Josef Shlovsky even proposed that they were spacecraft placed in orbit by Martians.

Our astronomer-detective pursues his investigation and discovers that the Moon's average density is only 3.3 times that of water, significantly smaller than the average density of Earth, which is 5.5 times that of water. The difference is due to the fact that our planet has an iron core, while the Moon does not. But—and it is an intriguing coincidence—the lunar density happens to be the same as that of the granitic rocks that make up Earth's crust. Could there be a connection there?

Progress in the investigation received a boost from the 382 kilograms of lunar rocks picked up at six different sites on the Moon between 1969 and 1972 and brought back to Earth by Neil Armstrong and eleven other American astronauts who walked on the surface of our satellite during the Apollo program. Laboratory analyses of these rocks revealed that they were uncharacteristically dry. Not the slightest trace of water molecules could ever be detected, in contrast to Earth rocks, which always contain some small amount of water, even in the most arid deserts. Another tantalizing difference is that, when compared to materials of terrestrial origin, lunar rocks lack volatile elements such as potassium and sodium, which vaporize at relatively low temperatures. They are, however, rich in refractory elements, such as calcium and aluminum, that require much higher temperatures to vaporize.

Our detective comes to the conclusion that the Moon is composed of a material that was once heated to a higher temperature (in excess of 1000°C) than were terrestrial rocks. The volatile elements must have escaped into space, leaving the Moon richer in refractory elements.

EARTH GIVES BIRTH TO THE MOON

At the end of a detective novel by Agatha Christie comes a fateful scene in which the sleuth gathers everyone in the living room to reveal the solution to the puzzle, including the name of the culprit. The investigation was not all that easy, and in the purest Hercule Poirot tradition, our astronomer begins by telling how he went down several false trails.

The first hypothesis is that the Moon formed the same way as all the planets and satellites in the solar system did some 4.6 billion years ago—by accretion of planetesimals. The process produced the Sun and its retinue of eight planets (Pluto is believed to be an unusually large asteroid orbiting in the outer reaches of the solar system), including the "giant planets" with their parade of moons—Jupiter is known to have sixteen of them, and Saturn nearly two dozen. According to this scenario, the nucleus of the nascent Moon would have condensed from a ring of material gravitating around Earth. However, this theory fails to explain why the Moon is so large relative to Earth, while the satellites of Jupiter and Saturn are much smaller compared to their mother planet. It sheds no light on why Earth is the only one of the terrestrial planets to have such a large moon. Furthermore, the facts that the Moon is made of drier material containing fewer volatile elements and more refractories, and that it has practically no iron core, remain total mysteries.

Forced back to square one, the astronomer-detective explores a new avenue. What if the Moon is a foreign celestial body that happened to wander by the solar system and became captured by Earth's gravity, never to leave again? It might explain why Earth is the only planet with such a large satellite. It would simply have been lucky enough to snare a massive object to form with it an eternal couple. Serendipity would have presented Earth with an ideal partner, providing poets with an inexhaustible source of inspiration and mankind with the magic spectacle of solar eclipses. What is more, if the Moon were a visitor from outside the solar system, its slightly different chemical composition would make sense.

The theory certainly has its appeal, but it cannot stand upon closer examination. The capture of celestial bodies approaching Earth is highly unlikely. Large numbers of asteroids routinely visit Earth's neighborhood. Yet they never go into orbit around it, always returning to the fringes of the solar system where they came from. Capturing the Moon is theoretically possible, but conditions would have to be tuned so precisely that no one takes the hypothesis seriously. Computer simulations show that the Moon would have to approach Earth to within less than 50,000 km (that is one-hundredth of a thousandth of the size of the solar system, which corresponds to the thickness of a hair compared to the length of a football field). It would have to do so without crashing into Earth, and within a very tight velocity window. Just a shade too fast, and it would escape Earth's gravitational pull and go back out, as most asteroids do. Besides, if it were captured,

Figure 12. *The Earth and its large Moon*. Ours is unique among the terrestrial planets in that it has a large moon, roughly one quarter the size of Earth. This picture shows on the same scale the 4 terrestrial planets (Mercury, Venus, Earth, and Mars), together with the 6 largest moons in the solar system: our own Moon, Jupiter's four Galilean satellites (Io, Europa, Ganymede, and Callisto), and Titan, Saturn's largest moon. Among the other terrestrial planets, Mercury and Venus have no satellite at all, while Mars has two tiny moons: Phobos and Deimos, whose sizes are 28 and 16 km, respectively, more typical of asteroids (see Figure 22). Our Moon is nearly as large as those orbiting the giant planets Jupiter and Saturn. Its unusually large size is believed to be explained by the fact that a huge asteroid collided with the Earth and tore loose a large chunk of Earth's crust, which subsequently condensed to form the Moon. (Photo courtesy of NASA)

the Moon would have a highly elliptical orbit, rather than the nearly perfectly circular one we observe today. Our astronomer-detective is forced to discard this theory and look for a different one.

A somewhat original explanation was suggested by George Darwin (1845–1912), the son of Charles Darwin, the inventor of the theory of the evolution of species. Darwin asserted that the Moon came from Earth itself. Our planet gave birth to its satellite by ejecting it as a result of the centrifugal force due to its own rotation. That is the same force that pushes you against the seat of a car when the driver rounds a curve at excessive speed. Earth was supposedly spinning much faster early on. If so, the enhanced centrifugal force would have been capable of ripping apart an entire region of Earth's mantle and propelling it into space, carving a gigantic hole in what is now the Pacific basin. The chunks of matter expelled from Earth would have subsequently condensed to form the Moon.

Can this theory explain all the evidence? It can, of course, account for the similar densities of the Moon and Earth's crusts, but it remains silent about the differences in chemical composition. The fatal blow to Darwin's theory came when people began to calculate how fast the still-pregnant Earth would have had to rotate to generate enough centrifugal force to eject the Moon. The verdict was unanimous: The numbers just did not add up. Earth would have had to spin ten times faster than it does today, completing a revolution in just under two and a half hours. It is difficult to conceive how the accretion process could have gradually built up an Earth rotating this fast. Moreover, assuming that it did, why would Earth have slowed down by a factor of ten after delivering the moon? Darwin desperately tried to salvage his theory by invoking resonance phenomena that might have amplified the ejection of material without requiring such large rotation speeds and centrifugal forces, but not many people really believed him. We know today that oceans in general, and the Pacific in particular, are created not by ejection of pieces of Earth's crust, but by drifting continental plates that form huge rifts where water rushes in.

THE GIANT IMPACTOR

Our astronomer-detective rubs his hands in delight; his eyes sparkle with glee. He is done with his list of false trails. As a true disciple of Sherlock Holmes and Hercule Poirot, he is about to unveil the one theory that accounts for all the pieces of evidence and will expose the guilty party.

He begins by reminding us of that epoch of great turmoil when planets were forming by accretion of planetesimals. Here and there, large bolides called asteroids roam across space at speeds of several tens of kilometers per second. Once in a while, an asteroid collides with a planet with incredible force, spraying large amounts of matter up in the air. In the less powerful collisions, the dispersed matter simply falls back down onto the planet (Figure 7). But in the most violent ones, the energy imparted is so huge that matter escapes the planet's gravity and is ejected out into space. We have seen that such a collision caused Earth to tilt, giving us the round of seasons and the glowing spectacle of fields of blooming lavender. Our astronomer-detective proceeds to argue that it was in fact a gigantic collision of the same type that ripped the Moon loose from Earth's crust. He goes on to describe the Giant Impactor: an enormous rocky asteroid with about one-tenth the mass of Earth, and roughly the size of Mars. It came crashing to Earth 4.6 billion years ago. The violence of the impact was so great that gobs of matter from both Earth and the Giant Impactor were thrown out into space. Part of the enormous energy was converted to heat, which liquefied and vaporized the ejecta. Water and other volatile elements evaporated and were lost in space.[3] This left the unvaporized part of the ejected material enriched in refractory elements. It eventually coalesced to form a moon depleted of volatile elements and rich in refractories (Figure 13).

The theory of the Giant Impactor is consistent with a host of other

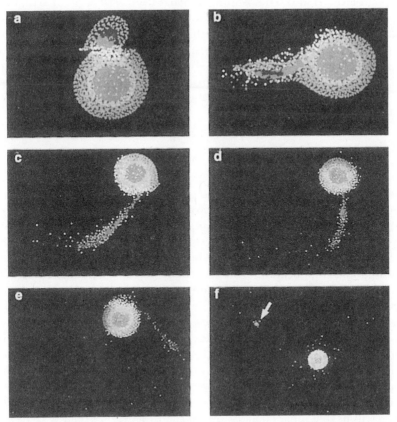

Figure 13. *The formation of the Moon.* The Moon was very likely formed when a giant asteroid collided with the young Earth during the first 100 million years of the solar system. The above sequence of pictures, from *(a)* to *(f)*, shows a computer simulation of the impact of a massive asteroid crashing into Earth. The simulation assumed an asteroid mass equal to one-tenth that of Earth. The impact rips a chunk of material away from Earth's crust *(a)*. That material is ejected into space, where it gradually coalesces into a single body under the influence of gravity. Frames *(b)* to *(e)* show the evolution of the process seen when one steps back and looks at an increasingly larger volume of space around Earth. The final result is the Moon, indicated by an arrow in *(f)*.

details. It accounts for the Moon's density being the same as that of Earth's crust. It also explains why the Moon is iron-poor: Because the heavy central core of the asteroid, rich in iron, sank into Earth.

The great power of modern computers has enabled us to confirm the plausibility of this hypothesis. The Giant Impactor theory has become the best candidate to explain the origin of the Moon. It is gaining wide acceptance because it accounts better than any rival theory for the available evidence. In this picture, Earth is the only one of the terrestrial planets to possess a large moon simply because it is the only one to have collided with such a large asteroid—a collision that could have shattered our beloved Earth into a thousand pieces if that asteroid had been just a little bigger. Once again, happenstance and contingency intervened to craft reality at its most profound level. Had the devastating asteroid been a bit larger, Earth would not exist today and we would not be around to talk about it. A random event that occurred 4.6 billion years ago is thus responsible not just for the soothing glow bathing the countryside during nights of full Moon but quite literally for our very existence. Indeed, the Moon plays a far more important role than just that of nocturnal beacon or muse for poets. It was crucial to the emergence of life on Earth because it regulated Earth's climate. Let us explore this point in more detail.

ICE INVADES EARTH

We sweat in the summer heat and shiver in the wintry winds because, as explained earlier, Earth is tilted and successively moves closer to or away from the Sun during its annual trek around its elliptical orbit. The variation of the Earth–Sun distance is quite modest—is less than 1 percent of the average value. But that is enough to give Chileans hotter summers and colder winters than those experienced by Americans, who live above the equator. An additional complication is the fact that neither the tilt of Earth nor the shape of its orbit are constant. They are perturbed by the gravitational influence of the Moon, the Sun, and the other planets, and change slowly over the course of time. As a result, the amount of solar heat that gives us a suntan in the summer and warms our faces in the winter also changes as the years go by. No need to panic, though, because these changes are quite tiny on the scale of a human lifetime. But they are noticeable over tens or hundreds of thousands of years.

Let us examine the fluctuations of Earth's rotation axis. Geological evidence tells us that it has changed by less than 1.3 degrees relative to its average inclination of 23.5 degrees over the last million years. This suggests that it has been fairly well-behaved over Earth's entire past history. Indeed, this small variation has provided the necessary climatic stability needed to ensure the emergence and evolution of life on Earth. That is not to say that a 1.3-degree shift is inconsequential as far as life on Earth is concerned. If an increase of that magnitude were to occur overnight, a Swede who lives at a latitude of 65 degrees north of the equator would suddenly receive 20 percent more solar heat during the summertime. But perhaps more important than the deeper tan among Nordic people, the extra heat has played and continues to play a determining role in the equilibrium of Earth's climate, because it melts the ice that has accumulated during the previous winter and prevents it from taking over entire continents.

Yet Earth did experience several episodes of glaciation in its past. The sediments accumulating at the bottom of lakes and oceans prove it unambiguously. The clay and sand that erosion removed from continents, the calcified skeletons of microscopic algae, ice-core samples—in short, everything that retains a memory of ancient climates—all tell us that the last Ice Age occurred some 20,000 years ago. All of Canada, the northern part of the United States, Greenland, and northern Europe were buried under roughly

4 kilometers of ice. According to the Yugoslav astronomer Milutin Milankovitch, glacial periods such as the quaternary era got started owing to a very special combination of circumstances. In particular, Earth must be at its closest point to the Sun in its orbit when it is wintertime in the Northern Hemisphere. In that case, winter is relatively mild (when Earth is closer to the Sun, it receives more heat) and shorter (the stronger gravity of the Sun speeds Earth along its orbit, shortening the cold season). By contrast, summertime is longer and cooler. Because of below-normal temperatures, the snows of the previous winter do not have a chance to melt completely. What fails to melt keeps on piling up year after year in the form of ice. Acting like a mirror, it reflects solar heat off into space, further accentuating the cooling effect. If, in addition, Earth's inclination is less than normal—say, 22 degrees—summertime, already cool for the previous reason, will be even more so, and Earth is then likely to enter an Ice Age.

Milankovitch's theory was not immediately accepted at first. But over the last two decades, evidence has been accumulating in its favor. The analysis of marine sediments allows us to reconstruct fairly accurately the climatic evolution during the past 3 million years. Other, somewhat less precise, geological studies can reach back some 200 million years. All the indications are that the mean temperature of oceans in the summers immediately preceding glaciation periods were indeed cooler, just as Milankovitch suggested.

WHAT IF SOMEONE TOOK AWAY THE MOON?

Even with a remarkably well-behaved rotation axis, Earth is still subject to severe episodes of glaciation. What would happen if the axis were to behave more erratically? Mars, our nearest neighbor, provides a good example of the possible consequences. Its tilt angle is now 25.2 degrees, but it is believed to have varied by a full 10 degrees in the past. The climate on Mars has almost certainly gone through extreme phases. In all likelihood, there was a time when the planet was listing a little too much toward the Sun; the resulting torrid summer heat probably caused the rivers that once flowed on its surface to evaporate, leaving behind only dried-up beds hinting at Mars's past splendor.

Why is the spin axis so fickle on the "red planet"? Could it be related to the fact that Mars lacks a large moon? As pointed out earlier, its two satellites, Phobos and Deimos, are about the size of large asteroids. This brings up an interesting question: How would Earth behave if, like Mars, it had no massive moon? To answer this question, we obviously do not enjoy the luxury of experimenting by changing the solar system the way we want it; and it is certainly not our intention to deprive poets of their muse. But computers are here to come to our rescue. Thanks to them, we can play God and reconstruct the history of the solar system under different conditions—for instance, by eliminating the Moon. The French astronomer Jacques Laskar at the Bureau of Longitudes, in Paris, and his coworkers have done just that.[4] What they found is that, absent the Moon, Earth's rotation axis would exhibit rather wild swings, from almost perpendicular to the ecliptic plane all the way to being practically parallel to it with an inclination of 85 degrees. The model suggests that these large variations could occur over periods as short as a few million years, a mere eyeblink in geological terms. Physicists would describe the behavior of Earth's rotation axis as "chaotic." Such behavior (we will explore chaos in all its glory in the next chapter) would provoke climatic changes that would spell disaster for life on Earth. While the rotation axis points straight up, the amount of heat received by each point on the globe would remain constant throughout the year. At the other extreme, when the axis lies in the ecliptic plane, as it does on Uranus, earthlings would endure traumatic climate changes. For six months, half of Earth would be plunged in the total darkness and frigid temperatures of an endless winter, while the same half would be bathed in the Sun's dazzling light and torrid heat for the next six months. With such extremes, which would come about without warning—chaotic behavior is inherently unpredictable—life would have an extremely difficult time gaining a foothold on Earth. And so, by keeping extreme changes in check, the Moon was crucial in making the advent of man possible. Once again, we can only marvel at the fundamental role of contingency in fashioning reality. The accidental collision of a huge asteroid with Earth, causing it to expel the Moon, enabled life to emerge.

THE MOON LIFTS THE OCEANS

A group of children are frolicking on a beach. Others are building sand castles under the watchful eye of a lifeguard perched atop his high chair. Dusk is approaching. Bathers gather their belongings and hurry home. The beach gradually empties. Water starts rising and the sand castles crumble in the waves. The children's play area gets flooded. From time to time, a bolder wave laps at the now-empty lifeguard's chair. The Moon has risen and appears to float above the horizon, casting a soft radiance on the water. It is an eerie feeling to realize that this full Moon, seemingly so fragile in the veil of the night, was able to move all this mass of water, flood the shore, and topple the children's sand castles.

The Moon is indeed responsible for the tides. The ebb and flow of the oceans is due to the forces of gravitation exerted by the Moon at two diametrically opposed places on Earth. The tide is high on the side of Earth facing the moon, because that is where gravitational attraction is the strongest. Somewhat paradoxically, the tide is also high on the opposite side of the globe, at the location on Earth farthest away from the Moon (Figure 14). There the Moon's gravitation is the weakest, and it pulls relatively less water from this area than from other parts of the globe, which results in a local high tide there as well. Thus any spot on Earth sees two high tides in a day, separated by about 12 hours. The first occurs when the Moon is directly overhead; the second comes about half an Earth turn later, when the same spot reaches its greatest distance to the moon.[5]

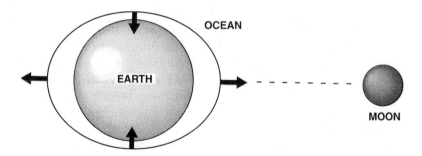

Figure 14. *The Moon is responsible for the tides on Earth.* The gravitational forces exerted by the Moon on Earth distort the surface of the oceans. They produce a rise in the water level (corresponding to high tides) along the direction of the Earth–Moon axis, and a drop of the water level (corresponding to low tides) along the perpendicular direction. Because a high tide occurs not only on the side of Earth closest to the moon but on the diametrically opposite side as well, there are two high and two low tides per day at any location on the planet, separated by about 12.5 hours (the interval is not exactly 12 hours because of the movement of the Moon around Earth). The Sun also exerts tidal forces on Earth's oceans. However, it is much farther away from Earth, and despite its much greater mass, its effect on tides is only about half that of the Moon.

If the Moon is capable of lifting oceans, the Sun is not idly standing by. Newton taught us that the forces exerted by a celestial body are proportional to its mass and inversely proportional to the square of its distance. This translates into tide-generating forces that depend inversely on the cube of the distance. The Sun is far more massive than the Moon, actually about 27 million times more. But it is also much farther away, the Earth–Sun distance being 389 times greater than that between Earth and the Moon. In the end, the Sun has only about half as much power to lift the oceans as the Moon does. Depending on the positions of Earth and the Moon relative to the Sun, the Sun can reinforce the tide-generating action of the Moon or oppose it. The relative positions of Earth, the Moon, and the Sun also determine the phases of the Moon, and indeed how strong the tides are goes hand in hand with the appearance of our satellite. During new Moon and full Moon, the Sun and the Moon line up with Earth. Their abilities to lift the oceans add up and the tides are particularly strong. At such times, the waves literally rush over the children's play area, and their sand castles do not stand a chance. At first and last quarters, on the other hand, the Moon, the Sun, and Earth are at right angles to each other, and the Sun offsets half the Moon's lifting power. This is when the tides are the weakest. The waves barely make it up the beach, and the sand castles seem to dare the ocean to knock them over.

THE MOON HAS A SECRET SIDE

The Moon plays a game of hide-and-seek with us. It conceals half its surface, because it always presents the same side to us (Figure 15). Yet it does not stand still. It spins about itself as it goes about its monthly journey around Earth. How, then, does it manage to always keep one side hidden? Its trick is to have synchronized its spin with its orbital motion. It rotates once about itself in exactly the same time (29.5 days) it takes to complete one revolution around Earth, so that the same side is always facing us. If you are not convinced, seat a friend on a chair in the middle of a room and start walking around him while constantly staring at him, never turning your back. You will realize that when you return to your starting point, you will have had to turn your body once completely around.

This perfect synchronism of the two motions of the Moon is no accident. Just as the Moon can lift the oceans on Earth, so does Earth exert tidal forces on the Moon. Our planet returns the favor to the Moon, as it were, and many times over at that, because Earth is some 80 times more massive than our satellite. Since there are no oceans on the Moon, Earth lifts the rocky lunar surface itself, which causes the Moon not to be perfectly round. Its diameter at the equator is 2 to 3 kilometers larger than its mean value of 3,476 km. This distortion into a slightly oblong sphere creates enormous stresses in the lunar crust. The only way to relieve the accumulated stress is through moonquakes. Seismometers left on the lunar surface by the Apollo astronauts record peaks in activity at each full and new Moon, when Earth, the Sun, and the Moon line up and the tidal forces of the first two combine. The same tidal forces that create a bulge at the Moon's equator have also slowed down the lunar rotation rate and synchronized its spin with its cyclical revolution around Earth. The end result was that the Moon shows only half its surface to Earth's inhabitants.[6]

Curiosity got the better of space scientists. In the 1960s, they sent space probes, part of the Lunar Orbiter series, to wrest the secrets of Moon's hidden side. They were in for a big surprise. The landscape they discovered was rather different from what they can see from Earth. Everyone expected craters separated by large seas of solidified lava. The dense jumble of craters was indeed there, silent witnesses to the era of intense asteroid pelting during the formation of the solar system, but there was no sign of extended solid lava seas, and that remains something of a puzzle. Some astronomers speculate that the Moon's mantle may be thicker on the far side—so thick, in fact, that rocky bolides crashing in were unable to pierce through it, which prevented lava in the interior of the young Moon from making its way back up to the surface.

THE NAUTILUS AND THE MOON

The tidal forces between the Moon and Earth have not only distorted the lunar shape, they have also gradually slowed down the rate at which the Moon revolves around Earth. Compelling evidence for this is provided by a marine organism called "nautilus." This mollusk is known for the elegant design of its shell, shaped like a perfect spiral divided into compartments by transverse partitions. This remarkable creature does not occupy all of the

Figure 15. *Earth viewed from the Moon.* Our white and blue-colored planet is far more hospitable than the arid and desolate lunar surface. Because the Moon takes exactly the same amount of time to spin once around itself as it does to complete one orbit around the Earth (gravitational forces between the Earth and the Moon are responsible for this synchronism), the Moon always turns the same side toward the Earth. From the vantage point of their planet, earthlings get to see only 59 percent of the lunar surface. (Photo courtesy of NASA)

shell, only the outermost portion. The shell has a very unusual growth pattern. Like a mason laying a new row of bricks per day, the nautilus adds each and every day a new layer to its shell and marks it by a striation. At the end of each month, when the Moon has completed one revolution around Earth, the nautilus has secreted thirty striations. At that point, it abandons its current compartment, closes it up with a partition, and moves into a new one. As such, the nautilus's shell provides a detailed record that allows us to reconstruct the evolution of the Moon's cycle through the course of time. If one studies ancient fossilized nautiluses, a stunning fact comes to light: The number of striations decreases in proportion to the age of the fossil. Instead of the thirty bands found on present-day nautiluses thriving in the deep waters of the South Pacific, there are only seventeen of them in 2.8-billion-year-old fossils. These ancient animals tell us that the Moon used to complete a revolution around Earth faster than it does today. Instead of the 29.5 days needed at present, it took only 29.1 days 45 million years earlier, and a scant seventeen days 2.8 billion years ago.

As the Moon's orbital motion slows down, the orbit itself becomes wider and the Moon gradually increases its distance from the earth. That has been confirmed by aiming laser beams fired from Earth at retroreflectors deployed on the lunar surface by astronauts. Such beams allow us to measure the Earth–Moon distance with extraordinary accuracy. All it takes is to time the round-trip duration, multiply it by the speed of light, and divide the result by two. Such measurements show that the Moon is spiraling away at a rate of 3.5 centimeters (cm) per year. By playing the film backward, we conclude that the Moon started out much closer to Earth when it originally formed.

As it gradually moves away, the Moon will look increasingly smaller, since its angular size is inversely proportional to its distance from Earth. At the present time, its angular size happens, by a curious coincidence, to match that of the Sun (roughly 0.5 degree), which gives rise to the magical sight of total solar eclipses when the Moon slips between Earth and the Sun and blocks the Sun's rays. Future generations will be deprived of the awesome spectacle of darkness descending in the middle of the day. Since the Moon will eventually become too small to obscure the entire solar disk, our descendants will have to content themselves with partial solar eclipses.

THE DAYS ARE GETTING LONGER

As Earth slows the Moon's orbital motion, so does the Moon slow down Earth as well. The days have been getting longer ever since Earth was born. The Moon accomplishes its braking action through the tides it creates. The ebb and flow of the oceans lifted by the Moon result in friction with Earth's crust. Whenever there is friction, heat is being generated and energy is being lost. To appreciate that, you need only touch the brakes of your bicycle just after you have had to stop in a panic to avoid colliding with a car. As a result, Earth gradually loses its rotation energy, it spins more slowly, and the time required for a full turn becomes longer and longer. But this is hardly cause for celebration for the hyperactive among us who keep complaining that there just isn't enough time in a day to accomplish everything. True, the days are getting longer, but they do so at a snail's pace. Someone fortunate enough to live 100 years will gain no more than 0.002 second during his or her lifetime. But on geological time scales, the cumulative effect becomes quite noticeable. For instance, 350 million years ago, a day was only 22 hours long. A few billion years farther back still, Earth used to spin four times faster than today. The Sun would hurry through its daily routine in a dizzying 3 hours. Our descendants, on the other hand, will experience increasingly longer days. The monthly cycle will also lengthen. As the Moon continues to drift away from Earth, it will take more and more time to accomplish one revolution around Earth. Days will lengthen relatively faster than months, and in about 10 billion years from now—5.5 billion years after the Sun will have exhausted its hydrogen fuel—they will have caught up to each other. Days and months will then be equal to 47 of our present days. At that point, the Moon will stop moving away from Earth. It will take exactly the same amount of time for Earth to rotate about itself as for the Moon to complete one revolution around Earth. The conditions will be very much the same as those experienced by the Moon today, as it turns around itself in precisely the same time it needs to revolve once around Earth. Just as the Moon now shows only about half its surface to Earth, Earth will eventually always present the same side toward the craters on the Moon.

A KILLER ASTEROID AND THE DINOSAURS

Each year, the sky is filled with swarms of birds migrating along mysterious but immutable routes. The soft glow of the Moon suffuses sleepy farmlands. Waves come topple the sand castles lovingly erected by children earlier in the day. As we explained before, all these events became possible because of enormously powerful collisions between the young Earth and massive objects that abounded in the early stages of the solar system, 4.6 billion years ago. Such collisions fall squarely in the domain of contingency and chance. Slight changes in orbit, and none of this would have happened. In that case, there would be no Moon, no tides, and no seasons. But the role of chance does not stop there. Modern science tells us that man's emergence is itself an accidental event.

Some 165 million years ago, dinosaurs were masters of Earth (Figure 16). Mammals, our direct ancestors, subsisted as best they could as small animals hiding in nooks and crannies to keep out of sight of voracious dinosaurs and assorted carnivorous monsters. But something happened some 65 million years ago, between the Cretaceous period and the Tertiary era, that would change everything. A celestial bolide, between 6 and 14 kilometers in diameter, and with a mass of some 10,000 billion tons, appeared in the sky. Larger than a mountain and slicing through the air at 25 kilometers per second, a hundred times faster than a bullet, the object plunged into the ocean on the edge of the Gulf of Mexico, disintegrating a second later against Earth's crust. The impact had the explosive force of a billion megatons of TNT, or five billion times the power of the bomb that destroyed Hiroshima. It

Figure 16. *The death of the dinosaurs.* Dinosaurs thrived on Earth about 100 million years ago. Toward the end of the Cretaceous period, 65 million years ago, they suddenly disappeared, together with about two-thirds of living species on Earth. Scientists believe that this massive slaughter was caused by the impact of a huge asteroid, weighing some 10,000 billion tons and about 10 km in size, which crashed into the Yucatán peninsula, in Mexico.

amounted to a million times the combined power of all the nuclear arsenals on the planet. Earth shook heavily from the violence of this cosmic impact. A 100-meter-high tidal wave raced across the Caribbean sea, ravaging Cuba, Florida, and the eastern coast of Mexico. The impact propelled more than a hundred thousand billion tons of vaporized rocks high into the atmosphere, leaving a giant crater 180 kilometers in diameter and more than 20 kilometers deep in Earth's crust. The vaporized rocky material cooled off in the atmosphere and began falling back down to Earth as hundreds of millions of small stones. A rain of gravel pelted Earth for the following hour. Atmospheric friction was such that the air became red hot. The nitrogen in the atmosphere began to combine with oxygen to form nitric acid, creating acid rain. Fires started to consume forests and spread rapidly over the entire planet. Most of the ejected rocks fell back down near the huge crater, but about 1 percent of the total material, amounting to tens of thousands of billion tons, remained suspended in the air for months on end in the form of a very fine dust cloud. The winds spread the dust particles all around the globe and a huge black cloud soon covered the entire Earth, blocking out the Sun's light and preventing its heat from warming the planet. The entire globe then entered a period of glacial winterlike night that lasted for several years. The temperature of entire continents plunged below freezing.[7] Photosynthesis, which nourishes plant life, came to a halt. The consequences of this severe darkening of the skies and this deluge of acid rain were devastating to both plants and animals. A total of 30 to 80 percent of plant species were wiped out. The disappearance of plants and trees in turn triggered the demise of two-thirds of living species, including the dinosaurs, which literally starved to death.

What is a catastrophe for some can turn out to be a blessing for others. The exit of the dinosaurs was a gift from the gods to our ancestral mammals. They managed to survive the disaster by foraging for grains and nuts buried underground. But now that their chief predators were no longer around, mammals began to proliferate and branch off into many new species. In just a few tens of millions years—an extremely short time in geological terms—cats, dogs, horses, and whales all made their entry onto the stage. More important, monkeys appeared, which would ultimately pave the way for *Homo sapiens*. In short, without an asteroid whose path accidentally crossed that of Earth some 65 million years ago, the dinosaurs might still be masters of the world, and mammals would probably have remained small nocturnal creatures hiding in unobtrusive corners. Under these circumstances, we would never have emerged. This once again illustrates vividly how chance and contingency can profoundly influence reality.

AN EXTRATERRESTRIAL METAL

This scenario, based on a rogue asteroid, is currently the most plausible hypothesis to account for the sudden disappearance of the dinosaurs. To be sure, other explanations have been proposed. Some invoke a period of intense volcanic activity that would have spewed so much soot and ash into the atmosphere that the Sun's light would have been blocked out and could no longer have reached the surface of the globe. The consequences would have been the same as in the case of the killer asteroid: A frigid cold would have settled worldwide and triggered mass extinctions. Others blame the explosion of a supernova, the cataclysmic death of a massive star in the vicinity of the solar system, which would have showered Earth with hazardous radiation and deadly particles. These alternate theories lack hard evidence, and none proved compelling enough to win the day.

By contrast, some important discoveries have buttressed the theory of the killer asteroid. In the process of looking for what wiped out the dinosaurs, the American physicist Luis Alvarez (1911–1988), his son Walter, a geologist, and their colleagues at the University of California at Berkeley were able to go back to the time the dinosaurs disappeared, 65 million years ago, by studying Earth's geological strata. The chronology of Earth's history is recorded in successive mineral layers deposited on its surface. The upper layers tell us the most recent history, while the deeper ones reflect the most distant past. The Alvarez team noticed that a particular layer of clay corresponding to the period when the dinosaurs became extinct, between the Cretaceous age and the Tertiary era, contains thin layers (a few centimeters thick) of particular dust and mineral grains that are known to form only during powerful explosions. Moreover, these same layers contain abnormally large amounts of a rare metal called iridium, which is not found naturally in Earth's crust but is present in high concentrations in asteroids and comets. This rare metal was detected not just near the presumed point of impact but over the entire planet. The team concluded that these deposits were of extraterrestrial origin, and that whatever event created them must have had global repercussions. That led to the idea of a deadly asteroid that came crashing into Earth and caused a huge explosion.

Another discovery came along to clear up what had long remained a mystery: Where did the asteroid hit? An object of that size striking Earth with such incredible force should have left a crater at least 180 km wide. Such a feature is believed to have been located off Mexico's Yucatán peninsula, between the Gulf of Mexico and the Caribbean sea. There indeed lies an enormous, half-buried crater named Chicxulub.

IS THE SKY ABOUT TO FALL IN?

The demise of the dinosaurs reminds us that bolides coming from deep in the sky can bring us unpleasant surprises. Could the calamity that befell the dinosaurs happen to us too? "By Toutatis, the sky is about to hit us over the head," keeps exclaiming Abraracourcix, the village chief in the popular French cartoon *Asterix*. Are such fears justified in light of what we know today?

The two greatest threats lurking out there in the skies are asteroids and comets. Asteroids are large chunks of rock generally orbiting the Sun between Mars and Jupiter, in a region astronomers call the "asteroid belt," 2 to 3.5 times farther out from the Sun than Earth is (Figure 1). With the exception of three asteroids whose diameters are known to exceed 300 km (roughly the distance between New York and Boston), and about a hundred others larger than 100 km (about half the distance between New York and Philadelphia), the vast majority of the hundreds of thousands of asteroids have a diameter less than 1 kilometer. If all objects in the asteroid belt were gathered into a single body, its diameter would barely reach 1,500 km, which is 2.3 times smaller than that of the Moon, and certainly much smaller than a planet. It is therefore highly unlikely that the asteroid belt is formed of the shattered remains of what used to be a planet. Rather, it is believed to be made of residual debris left over from the time of the formation of the solar system. Most asteroids remain safely confined to their normal orbits in the asteroid belt and do not come bother us. But once in a while, chaotic events (discussed in the next chapter) knock them off their usual path, out of the belt and toward Earth. Some of them will have trajectories crossing that of Earth, creating a definite risk of collision.

The other possible threat in the heavens comes from comets (Figure 17). These are enormous chunks of rock a few kilometers in size, covered

a

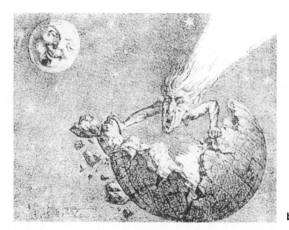

Figure 17. *Comets*. From time to time, they wander into the inner regions of the solar system and treat earthlings to a magnificent spectacle. A comet is a rocky body, a few kilometers in size, covered with ice. It can be compared to a large dirty snowball with a solid core and a large amount of dust. Originating either in the far reaches of the solar system (in Oort's comet cloud) or in closer regions (in Kuiper's belt), a comet is sent on a hyperbolic or parabolic orbit around the Sun by the gravitational nudge of a nearby star. As the comet approaches the Sun, its ice melts and vaporizes; the gaseous matter (primarily water vapor, with a small amount of carbon dioxide and monoxide), mixed in with dust grains, is pushed back by the wind of solar radiation and particles, creating those stunning comet tails that make for awesome sights. The comet grows larger and more brilliant as it gets closer to the Sun. In fact, as shown in the lithograph *(a)*, which depicts comet Donati above the Conciergerie in Paris on October 5, 1958, a comet typically develops two tails: a thin, straight tail composed of ionized atoms (atoms that have lost a few electrons and are positively charged), and a more diffused, slightly curved tail, composed primarily of neutral atoms mixed in with dust grains.

Before their nature came to be understood, comets used to inspire fright and superstitious beliefs among men; they were considered harbingers of great events and destruction. The cartoon *(b)*, published in 1857, depicts the destruction of Earth by a comet, while the Moon grins. The reproduction *(c)* shows a detail of the Bayeux Tapestry, made between 1073 and 1083, representing comet Halley during its 1066 visit, at the time of the conquest of England by William the Conqueror, Duke of Normandy.

with dust and dirty ice. They are sometimes described as "dirty snowballs" or "icy mudballs." For the most part, they remain safely parked in two regions. The first is located well beyond the solar system, at a distance roughly 50,000 times greater than the distance between Earth and the Sun. This region, containing an impressive 1,000 billion comets, is called the Oort cloud, after Jan Oort (1900–1992), a Dutch astronomer who postulated its existence. Much closer to us is another region of comets called the Kuiper belt, named after another Dutch astronomer, Gerard Kuiper (1905–1973), who discovered it in 1951. It is located near the outskirts of the solar system, at only 35 to 45 times Earth–Sun distance. It contains at least 100 million comets.

Most of the time, comets remained uneventfully in their places of origin. Occasionally, though, tranquillity is upset by the passage of one of the many clouds of gas and dust roaming intergalactic space or a nearby star that gives a few comets just enough of a gravitational nudge to send them on their way toward the solar system. As these comets approach the Sun, the increasing heat vaporizes the ice covering the rocky nucleus and creates spectacular comet tails that can stretch over millions of kilometers, and fill Earth's inhabitants with wonder.[8] The Hale-Bopp comet, which dazzled us with its brilliance in 1997, is a perfect example of such happenings. Nearly 200 comets have been identified that regularly cross the path of spaceship Earth. An eventual collision cannot be ruled out.

STONES FROM THE HEAVENS

The Earth has been struck numerous times in the past by these rocky monsters from space. Even today, about 300 tons of rocks and dust fall each day onto Earth's surface. This constant celestial rain is what causes abandoned towns to eventually be buried and disappear forever, at least until archaeologists excavate them to catch a glimpse of their ancient splendor. Fortunately for our health, most of the objects falling from the sky are no larger than a grain of sand by the time they hit the ground. In fact, Earth's atmosphere constitutes a shield that protects us from all rocky space debris with a mass smaller than 100,000 tons and a diameter less than a few tens of meters. Their friction with the atmosphere is so intense and the braking action of the air so extreme that these large bolides generally disintegrate into a shower of small pieces. These become red hot and actually begin to burn up. Those fiery asteroids, also called meteors, create spectacular bright streaks of light across the starry night sky that we call "shooting stars" (Figure 18). When the meteorite is a piece of comet debris made of ice and dirt, it disintegrates relatively quickly as it enters the upper atmosphere, more than 50 km up. On the other hand, a meteorite made of rock is better able to withstand the fiery reentry and can make it all the way down to the ground, where it can be recovered in the form of a charred rock (Figure 19). Each year in mid-August, clear summer nights are graced by a festival of shooting

Figure 18. *Comets, asteroids, and meteors: interplanetary wanderers.* Planets are not the only celestial objects populating the solar system. There are also asteroids, meteors, and comets, small bodies that once in a while come visit our corner of the solar system and bring us precious information about its early stages.

Earth gets bombarded by some 300 tons of celestial rocks and dust every day. When small asteroids, called meteors, penetrate the Earth's atmosphere, the friction against air molecules heats them, causing them to burn up and trace fiery lines across the starry sky, providing us with the magnificent spectacle of shooting stars. If they are large enough not to completely disintegrate, they fall to Earth in the form of charred rocks called "meteorites" (see Figure 19).

stars. The reason is that Earth then crosses the orbit of a comet whose rocky nucleus is slowly coming apart. It is the fiery death of the swarm of cometary debris that produces this marvelous light show.

For a long time the scientific community scoffed at the notion of stones falling from the sky. The venerable Academy of Sciences in Paris dismissed the possibility as late as the latter part of the eighteenth century. Only in the early nineteenth century did Jean-Baptiste Biot (1774–1862) succeed in proving the extraterrestrial origin of these scorched rocks. In 1803, rumors spread that a rain of stones had fallen on the village of L'Aigle in northwestern France. Biot was dispatched by the Academy to investigate. He examined hundreds of rock fragments scattered over tens of square kilometers and interviewed dozens of local peasants, carefully cross-checking their testimony. By conducting a genuine and rigorous scientific inquiry, he managed to assuage the skepticism of his colleagues and convinced them that these "stones falling from the sky" were for real.

None of these meteorites ever harmed anyone. There is not a single documented case of someone being wounded, let alone killed, by a rock from the sky. True, there are a few isolated reports of car tops transpierced, garage roofs damaged, or mailboxes punched through, but they do not amount to much. These events are called "encounters of the first kind." They involve celestial rocks smaller than 10 meters that are mostly consumed upon entry into Earth's atmosphere and rarely reach the ground. Even when they do, the impacts are benign and damages quite modest.

Figure 19. *Rocks falling from the sky*. This snapshot shows a variety of meteorites. In the upper left corner is a piece of the carbonaceous meteorite Allende, featuring white inclusions; it fell near Chihuahua, Mexico, in February 1969. Radioactive dating indicated an age of 4.6 billion years, which means that this meteorite constitutes a specimen of primitive planetary matter. In the upper right corner is a fragment of the iron meteorite responsible for the formation of Meteor Crater, in Arizona. The lower images show examples of stony meteorites. The one at left contains pretty crystals and iron inclusions with lighter colors.

EARTH'S SCARS

About 2 percent of asteroids measure between 10 and 100 meters. They are responsible for "encounters of the second kind." Fortunately for life on Earth, they manifest themselves much more seldom, at the rate of about one visit every few centuries. These asteroids are composed of stone or iron. A stone asteroid plunging into the atmosphere at high velocity (around 20 km per second) experiences such enormous stresses that it breaks apart and gets flattened like a pancake. It generally explodes before reaching ground level. The explosion is so powerful that a shock wave blows everything down over a radius of several kilometers. An iron asteroid, by contrast, generally does not disintegrate into small pieces as it traverses the atmosphere. It typically reaches the ground almost intact and causes considerably more damage. Earth displays several scars inflicted by such visitors from space. Over 150 impact craters have been recorded, most of them formed in the last 200 million years. That is not to say that stones were not falling from the sky before that. Earth was being bombarded just as regularly then, but erosion caused by rain, rivers, and oceans, as well as the movement of tectonic plates and changes of Earth's surface due to human activity, have had ample time to erase wounds older than 200 million years. The majority of impact craters have been located in North America, eastern Europe, and Australia. It is not that asteroids have a particular preference for these places, but that their surface area is large, they are geologically stable, and that is where searches happen to have been conducted the most systematically. Thousands of craters remain to be discovered in the depths of oceans.

Of all the asteroids that have struck Earth, the one that exploded over Tunguska, in the Siberian tundra, has generated by far the most controversy because of the mystery shrouding it. On the morning of June 30, 1908, a rocky object weighing 100,000 tons and measuring some 50 meters across plunged into the atmosphere above the Tunguska River before totally disintegrating at an altitude of about 10 km. The gigantic explosion was heard for thousands of kilometers around, decimated entire herds of deer, and flattened the forest in a radius of 30 kilometers. People living in nearby villages were suddenly awakened by a blast that must have sounded like the end of the world. Running to their windows, they saw an immense fireball. The energy released was equivalent to 15 megatons of TNT, or 75 times more than the bomb dropped on Hiroshima. In the ensuing days, the sky glowed so red that, according to some accounts, people as far away as western Europe could read a newspaper at night without any supplemental lighting. The effect was as spectacular as a volcanic eruption. Because the asteroid exploded before hitting the ground, there were neither impact crater nor asteroid fragments. The only sign that something unusually violent had happened was the devastated forest, with flattened and charred tree trunks as far as the eye could see (Figure 20).

This scene of destruction, combined with the absence of any visible impact crater, inspired the most outlandish hypotheses. Believers in extraterrestrial civilizations insisted that a flying saucer had landed and that its engines had scorched the forest. Today, the theory of an exploding asteroid is widely accepted. Scientists have examined the trees around Tunguska, silent witnesses to the cataclysm, as they try to slowly recover from the event. Their bark appears to contain numerous microscopic particles with the unusually high concentration of copper and/or nickel typically found in asteroids. It seems, then, that the culprit was indeed an asteroid composed of either stone or iron. Iron appears unlikely in this case, because it is too dense and resistant a material, and does not break apart readily. An iron asteroid would almost certainly have made it to the ground in one piece, which would have caused even more serious damage than charred forests and wiped-out deer herds. The best indications are that it was in fact a stone asteroid, which entered the atmosphere at a 45-degree angle and detonated about 10 kilometers above ground level. The explosion would have projected vast quantities of dust (estimated in the millions of tons) to such a high altitude that it could reflect the light from the Sun long after it had set below the horizon. That would explain the red glow of the sky (dust grains

Figure 20. *The Tunguska event.* On June 30, 1908, a spectacular explosion occurred in the region of Tunguska, in Siberia. Heard at distances of more than 1,000 km, it ejected millions of tons of dust into Earth's atmosphere. Every tree was flattened within a radius of 30 km. Decades later, the forest had not yet recovered, as shown by this photograph taken twenty-one years after the event. Scientists believe that the destruction was caused by a stony asteroid, 50 meters in diameter, exploding 10 kilometers above ground.

have the ability to absorb the blue portion of the solar spectrum and transmit the red), which reportedly enabled western Europeans to read at night. In the end, the Tunguska asteroid perhaps caused more fear than harm. It was, nonetheless, a spectacular explosion in a remote place of the world, and it left little trace besides a razed and scorched forest.

Another asteroid of the second kind, which left a far more visible scar on the ground, is the one that fell at Meteor Crater in the Arizona desert (Figure 21). This one was definitely made of iron, a much tougher material. As a result, it made it through the entire atmosphere practically intact, retaining much of its original size of roughly 50 meters across. It crashed some 50,000 years ago with an energy of 15 megatons (similar to the Tunguska asteroid), creating an enormous cavity 1.5 kilometers in diameter. Earth trembled and debris was thrown as far as several tens of kilometers from the location of the impact.

Can we sleep in peace knowing that the sky is full of such threats? Statistics tell us not to worry too much. On average, a collision with a rocky asteroid of the Tunguska type occurs somewhere on Earth only once every few centuries. Even if such an event did take place, the asteroid would very likely touch down in one of the oceans that cover two-thirds of Earth's surface, or it might hit one of the vast uninhabited regions of the continents. An asteroid larger than 100 meters falling in the ocean would trigger a huge tidal wave hundreds of meters high that would come crashing on land with unimaginable force, destroying coastal cities. Damage would be consider-

Figure 21. *Meteor Crater, in Arizona.* This is one the best-preserved impact craters on Earth (erosion tends to erase craters on Earth, unlike the situation on the Moon and Mercury, where there is no atmosphere and, therefore, no rain and erosion). Meteor Crater is more than one kilometer in diameter; it was formed 50,000 years ago when a roughly 50-meter-wide iron asteroid hit the ground, releasing an energy of 15 megatons. (Photo courtesy of NASA)

able, but it would remain largely localized and would not affect the entire planet. Even if by a stroke of bad luck the asteroid were to hit a populated area, the repercussions would be confined to a radius of a few tens of kilometers. There would be no climatic or environmental consequences on a global scale. It may not be much consolation if you happen to live in the area that is hit, but 99.999 percent of the world's population would remain unaffected. In any event, there is no mention of such an event anywhere since history began to be recorded. Your insurance agent would come out on top if he sold you a policy protecting you against being hit over the head by a Tunguska-type asteroid. The chances of being killed in such an accident are less than one in ten million per year.

Iron objects of the Meteor Crater type can cause far more damage, but they are much rarer still. Ferrous meteorites constitute less than 5 percent of all matter falling to Earth.[9] They strike Earth no more often than once every few tens of thousands of years. The annual probability of getting hit by one over the head is smaller than one in a billion.

WILL WE SUFFER THE SAME FATE AS THE DINOSAURS?

We have seen that life on Earth is not threatened by asteroids of the second kind, whose size is less than 100 meters, such as the ones that hit at Tunguska and Meteor Crater. But what does the future hold when it comes to "encounters of the third kind," involving objects the size of a mountain, between 1 and 10 kilometers across? Unlike their smaller counterparts, asteroids this big can cause damage on a global scale. The whole planet would be affected. The energy released by a space rock 1 kilometer in diameter is comparable to the explosion of a 1-million-megaton bomb, which is 10,000 times more than the most powerful thermonuclear devices. It would take the simultaneous explosion of all the bombs stored in every nation's arsenal to cause the same amount of damage. In the early 1980s, the topic of "nuclear winter" made the front pages of newspapers. The global effects on our civilization of a worldwide nuclear war were described in their gory details in the press. The scenario described then was not too different from what would happen following an encounter of the third kind with a 1-kilometer-wide asteroid. On average, one such asteroid hits our planet once every 250,000 years. The impact would spew vast quantities of dust into the upper atmosphere. Smoke from innumerable forest fires ignited by the collision would compound the problem. The mixture would act as an opaque screen blocking the Sun's light and heat for many months. The globe would cool by a few degrees per month over a period of several years. A long cold night would descend on Earth. Photosynthesis by plants and trees would come to a stop. The food chain would be disrupted. Worldwide production of grains and rice would plummet. Famine and epidemics would become rampant. A billion people or more would die of starvation and disease. Few nations could cope with such a calamity. Health, political, and economic infrastructures painstakingly put in place by society would collapse and our entire civilization would teeter on the edge of destruction. The probability of such an event occurring in any given year is of the order of one in a million. Since our life expectancy is roughly 100 years, we are faced with a probability of 1 in 10,000 over a lifetime. Your insurance agent will tell you that this is 100 times less than the chances of dying in an automobile accident, 200 times less than the chances of being killed by a firearm in the United States, but more than the combined probability of becoming the victim of an earthquake, a hurricane, and a volcanic eruption.

We now come to the most devastating encounters, those of the "fourth kind." They are caused by celestial bolides with a size of 10 kilometers or more (Figure 22). We know that such massive monsters visit Earth from time to time. A 15-kilometer asteroid did in the dinosaurs some 65 million years ago. Luckily for us, bolides of this size are far less common in space and their encounters with Earth much less frequent. On average, an asteroid or comet 10 kilometers in size would hit Earth only once every 10 million years. Increase the size to 15 kilometers, and the frequency drops to once in 100 million years. Improbable as it may be, an encounter of the fourth kind would have catastrophic results on the evolution of life on Earth. Civilization would not teeter anymore. It would simply vanish.

A 10-kilometer-wide bolide streaking at a speed of 20 km/s would not even notice the atmosphere. It would go right through it in one or two seconds and explode on impact with a power of 1 billion megatons, or a thou-

Figure 22. *Three asteroids.* They were photographed in 1995 by the space probe Galileo on its way to Jupiter. At top is a picture of Gaspra, an object in the asteroid belt, photographed from a distance of 1,600 km with a resolution of 100 m. Gaspra is shaped like a potato measuring 16 km x 11 km x 10 km; it probably is the remnant of a catastrophic collision between two large asteroids in the belt. The lower photographs show, on the same scale, the two moons of Mars: Deimos *(at left),* measuring 16 x 12 x 10 km, and Phobos *(at right),* measuring 28 x 23 x 20 km. These moons are probably asteroids captured by Mars's gravity. These three bodies are not massive enough for gravity to sculpt them into spheres. They are peppered with impact craters produced by collisions with smaller asteroids. If an asteroid similar in size to these three were to collide with Earth, it would have catastrophic consequences for life on our planet. (Photo courtesy of NASA)

sand times the combined power of all the nuclear weaponry stockpiled across the globe. A crater 100 kilometer in diameter (roughly half the distance from New York to Philadelphia) would be gouged in Earth's crust. Red-hot material would be thrust so high in the air as to go into orbit. Incandescent debris would rain down on the planet, starting destructive fires everywhere. Entire continents would be set ablaze. Smoke and soot would mix with dust in the atmosphere, obscuring the Sun for years. Temperatures would drop 10 to 20 degrees. It would be freezing in summertime, and a long dark winter would spread worldwide. Toxic and acid rains (fires produce enormous amounts of carbon that react with the atmosphere to form toxic substances) would flood the planet. The initial period of cold and darkness would be followed by one of torrid heat caused by the greenhouse effect of carbon dioxide in the atmosphere.

A collision of the fourth kind would have truly global consequences. We have seen how a 15-kilometer-wide asteroid triggered a long winterlike night and snuffed out two-thirds of the species living during the transition from the Cretaceous to the Tertiary. Dinosaurs as well as ammonites and belemnites "suddenly" vanished. This event was not unique. Studying fossils encrusted in geological deposits of various ages, paleontologists have found evidence of other abrupt changes in the landscape of terrestrial life. Mass extinctions appear to have occurred roughly every 40 million years. In the 250 million years since the end of the Permian epoch, there might have been as many as six large-scale cataclysms. The saga of life on Earth has had to be restarted many times, only to be interrupted again. Could it be that these interruptions were all caused by deadly bolides of the fourth kind?

PROTECTING EARTH

Is this too pessimistic a view of the long-term evolution of life on planet Earth? Perhaps it is. At any rate, we know of the existence of at least a hundred asteroids with a diameter larger than 1 kilometer whose orbits cross that of Earth (Figure 23). There are probably two thousand more such asteroids that we have never detected. The risks of collision are therefore quite real. In January 1989, a rocky object roughly 5 kilometers in size and named,

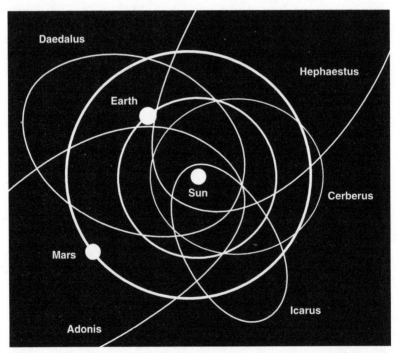

Figure 23. *Asteroids crossing Earth's orbit.* The orbits of some asteroids cross that of Earth, creating risks of collisions. This figure depicts five of them, named after five characters in Greek mythology: Adonis, Cerberus, Deadalus, Hephaestus, and Icarus. Approximately one hundred such high-risk objects have been identified.

appropriately enough, Toutatis, came within 15 million kilometers of Earth—almost nothing on an astronomical scale. Toutatis is due back in our neighborhood in September 2000. Another scheduled visitor is the comet Swift-Tuttle, whose solid core is at least 5 kilometers wide. Its orbit brings it so close to Earth that there is a non-zero probability that it will hit us broadside during its next visit in August 2126.

Is there anything we could do if an astronomer were to announce tomorrow that he had just spotted an asteroid or a comet heading straight toward Earth? Unlike other natural disasters such as earthquakes, hurricanes, and volcano eruptions, against which we are totally helpless, mankind now has the necessary technology to prevent the sky from falling in on us. If we decided to do so, we could change the trajectory of a bolide as large as a mountain. A threatening asteroid can be detected decades in advance. That would leave us plenty of time to prepare to send a nuclear bomb aboard a rocket. The idea would be to detonate it in the immediate vicinity of the threatening body and knock it slightly off course. A change in the velocity of the bolide of just a few centimeters per second, which amounts to less than one part in a million, would be enough to do the job. The trick would be not to shatter the threatening asteroid in a thousand fragments, in which case we would have to deal with a multitude of objects aimed at Earth instead of just one.

The case of a comet speeding toward us is more complicated. Because it comes from so far away—beyond the solar system (we have seen that a comet nursery called the Oort cloud lives at a distance some 50,000 times Earth–Sun distance)—and because it can be seen only when it is close to the Sun, when enough ice has been vaporized, a potentially dangerous comet cannot be detected much more than a year before it might crash into us. Any corrective action could then take place only when the comet is already much closer to Earth, which would require a correspondingly larger change in velocity, and hence a far more powerful nuclear bomb. To make matters worse, a comet, unlike an asteroid, is not subjected to just the gravitation forces exerted by the Sun and the nine planets. As ice is vaporized off the comet's nucleus by the solar heat, the comet changes course ever so slightly, which makes it extremely difficult to calculate its trajectory with any accuracy. What this means is that, by the time we should decide to go ahead and intercept the comet to deflect it, we still would not be certain whether it presents a real danger or not.

Fortunately, threatening comets are far less numerous than potentially dangerous asteroids. They make up less than 10 percent of the total. Needless to say, monitoring bolides, evaluating the risk they pose, and mounting a mission to deflect them if necessary, requires extensive funding. As always, it is the politicians who control the purse strings. They tend to drag their feet, unconvinced that the danger is serious. And who can blame them? When problems of crime, drugs, and poverty keep nipping at you all day long, whether the sky is about to fall on your head seems rather esoteric. Yet things changed dramatically in July 1994, when the entire world was able to watch live on television an encounter of the third kind. It happened not on Earth but on Jupiter, the giant planet of the solar system.

JUPITER'S WOUNDS

Earth is not the only planet in the solar system to be the target of murderous bolides. With its large mass and enormous gravity, Jupiter, too, attracts many undesirables. The comet Shoemaker-Levy 9 plunged into the dense Jovian atmosphere in July 1994. It did not show up as a single piece. During its previous approach, in July 1992, it came dangerously close to the giant planet without actually hitting it. But Jupiter's tremendous gravitational forces tore the rocky nucleus of the comet apart into some twenty pieces, the largest one about 1 kilometer across, which then continued on in a procession following the same orbit (Figure 24). The broken pieces of the comet lined up like a pearl necklace stretching over one million kilometers, or three times Earth–Moon distance. They crashed into the atmosphere of hydrogen and helium, one after the other, at a speed of 60 km/s over a period of several days. The impact of Shoemaker-Levy 9 with Jupiter was the first planetary catastrophe to be seen live on television screens. Both ground- and space-based telescopes (such as the Hubble) were all trained on this extraordinary encounter of the third kind. They captured images that were almost instantly relayed across the entire world.

In just a few days, Earth's inhabitants were able to see the damage caused by some twenty crashes into the layer of thick clouds surrounding the giant planet. Lightning, explosive mushrooms, convection patterns, and gaseous jets attested to the violence of each impact. Within a few minutes, a bright fireball emerged from each impact zone, heating the surrounding gas to thousands of degrees. The energy released was equivalent to that of five billion Hiroshima-type bombs, for a total of 1 billion megatons. Each individual hit was as powerful as the one that caused the extinction of the dinosaurs 65 million years ago. The planet shook and its interior kept on ringing for days after each collision. Giant wounds, as large as Earth, formed in Jupiter's upper gas layers. They were dark colored and looked much like a string of black eyes. The planet eerily resembled the battered face of a boxer who has just been pummeled by too many blows from his opponent (Figure 25). Multiple sprays of comet debris were ejected up into space before they fell back down into Jupiter's atmosphere. All these wounds will take a long time to heal. Months, perhaps years, will pass before the scars fade away and the planet's atmosphere returns to normal. Some of the comet fragments will be carried around Jupiter's atmosphere by the fierce Jovian winds for years before they finally dissipate.

A COSMIC WATCH

Seeing this planetary collision live awoke politicians to the reality of threats lurking in space. In August 1994, only weeks after the last piece of the comet

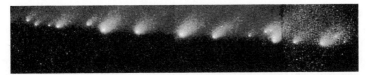

Figure 24. *Fragments of comet Shoemaker-Levy 9.* A comet loses matter every time it swings by the Sun because the ice covering its core evaporates. After hundreds of passages, the comet has lost so much matter that its rocky core can no longer withstand the gravitational forces of the Sun and the planets and disintegrates into several pieces. That is what happened to comet Shoemaker-Levy 9, which split into 21 fragments in 1992 as it flew by Jupiter. This photograph, taken in May 1994 by the Hubble space telescope, shows the 21 fragments following each other along the comet's orbit, and stretching over a distance of 1.1 million kilometers. (Photo courtesy of NASA)

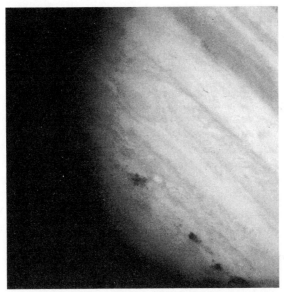

Figure 25. *Pieces of comet Shoemaker-Levy 9 plunging into the Jovian atmosphere.* In July 1994, the whole world was able to watch live on television and on the Internet the separate pieces of the comet crashing into Jupiter, releasing an energy equivalent to millions of megatons of TNT. Each fragment plunged into the Jovian atmosphere at a velocity of about 60 km/s and disintegrated in a violent explosion that ejected huge cometary dust clouds into the atmosphere above. This photograph, taken by the Hubble space telescope, shows three earth-size black spots in Jupiter's atmosphere, indicating the impact of three separate fragments. The black color is probably due to organic molecules in the comet's fragments, or to shock waves created by the impact. (Photo courtesy of NASA)

Shoemaker-Levy 9 had crashed into Jupiter, the U.S. Congress asked NASA to come up with ways to identify objects likely to cross Earth's orbit. The specific mission was to conceive a program of cosmic watch, dubbed Spaceguard, using a worldwide network of telescopes monitoring the skies and capable of cataloguing all celestial bodies posing a potential threat to Earth. The network is supposed to track any object with a diameter greater than 1 kilometer at risk of crossing our own orbit within the next ten years.

For the time being, the Spaceguard program remains a proposal. Some scientists are of the opinion that Congress erred on the side of optimism. With a network of six telescopes 2 to 3 meters in diameter, spread evenly around the globe, and equipped with the most advanced and sensitive light detectors, it will take not ten, but twenty-five years to complete a reasonably accurate census of potentially threatening objects. The network would scan the entire sky twenty-four hours a day, looking for nearby celestial objects moving rapidly against the background of the more distant fixed stars.

Several problems stand in the way, not only technological or scientific, but political and military as well. The political aspect of the issue involves the financing and the need for international cooperation. It would be unthinkable for a single country to deal with a problem that clearly affects all nations of the globe. The fate of human civilization should be a matter of concern for everyone. The military aspect has to do with necessary checks to ensure that an effort intended to protect Earth does not turn into a pretext to develop war-fighting technology.

COMETS BRING LIFE

We have seen how celestial rocks colliding with Earth have fundamentally altered the fabric of reality by introducing sudden changes into the biosphere. To fashion complexity, Nature resorts to every means at her disposal. Through the bias of rocky bolides, she used contingency to give us the cycle of seasons, the soft glow of the Moon, and to allow our ancestral mammals to flourish. She may even have taken advantage of these stones fallen from the sky to introduce life itself on Earth.

To see how this came about, we must once again go back 4.6 billion years, to the time when the solar system was forming—more precisely, to when the planetesimals had just completed their accretion phase and given birth to the brand-new Sun and its court of planets. Early Earth did not include a single drop of water. Yet today the oceans cover two-thirds of the planet's surface. How can we explain this mystery? Comets may hold the answer. Made of a mixture of ice and dust, they are chock-full of water. During the first billion years of the solar system, the newly formed planets were being constantly bombarded by comets and asteroids. The cratered surfaces of the Moon (Figure 8) and Mercury (Figure 9) provide vivid testimony of this process. During this period of great turmoil, comets are believed to have ferried to Earth many times the amount of water contained in today's oceans. As a bonus, they also brought quantities of silicates from their rocky nuclei that built up Earth's crust, as well as a second-generation atmosphere.[10]

Comets and asteroids might have played an even more important role. These objects from outer space may well have started life on Earth. Some scientists believe that they seeded the oceans with organic substances, such as amino acids; these can readily assemble themselves in long chains, producing proteins and DNA—the "building blocks" of life. These organic materials may have been dumped from the sky at a rate as high as 10,000 tons per year, accumulating in a layer more than 1 kilometer thick. This hypothesis was inspired by the discovery of many organic substances in recovered celestial rocky objects. The meteorite that fell in 1969 near Murchison, Australia, contains more than 400 such substances, including several organic compounds not found on Earth. The analysis of comet nuclei shows that they are made half of dust and ice, and half of silicate rocks and organic matter. The organic compounds are believed to have formed out of the many molecules (numbering nearly 90) existing in interstellar space. Astronomers began to discover them in the latter part of the 1960s and at first were completely stunned. No one had thought it possible that the frigid temperatures (-260°C) and nearly total vacuum of interstellar space could be conducive to the formation of a variety of molecules such as carbon monoxide (CO), water (H_2O), methane (CH_4), ammonia (NH_3), and others mixed in with an abundance of hydrogen. But Nature proved herself quite creative in such an inhospitable environment. True, no one has ever found amino acids in interstellar space, but plenty of precursors have been detected, including hydrogen cyanide (HCN) and ammonia (NH_3). By reacting with water coming from comets and asteroids, these substances can produce amino acids.

Thus, it looks like objects from the sky did far more than enable the proliferation of our early mammal ancestors by causing the extinction of their fierce predators, the dinosaurs. By delivering the water and gases necessary to our oceans and our atmosphere, they are also responsible for the crucial ingredients of our biosphere. They are the link between the immensity of the cosmos and spaceship Earth. We are all made of atoms manufactured in the nuclear alchemy taking place inside massive stars, and ejected into interstellar space when these stars exploded in their violent death. As true messengers from space, comets and asteroids picked up this star dust and deposited the seed of life on our beloved planet.

DETERMINISM AND FORTUITY

We have seen that completely fortuitous and unpredictable events in the heavens can profoundly influence our everyday lives. Contrary to the laws of physics, such events are dictated not by necessity but by pure chance and serendipity. On every level, reality is built through the combined actions of what is determined and what is undetermined, through the interplay between chance and necessity. In the context of the solar system, physical theory—embodied in Newton's universal gravitation law and the laws governing the behavior of gases—could have predicted the formation of the Sun as a consequence of the collapse of the protosolar nebula, and the building of planets as a result of the accretion of planetesimals. It could have told us ahead of time that most planets would be orbiting in the Sun's equatorial plane. Theory alone could foretell that planets would rotate about themselves and around the Sun in the same direction as the Sun itself, from west to east. That was preordained by the original rotation of the primordial solar nebula. But theory could never have predicted the ultimate number of planets (why nine rather than five?). No physical law fixed a priori what the inclination angle of Earth should be. The fact that a rocky bolide tore the Moon loose from Earth's crust, giving poets and lovers their muse, was purely accidental and in no way predetermined. Chance exists in its own right at every level of reality.

Fortuity and necessity are both indispensable tools in Nature's panoply. They are complementary colors in her palette. The laws and constants of physics were fixed since the early moments of the universe, and steer it toward ever-increasing complexity. In the course of 15 billion years, Nature began with an energy-filled vacuum and successively produced elementary particles, atoms, molecules, DNA chains, bacteria, and all living beings, including man. On this huge canvas determined by physical laws, Nature learned to take advantage of fortuity to invent and create complexity. Randomness provided her with the freedom required to innovate, to broaden the realm of the possible defined by the often too restrictive laws of physics, and to satisfy her yearning for variety and complexity. Everything is put to good use: chance and necessity, random events and deterministic laws. That is why reality will never be completely described in terms of just the laws of physics. Contingency and accidents will forever preclude a comprehensive explanation of the world. To account for the advent of man, we may invoke the asteroid that appeared in the sky 65 million years ago and struck Earth, killing all the dinosaurs. But we will never be able to explain why it showed up at that precise moment. To account for the beauty of spring flowers, we may invoke a collision of Earth with an asteroid that tilted our planet, but we would be at a loss to predict that the impact would tilt it by just 23.5 degrees, rather than cause it to lean completely on its side like Uranus, which would have given us 6-month-long instead of 12-hour-long days and nights.

Chance is not the only agent that contributed to freeing Nature. The laws of physics themselves lost much of their determinism. With the development of chaos theory, randomness entered the macroscopic world in force.

« 3 »

Chaos in the Cosmic Machinery, and Uncertainty in Determinism

THE END OF CERTAINTY

The seemingly impregnable walls of certainty that had protected the fortress of Newtonian physics began to show cracks early in the twentieth century. They would eventually crumble one after the other. Einstein introduced relativity theory in 1905, which tossed out the certainty of absolute space and time. In the 1920s and 1930s, quantum mechanics destroyed any hope of measuring everything with any arbitrary accuracy. The velocity and position of an elementary particle could no longer be determined simultaneously with unlimited precision. The last bastion of certainty collapsed at the end of the century: The emerging field of chaos eliminated once and for all the Newtonian and Laplacian tenet of Nature's unconditional determinism. Before the advent of chaos, the operative word was *order*. The word *disorder* was anathema and banned from the language of science. Anything apt to exhibit irregularity or disorder was considered a monstrosity. The science of chaos changed all that. It introduced irregularity in regularity, disorder in order. It captured the imagination not only of scientists but also of the public at large, because chaos theory deals with objects on a human scale and speaks to everyday experiences.

Relativity theory applied to the world of the infinitely large, the world of "black holes," galaxies—indeed of the entire universe. Quantum mechanics concerned itself with the other extreme, the world of the infinitely small, things like electrons, atoms, and molecules. Chaos, on the other hand, projects an air of familiarity. Who among us has never complained of "chaos" in our lives? Chaos deals with daily experiences: smoke volutes rising from a cigarette, a flag flapping in the wind, endless traffic jams on a highway, or even water drops dripping from a leaky faucet. Chaos theory turns ordinary things into legitimate objects of study.

The science of chaos is seductive because it is global and topples barriers between various disciplines. It unites researchers in different fields and works against the tendency toward excessive specialization that seems so pervasive in modern research. It is appealing because it eradicates the last bastion of determinism and restores free will to its rightful dominance. Besides, it is a "holistic" science that considers the whole and forces reductionism into retreat. The world can no longer be explained only in terms of its constituents (quarks, chromosomes, or neurons); it must be apprehended in its totality.

Yet, for all its allure, the field of chaos did not really take off until the 1970s. It did so thanks to an unexpected ally—the computer. That tool came to be just as essential to the study of chaotic systems as the microscope was to microbes, the particle accelerator to the subatomic world, and the telescope to the exploration of deep space.

Late as it was in emerging as a bona fide topic of study, the science of chaos counted some illustrious pioneers. One of them was the French mathematician Henri Poincaré (1854–1912), who began rebelling against the dictatorship of Newtonian determinism as early as the end of the nineteenth century.

DETERMINISM AND THE TENNIS BALL

Picture yourself playing tennis. You are about to return the ball speeding toward you. One stroke of the racket, and the ball flies back across the net. A physicist could tell you exactly where the ball will land on the court if he has access to two pieces of information. First he needs to know the initial conditions—that is, the precise spot where the ball was struck, and the speed with which it began its return trip. Second, he must have a knowledge of the laws of physics, in this case Newton's law of universal gravitation. The law in question is quite well known and has been verified in great detail in the laboratory, both in the context of ordinary life (the way a ripe apple falls to the ground) and on the scale of the entire solar system (through the motions of planets). In principle, it is possible to determine the initial conditions fairly accurately—for instance, by filming the action with a video camera. The tape would show where in space the racket struck the ball. To determine the moment of impact, all that is needed is a good stopwatch. Finally, the velocity can be calculated by looking at two successive frames of the tape: One can measure the distance the ball has traveled and divide it by the time interval between the two frames. All these initial conditions can be known only with varying degrees of accuracy. For instance, the ball's position may be determined with an error of a few centimeters. Likewise, the instant when the ball is struck may suffer an uncertainty of a fraction of a second. Does this lack of precision mean that the future behavior of the tennis ball cannot be specified—in other words, that it is unpredictable? Not in the least: It simply implies similar uncertainties in our ability to predict where and when the ball is going to bounce off the ground. The motion of the tennis ball does not depend very sensitively on the initial conditions. That is what makes the game of tennis possible in the first place. If the path of the ball were unpredictable, if small changes in the initial conditions led to erratic trajectories, the ball might end up anywhere on the court and no rally could take place.

SMALL CAUSE AND LARGE EFFECT

As it turns out, there are many situations in Nature and in ordinary life that depend extremely critically on initial conditions. A minute change in the initial state of the system can produce huge changes in subsequent behavior. Mathematicians describe these situations by saying that the changes grow exponentially with time. That describes, for instance, any quantity that doubles in a given time interval, doubles again in the next, and so on. Let us take a specific example. Suppose that inflation goes out of control and reaches a rate of 100 percent, which means that money loses half its value and prices double every year. A loaf of bread costing $1 today will cost $2 next year, $4 the year after that, and the process keeps going. After ten years, the cost will have soared to $1,024, and to a shocking $1,048,576 after twenty years. This example helps to understand why countries with large inflation rates undergo social upheavals. That is precisely what happened in Germany in the 1920s, when suitcases chock-full of money were needed to buy a simple loaf of bread. We benefit from the same exponential growth when we deposit our money in a savings account. Assuming a constant interest rate of 5 percent, we will see our capital double in slightly over 14 years, quadruple in a little more than 28 years, and grow almost eightfold in a little more than 42 years. Needless to say, this neglects the IRS, which will go out of its way to reduce your gain by as much as it can.

Likewise, there exist physical systems that are extremely sensitive to initial conditions. A tiny perturbation completely alters the result. Unlike the case of a tennis ball, if the initial position or velocity are changed ever so slightly, the trajectory of the perturbed object, which starts out very similar to its unperturbed counterpart, will gradually deviate from it exponentially until the two no longer have anything in common with each other. That is what is called "chaos."

Your trustworthy dictionary will tell you that the word *chaos* describes "disorder," "general confusion." In truth, to a scientist, it does not necessarily mean "lack of order." Rather, it denotes indeterminacy, or impossibility to make long-term predictions. Because the final state depends so critically on the initial conditions, the smallest perturbation can have drastic repercussions. This fundamentally limits our ability to predict what the final outcome will be. We should remember that our knowledge of the initial state suffers irrevocably from some degree of uncertainty, no matter how small. In chaotic systems, this unavoidable inaccuracy gets amplified exponentially, preventing us from predicting the final state.

A PROPHET OF CHAOS

Poincaré was the first scientist to reflect on the sensitivity of certain systems to initial conditions, and to realize that in such systems a minute change early on leads to repercussions of such large magnitude that predicting the future becomes hopeless. Laplace had proclaimed his famous deterministic credo, stating that "for an intelligence that could embrace in a single formula the movements of the largest bodies in the universe and those of the smallest atoms, nothing would remain ambiguous, and the future as well as the past would be in plain view before it." Poincaré adopted a decidedly opposite stance, issuing a premonitory warning in his book *Science and Method*, published in 1908:

> A cause so small as to escape us can have a considerable effect which we cannot see; we then declare that the effect is due to chance. If we knew exactly the laws of Nature and the situation of the universe at the initial instant, we could predict exactly the situation of that same universe at any subsequent time. But even assuming that natural laws no longer held any secrets for us, we still would know the initial conditions only approximately. If that enables us to predict any subsequent situation with an uncertainty of a certain order, we need look no further and we declare the phenomenon to be predictable and governed by laws. But that is not always the case. There are situations when small differences in the initial conditions can produce very large ones in the final result; a small error on the former can lead to a huge error on the latter. In those cases, predictions become impossible.

Despite this warning, the science of chaos did not catch on immediately. Poincaré was too far ahead of his time. Moreover, computers had not yet appeared on the scene to enable mathematicians to extrapolate the behavior of systems sensitive to initial conditions far into the future, thereby confirming Poincaré's brilliant intuition. There matters stood for over half a century. The American meteorologist Edward Lorenz (1917–) ultimately picked up the torch, almost serendipitously, in 1961.

THE BUTTERFLY EFFECT

While working at the prestigious MIT, Lorenz was using a computer to study weather patterns. In the early 1960s, computers were a jumble of wires and vacuum tubes. They were bulky and not very pretty to look at. They were also notoriously unreliable. It was something of a miracle if they kept working for more than a week at a time. Even though Lorenz's computer filled an entire room, it had neither the speed nor the memory required to model the atmosphere and oceans realistically. Lorenz was forced to simplify the laws of meteorology to the extreme by describing the movements of air and water with very simple equations; it is, after all, the interactions between these two elements that make rain and sunshine. The computer's task was to regurgitate weather bulletins. One such bulletin would predict westerly winds suddenly shifting to the north. Another would warn of the birth or death of hurricanes. In yet others, heat waves would develop and the air pressure would suddenly drop. The computer would communicate its verdicts by printing wiggly curves showing how a particular variable—for instance, the speed of the northerly wind—varied as a function of time. In calm weather, the curve would be shaped like a valley, and it would exhibit peaks when wind gusts picked up.

On a winter day in 1961, Lorenz needed to finish a computer run interrupted prematurely. To save time, he decided to resume in midstream rather than start all over again from the beginning. Lorenz got the run going and went to get a cup of coffee. When he returned, a completely startling result greeted him, which was to give birth to a new science—chaos.

Lorenz had expected the new curve, started halfway through the old run, to be an exact replica of the previous curve. To his amazement, that was not at all the case. The two curves indeed were fairly close together at first, but they soon diverged so rapidly that any similarity had completely vanished a few months into the model (Figure 26). Lorenz immediately suspected yet another computer failure, but a quick check proved him wrong. The reason for the discrepancy between the two curves had to do with the

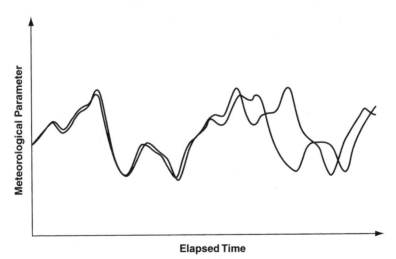

Elapsed Time

Figure 26. *The butterfly effect and meteorology.* The meteorologist Edward Lorenz discovered that, if he started with almost identical initial conditions, two computer simulations of the weather, depicted by the two curves in this diagram, diverged more and more as time went by, until they no longer had anything in common a few months later. This makes long-term meteorological predictions impossible. Such extreme dependence on the initial conditions is often referred to as the "butterfly effect": A butterfly beating its wings in Brazil can provoke tornadoes in Texas.

numbers he had entered into the computer as initial conditions for the new run. Just before the interruption of the previous run, the computer had returned the number 0.145237 for a particular variable, the memory of the machine being able to store only six decimal places. But when he typed in the starting value for the new run, Lorenz decided to round things off and entered 0.145 instead, convinced that a difference of less than one part in one thousand would be of no consequence. He was grievously mistaken: A small change in the initial condition had caused an enormous change in the final result.

Thus chaos imposes fundamental limits on our ability to predict the weather. That does not mean you should boo the weatherman every time he appears on your favorite local television news program. Short-term meteorological predictions, over a period of a day or two, and over a relatively limited area such as a state, are fairly reliable. Thanks to television, we have become accustomed to satellite pictures of white cloud masses swirling over continents. With the help of these pictures, and knowing the wind direction, meteorologists can relatively easily predict the weather in the next twenty-four to forty-eight hours. Longer-term forecasts require computer models of the overall circulation of air masses across Earth's atmosphere. Computers are fed detailed meteorological data such as pressure, temperature, and humidity, monitored by weather stations scattered all over the globe, geographical data such as the locations of oceans and mountains, as well as physical laws describing the behavior of air masses. They are then expected to crunch all these numbers and predict the weather a couple of weeks out. It does not take long to realize that after a few days, predictions have almost nothing in common with reality. In short, the chances that your picnic will be rained out tomorrow despite predictions of bright sunshine by the weatherman are low, although not entirely zero; beyond six or seven days, though, predictions become downright speculative, if not completely false. There is nothing we can do to get around this fundamental limitation. The causes of uncertainty are built into Nature. To predict its tempers and tantrums, you may imagine blanketing the entire globe wall-to-wall with weather stations. There would still be tiny fluctuations in the atmosphere, so minute as to elude detection, which can grow into massive wind systems and destructive hurricanes, and have the ability to modify the weather over the entire planet. That is why chaos is often described in terms of the "butterfly effect" metaphor: A butterfly flapping its wings in the Amazonian forest can trigger torrential rains in New York. Our knowledge is limited not only by the way Nature works but also by the very tool we use to try to unlock her secrets. Computers do not have unlimited memory to store numbers with endless decimal places. As Lorenz found out, we will always have to cope with round-off errors. No matter how we slice it, long-term weather prediction will remain a pipe dream.

RATIONAL HEAVENS AND IRRATIONAL MAN

It was thus the study of weather that drew the attention of the scientific community to the existence of chaotic phenomena. Yet, historically, chaos first manifested itself not in meteorological studies but in an area where it was least expected—the solar system. In order to comfort himself from the vicissitudes of life and to alleviate his anxiety about the future, man has always turned to the sky in the hope of finding there the certainty that is so sorely missing in human affairs. In the midst of the unsettling impermanence of life, celestial phenomena recur faithfully as a kind of reassurance against tomorrow's unknowns. The planets and the stars repeat their motions from east to west, night after night, with exquisite precision. The Sun marches inexorably across the sky during the day. The Moon changes her appearance on a regular schedule every month. To our forebears, the regularity of the celestial machinery was proof that the universe was rational, in spite of the irrationality all around them.

Not surprisingly, the heavens proved more subtle and more complicated on closer examination. Detailed observations revealed that the Sun, for one, does not rise in exactly the same spot on the horizon every day, but that it does so in a cyclical to-and-fro motion. So do the stars at night as the year progresses. Some even drop out of sight for entire seasons. Moreover, there are points of light in the sky that behave strangely. They do cross the firmament from east to west, but they change their positions relative to the stars, sometimes even retracing their steps for short periods of time (their motion is then called "retrograde"). We now know that these wayward shining objects are planets (the word means "wanderers" in Greek) in the solar system. Yet, in spite of all these complications, long-term observations have revealed subtle rhythms. The cyclical to-and-fro pattern of the Sun's rise on the horizon and the appearance of certain stars and constellations at night are all related to the passing seasons. The Moon has its own rhythm, which is much shorter than that of the Sun. It completes about thirteen cycles for every solar cycle. And so the heavens appear indeed rational and predictable, in contrast to the unpredictability of human affairs.

THE MENHIRS AT STONEHENGE

Visitors to the Salisbury plain, in southern England, are greeted by several concentric rows of dolmens and menhirs protruding out of the ground. Standing about 3 to 6 meters tall, they have long been a favorite destination of tourists. The impressive collection of megaliths at Stonehenge was probably erected during the Bronze age, between 2000 and 1500 B.C. It is a magnificent monument to the human spirit intent on conquering the fear of infinite spaces by focusing on rationality in the sky (Figure 27).

Stonehenge amounts to a huge cosmic clock designed to keep track of the motion of the Sun during the year. Four thousand years ago, the people living in what would ultimately become Great Britain had noticed the to-

Figure 27. *An ancient observatory*. Stonehenge is a megalithic astronomical observatory, erected some 4,000 years ago in the Salisbury plain of Southern England. This observatory was originally composed of thirty stone blocks, 4 m in diameter, arranged in a 30-m-diameter circle. Some stones are aligned in the direction where the Sun and Moon rise at certain special times of the year, such as the summer and winter solstices.

and-fro motion of the spot where the Sun rises on the horizon. They had learned that this phenomenon was cyclical and obeyed very precise rules. At the summer solstice (on the twenty-first of June, which at our latitudes is the longest day of the year and marks the beginning of summer), the Sun rises at the most northerly point on the horizon. Six months later, at the winter solstice (December 21, which for us is the shortest day of the year and ushers in winter), it peeks out at the most southerly point. Half a year later, it emerges again at the most northerly point, and so forth. In order to connect with the cosmos and celebrate the heavens, the men of the Bronze Age built a megalithic observatory whose central walkway was aligned with the most northerly sunrise spot. If you ever visit Stonehenge during summer solstice, you will be treated to a glorious view of the rising Sun framed by a stone gateway erected some forty centuries ago.

THE PUZZLE OF THE MOON

The people who built Stonehenge were also interested in the motion of the Moon. The many mounds and ditches dotting the solar observatory appear to be lined up with the most northerly point where the Moon rises above the horizon. This alignment is only approximate, though, because the Moon's apparent motion is much more complex than that of the Sun. It does not rise at the same location after each cycle, appearing instead at different spots during successive cycles. It does not return to its most northerly point for almost nineteen years. There are many different lunar cycles. For instance, 29.5 days elapse between two successive full moons, during which time the Moon goes through every one of its phases, from new moon to full moon, through crescent moon, quarter moon, and gibbous moon. But it takes only 27.3 days for the Moon to return to the same alignment relative to the backdrop of stars. So varied are the lunar cycles that they must have caused many headaches to the builders of Stonehenge and their predecessors. One might think that with the birth of modern science in the sixteenth century, the development of mathematics and physics in the ensuing centuries, and the advent of computers in the twentieth century, the problem of the movements of the Moon should have been solved a long time ago. As it turns out, that is not the case, and not for lack of talent and effort. Even though the greatest scientific minds have tackled the issue, the mystery of the Moon's movements endure. Still, those who have wrestled with her secrets have not come up empty-handed. They were even able to capture chaos, as we are about to see.

CIRCLE UPON CIRCLE

Because it is the closest celestial object to us, the Moon, which projects its soft glow on sleepy farmlands and inspires the amorous embraces of lovers in the safety of the night, has always captured the human imagination. In times past, priests in search of power and politicians hungry for fame have scrutinized its movements with great care. Ancient civilizations believed that during an eclipse the Moon (or the Sun) was devoured by a mythical dragon or some other celestial monster. Eclipses used to inspire awe and fright. By announcing them in advance, politicians and priests knew they could impress the masses with their prescience and knowledge.

Meanwhile, the scientists were not sitting on the sideline. In the second century of our era, the Greek astronomer Ptolemy (ca. 90–ca. 168) carried out a great synthesis of all the knowledge acquired by his predecessors. He proposed a lunar model that was to prevail for over a thousand years. In a geocentric universe, Earth occupied the center of the solar system, and all other celestial bodies, including the Sun, revolved around it. The Moon pursued her path around Earth in a plane inclined at 5 degrees relative to the plane containing Earth, the Sun, and all the other planets (the so-called zodiacal plane). Ptolemy faced two problems. First, the Moon did not move uniformly in its orbit. It picked up speed in certain places and slowed down in others, which challenged the doctrine of perfect and uniform movements of celestial objects so dear to Aristotle (384–322 B.C.). Moreover, the distance between the Moon and Earth was clearly variable, since the Moon's angular size was not constant. That was inconsistent with the notion that the Moon moves on a sphere centered on Earth. To resolve these problems, Ptolemy removed the Moon from its celestial sphere and placed it on a small circle called an epicycle, the center of which was itself on another larger circle. In this model, the net movement of the Moon resulted from the superposition of two independent movements: the uniform motion of the Moon on an epicycle whose center in turn moved uniformly along a circle centered on Earth.

THE MOON REFUSES TO SUBMIT TO CALCULATIONS

By the sixteenth century, calculating lunar positions and eclipses had turned into a veritable cottage industry. Spurred on by the needs of navigators who had to know precisely the positions of their ships and by the necessity to establish calendars for the purpose of scheduling fairs and celebrations, mathematicians and astrologers were kept busy calculating tables of the Moon's positions. Because Ptolemy's model was incorrect (Earth is in fact not at the center of the world), errors kept on accumulating with each passing year. The Moon in the sky deviated ever more from calculated positions, and it was necessary to periodically recalculate everything from scratch and generate new tables. But even with revised numbers, lunar eclipses would sometimes arrive one or two hours ahead of schedule. Worse than that, calculations would produce different results depending on the methods used.

The geocentric universe came to an end in 1543 when the Polish canon Nicolas Copernicus (1473–1543) dislodged Earth from its central position and installed the Sun in its place. Using a wealth of data on the positions of the planets measured with unprecedented accuracy by the Danish astronomer Tycho Brahe (1546–1601), the German Johannes Kepler (1571–1630) finally succeeded in 1609 in solving the secret of how the planets move. He concluded that they follow elliptical orbits rather than circular ones, as Aristotle and almost everybody else after him had believed. The planets accelerate as they get closer to the Sun, located at one of the focal points of the ellipse, and slow down as they move away from it. The myth of the uniformity of celestial movements had to be abandoned.

Kepler also worked on the problem of the Moon. He discarded Ptolemy's epicycle and placed the Moon on an elliptical orbit around Earth. His calculations being much more accurate, the tables he produced were used for decades after they were published in 1627. But the Moon still did not fully cooperate. It continued to deviate from predictions after a while.

While Kepler did elucidate the movements of planets, just what caused that movement remained a mystery. What is it that is pushing the planets along their orbits? Surely not an army of angels, as many people believed during the Middle Ages. The answer to that question was brilliantly provided by the Englishman Isaac Newton (1642–1727) in 1666. He had the revolutionary insight that the fall of an apple in an orchard was related to the movement of the Moon around Earth. The movements of both the apple and the Moon were dictated by one and the same law—the law of universal gravitation. The planets move along elliptical orbits around the Sun because they are subject to the Sun's gravitational influence.

THE THREE-BODY PROBLEM

Newton also became interested in the problem of the Moon's orbit. He introduced a novel element in the analysis. Instead of making Earth solely responsible for the movement of the Moon, he also included the gravitational influence of the Sun. In Newton's model, Earth, the Moon, and the Sun all attract each other through a force that depends only on their masses and distances. Newton's hope was to explain the Moon's complicated and periodic irregularities by means of the perturbation the Sun exerts on the elliptical orbit of the Moon around Earth.

At first sight, you would think that the problem of the orbit of a celestial body subjected to the gravitational influence of two others (mathematicians call this a "three-body problem") should present no great difficulty. After all, Newton himself had solved the two-body problem in his masterpiece *Mathematical Principles of Natural Philosophy*, published in 1687. The orbit of a body subjected to the gravity of just one other can only be an ellipse, a parabola, or a hyperbola. You might think that going from two to three bodies should be a breeze, but you would be completely wrong. The orbits of three interacting bodies cannot be described by a simple mathematical formula, contrary to the case of just two. The problem becomes even more daunting with four or more bodies.

Unable to find an exact solution to the problem of the Moon's movements, Newton had to resort to a technique mathematicians refer to as "perturbation method" to work out an approximate solution. The idea is to start with the dominant phenomenon—namely, the gravitational interaction between Earth and the Moon. That problem is easy, since it involves only two bodies. The influence of the third body is then added on as a perturbation to the idealized two-body simple case. Unfortunately, calculating this perturbation (or perturbations in the case of more than three bodies) is not easy. For all his genius, even the great Newton failed to work it out. As he would later reminisce, "Never did I get bigger headaches than when I was working on the problem of the Moon." Despite a full year of intense calculations, the positions he managed to calculate with his theory still differed from those observed in the sky by one-sixth of a degree. That is a significant error, since the angular size of the full Moon is itself only half a degree. Because the Moon is so close to us (it lies at a distance of 384,000 km and it takes barely more than a second for its light to reach us), such a difference is easily measurable. Newton considered his work on the Moon a great failure. After such a sustained but unrewarded effort, he moved away from science. Perhaps he felt he had lost the creative power of his youth. He was only twenty-three years old, when in 1666, within one year, he forever changed the face of the universe by discovering not only universal gravitation but also differential calculus and the nature of light. In 1696, Newton resigned from his professorship at the University of Cambridge and spent most of the remainder of his life in an administrative post at the Royal Mint, supervising the minting of England's new currency.

The Moon refused to give up her secret because the perturbation method used by Newton did not really apply to the real situation. The gravitation exerted by the Sun turns out to be a significant part of the total force acting on the Moon. As such, it could not be treated as a small perturbation. As is often the case in human interactions, the intrusion of a third party into the life of a couple can cause far more insidious damage than just mild annoyance.

Newton's failure did not discourage his successors, who tackled the problem of the Moon with renewed vigor. Promises of glory, prestige, and influence proved to have as much appeal as pure scientific interest. Results and counterresults were announced in rapid succession at the science academies of Europe, primarily in Paris, London, and Berlin. Rivalries developed. The greatest mathematicians of the eighteenth and nineteenth centuries took a crack at the problem, including the Swiss Leonhard Euler (1707–1783), and the Frenchmen Joseph-Louis Lagrange (1736–1813) and Pierre-Simon de Laplace (1749–1827). In the process, they discovered new aspects of mathematics, but the Moon continued to stubbornly refuse to surrender to calculations. Euler devoted his entire life to that pursuit, but finally had to concede defeat: "For the last forty years I have tried to devise a theory of the Moon's movements based on the principles of gravitation, but so many obstacles stood in the way that I had to stop. . . . I cannot see how such research about the Moon can succeed, nor how it could be used for any practical purpose." Laplace managed to reduce the discrepancy between calculated and observed positions to less than $1/120$ of a degree, but in the end he, too, failed to tame the Moon. As a matter of fact, all calculations done in the nineteenth century with the help of Newton's laws and a formidable mathematical arsenal were barely more accurate than the results obtained empirically two thousand years earlier by Greek and Babylonian astronomers by observing eclipses.

A brand-new approach was needed to make any headway, a radically different "paradigm," to use the word of the modern science historian Thomas Kuhn.[1] The young French mathematician Henri Poincaré was the one who found this approach. He proposed an extremely original method to attack problems of celestial mechanics, which ultimately would force him to confront—against all expectations—the notion of chaos. He proved that, in the case of three bodies mutually interacting by gravitation, Newton's equations contained within them not just regular and predictable phenomena but irregular and unpredictable ones as well. The Moon had not conformed to calculations because its behavior included an unpredictable component that Newton and his followers had never suspected. To put it succinctly, Newton's equations harbored the seed of chaos.

ARE PLANETARY ORBITS ETERNAL?

Jules-Henri Poincaré (Figure 28) was a brilliant research scientist with a highly original mind. He was only twenty-seven years old when he was

Figure 28. *Henri Poincaré (1854–1912)*. One of the greatest mathematicians of his time, he was also a pioneer of the theory of chaos. While working on the three-body problem, he was the first to realize that the evolution of certain physical systems depends so critically on the initial conditions that predicting their future behavior is impossible. He thus discovered that chaos resides at the very heart of Newton's deterministic equations.

appointed professor of mathematics at the University of Paris. He became interested in the three-body problem during a mathematics contest organized by the University of Stockholm to mark the sixtieth anniversary of Olaf II (1829–1907), king of Sweden and Norway. One of the topics of the contest had to do with the stability of the solar system. The goal was to resolve in favor of one of two possibilities: Either the planets are to follow faithfully their orbits around the Sun forever, endlessly retracing the same path, in which case the solar system is stable, or, because of the cumulative effects of gravitational perturbations, planetary orbits are destined to change radically in the distant future, completely altering the organization of the solar system, in which case the situation is unstable. The German mathematician Gustav Dirichlet (1805–1859) had claimed in 1858 that he had invented a new technique to solve Newton's equations and successfully proved that the solar system was stable. Planets would never deviate from their present orbits, which were fixed for all eternity. Unfortunately, Dirichlet died a year after he made his provocative statement without leaving any document supporting his claim. Because Dirichlet had a reputation for coming up with ingenious and incisive derivations, his assertion about the stability of the solar system was taken very seriously. Many of his contemporaries and successors tried their hand at the problem, without success. Hence the idea of a contest. The hope was that some genius might be able to rediscover Dirichlet's lost proof.

ENDLESS SERIES

Before Poincaré came along, mathematicians believed they could resolve the issue of the solar system's stability simply by working out the solutions to Newton's equations. These were rooted in the idea that the position and velocity of an object at any given instant completely determine all future positions and velocities of that object, to the same degree that the present characteristics are the result of past ones. In other words, the laws of physics relate the current state of the world to what it was immediately before or what it will be immediately after. Because they describe the differences in conditions between the present moment and the one just before or after, Newton's laws can be expressed in terms of what are called "differential equations."

The solutions to such differential equations take on the somewhat unwieldy form of an endless summation of algebraic terms—what mathematicians call a "series." Since most people have neither the patience nor, for that matter, the longevity required to calculate all the terms of an infinite series, the next-best strategy is to evaluate only its first few terms to at least get an approximate idea of how the planets are going to behave. The hope is that the neglected terms will be so small as to contribute virtually nothing to the sum. When that is the case, the series converges rapidly toward the final result—such series are said to be "converging." As it happens, not all series converge. There are some that stubbornly keep diverging. In those cases, additional terms do not become negligibly small. Instead, they keep on contributing to the sum, which never approaches any finite limit.

When it comes to the solar system, the sixty-four-thousand-dollar question then is: Does the series that describe the movements of planets converge, in which case the solar system is stable and the planets will just keep on orbiting the Sun endlessly, or does it diverge, in which case the orbits are irrevocably condemned to change?

If Dirichlet was right, it should be possible to prove that the series applicable to the solar system converges. That was, in a nutshell, the challenge issued by the organizers of the contest honoring the king of Sweden to the mathematicians of the world.

REALITY IN ITS TOTALITY

Poincaré met the challenge brilliantly. But he had no patience for endless series. What he wanted to do was embrace reality in its totality. As it happens, the traditional approach for solving differential equations isolates fragments of reality. The whole is then reconstructed by juxtaposing these fragments end-to-end. As Poincaré himself wrote: "Instead of considering the evolution of a phenomenon in its totality, one seeks only [with differential equations] to relate one instant to the one immediately before. The assumption is that the present state of the world depends only on the very recent past and is not at all influenced by any memory of the distant past." This decoupling from the past implicitly gives reality the appearance of a continuum from which disorder and chaos are precluded. Moreover, solving three-body differential equations with the method of infinite series is exceedingly difficult, if not downright impossible, even for the most accomplished practitioners of this art. Those who tried it often ended up restricting themselves to special cases that could be solved but gave only a simplified and partial view of reality. Poincaré refused to be forced into the position of an ant peering no farther than the next clump of grass. He wanted the bird's-eye view of an eagle soaring high above mountains and valleys. He felt that studying a single orbit corresponding to a particular set of initial conditions was far too restrictive a way to deal with the three-body problem. He was convinced that success could come only by taking a broader view of all possible orbits, including all conceivable combinations of initial conditions. To that end, he took advantage of his extraordinary geometrical intuition and invented a new technique, which to this day continues to be the primary tool for studying chaos.

AN ABSTRACT MULTIDIMENSIONAL SPACE

We all live in a space with three dimensions. We can move forward or backward, to the right or the left, and up or down. In this kind of space, the position of a tennis ball that has just cleared the net or of a Phileas Fogg who travels around the world in 80 days is defined by three spatial coordinates. To grasp things in their totality, Poincaré decided to discard the familiar space in which we operate every day. Through the sheer power of his imagination, he moved into an abstract space with multiple dimensions, something called "phase space" (Figure 29). There, the position of a tennis ball or that of Phileas Fogg is determined not just by the three familiar spatial coordinates but by three velocity coordinates as well, tracking the speed from top to bottom, from left to right, and from front to back (or vice versa). On the stage of this abstract space, velocity is treated on a par with position; both play equally important roles (Figure 30a). Three additional dimensions are needed to accommodate the new actor. Accordingly, it now takes a total of six dimensions to track a tennis ball, and another six to take care of Phileas Fogg. Likewise in a three-body problem: Six dimensions are needed to account for the Moon, another six for Earth, and six more for the Sun. In this approach, a total of eighteen dimensions are, therefore, required to get a global description of three interacting bodies. This proliferation of dimensions allowed Poincaré to rise from ant to eagle. Instead of crawling on the ground, he could soar high in the sky. Instead of being stuck with an isolated snapshot of just one player at a time, he enjoyed an overall view of the whole stage and all the actors at once. In this multidimensional space, the entire solar system is represented by just a single point, rather than by the ten different ones (one for the Sun and nine others for the planets) in conventional three-dimensional space. That is the power of this mathematical construct called "phase space." No matter how

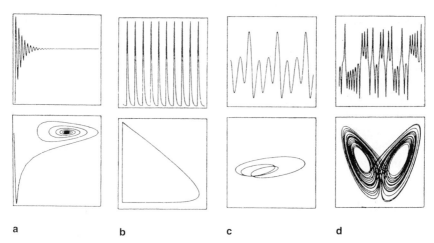

a   b   c   d

Figure 29. *Portraits in phase space*. The dynamical behavior of a system can be represented in two different ways. The standard way consists in describing the evolution of the system as a function of time *(top diagrams)*. The modern way relies on studying the trajectories of a point representing the dynamical state of the system in phase space *(bottom diagrams)*. For instance, system *(a)* converges toward an equilibrium state after many oscillations, which, in phase space, corresponds to loops collapsing into a single point. System *(b)* repeats itself at regular intervals, which corresponds to a closed cyclical orbit in phase space. System *(c)* also has a periodic motion, albeit more complicated. It repeats itself only after every third oscillation: it is said to have a period-3 cycle. This translates into more complicated loops in phase space. System *(d)* is chaotic; it is represented in phase space by Lorenz's "strange attractor," a figure shaped like a pair of butterfly wings (see Figure 37).

complex the situation, how convoluted the plot, how baroque the setting, regardless of the number of actors, a single point in that abstract space is sufficient to describe a system in its entirety.

CHAOS IN A LONG LAZY RIVER

What really piqued Poincaré's interest was not so much the static and frozen aspect of a system as its dynamic and evolving character. He wanted to understand not the Moon fixed in its orbit but how that orbit changes during the course of eons. As a system evolves, its representative point describes a curve in multidimensional phase space. Change the initial conditions slightly, and the point traces a different curve. The ensemble of solutions to the differential equations applicable to the system corresponds to a family of curves in phase space, like the streams of a river flowing toward the sea. Picture a leaf torn loose by the wind, twirling in the air before landing on the surface of this mathematical river that carries it away. It can either follow a straight path on the surface of a long, calm river, or it can run into rough waters and describe swirls and arabesques. How the leaf moves along this abstract mathematical river is of considerable interest, because it tells us something about the motion of the system in real space. Whether or not the evolution of the system depends critically on the initial conditions is

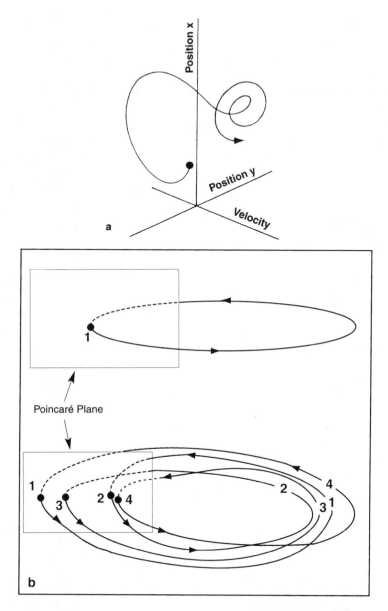

Figure 30. *(a) The abstract phase space.* In order to visualize the global behavior of a dynamical system, it is often useful to represent it by a point in an abstract multidimensional space called "phase space." This space has as many dimensions as necessary to characterize the complete system. For instance, the axes shown here represent spatial as well as velocity coordinates. As the system evolves, both its position and velocity change, and the corresponding point traces out a curve in phase space.

(b) Poincaré's plane. To follow the trajectory of a point in phase space, the mathematician Henri Poincaré imagined slicing through that space with a vertical plane, now called "Poincaré's plane." The points where the trajectory intersects the plane trace a pattern that makes it possible to keep track of the various behaviors of the system. For instance, a single point in this plane *(top)* corresponds to a periodic motion. By contrast, a cyclical movement repeating itself every fourth pass only would be represented by four distinct points *(bottom)*.

revealed immediately by watching two leaves whose paths initially follow each other closely in the mathematical river. If the two paths end up diverging, the system is quite sensitive to the initial conditions. If not, the paths remain similar, and the system does not depend sensitively on the initial conditions.

These mathematical rivers behave rather differently from a Hudson or a Mississipi. While real rivers always empty into the sea and never flow back toward their source, mathematical rivers have no problem doing just that. If the movement in real space is periodic—that is to say, repeats itself regularly, as is the case, for instance, of the motion of a pendulum—the corresponding mathematical river comes back to its starting point, and the loop closes on itself. Likewise, while a real river can divide into two or more branches, mathematical trajectories in phase space are forbidden to split. Nor are they allowed to intersect another trajectory, whereas the confluence of two rivers in real life is commonplace.

UNCERTAINTY IS LURKING IN THE SHADOWS

Having gained a global perspective through the notion of mathematical phase space, Poincaré was ready to tackle the three-body problem—the subject of the competition organized in honor of the king of Sweden. In the process, he would revolutionize the branch of mathematics that deals with the interrelations between forces and movements, a branch called "dynamics." More important, he was about to push out of the shadows an actor no one had been expecting, and which would turn out to play a pivotal role in the making of reality: chaos.

The dynamics of two bodies interacting gravitationally had been solved by Kepler and Newton. They taught us that a planet perpetually pursues its journey around the Sun along an elliptical orbit, with the Sun occupying one of the foci. The three-body problem, as we have mentioned, is a much tougher nut to crack. To analyze the mathematical flow in phase space, Poincaré imagined a fictitious screen that would not interfere with the rivulets, but recorded the precise spot where each would cross the screen. An ellipse in real space, to take one example, corresponds to a loop in phase space. If a planet endlessly retraces the same orbit in the sky, as Newton believed must happen, the same loop is repeated over and over in phase space. The loop traverses the screen—called "Poincaré plane" in technical parlance—at a single point (Figure 30b, top). Thus, a periodic movement in real space is represented by a single point in the Poincaré plane. A more complicated movement, but one that still repeats itself every fourth pass, will show up as four distinct points in that same plane (Figure 30b, bottom). A movement that never repeats itself at all will generate an infinity of points. Much as the colored dots painted by George Seurat transform a bare canvas into a bucolic Sunday scene on the shores of the "Grande Jatte" island, the points on Poincaré's plane can create patterns of amazing beauty and artistry, whose intricacy directly reflects the complexity of the corresponding movement in real space. Modern computers have been enlisted to become artists, generating with ease patterns of stunning loveliness. Something as innocuous as a trajectory meandering its way in phase space can end up tracing in Poincaré's plane marvelous landscapes reminiscent of mountains and valleys, of islands in the middle of wild rivers that empty out into the ocean. A remarkable fact is that the same pattern recurs ad infinitum. If you isolate a piece of that landscape and magnify it, you will see the same scene of mountains and valleys, islands and rivers, but on a smaller scale.

It was while studying these patterns that Poincaré came face-to-face with chaos. Obviously, in 1888, he did not have access to the powerful computers and laser graphics terminals that make our lives so much easier today. All he had was his uncanny ability to mentally maneuver in phase space. His genius was such that he was able to envision situations where a minute change in the initial position or velocity of one of the three bodies could completely change its orbit. A tiny change could be enough to switch from stability to chaos. Poincaré realized that regularity and chaos were closely intertwined, and that the unpredictable was never very far away from the predictable. In the ever-mesmerizing Poincaré plane, islands of turbulence stand out in the middle of well-behaved areas, just as orderly regions can pop out of otherwise chaotic regions. Poincaré came to the realization that a system seemingly as simple as the Moon, Earth, and the Sun, in spite of being governed by as precise and constraining a law as Newton's universal gravitation, can give rise to unpredictable and indeterminate behaviors. Uncertainty was muscling its way onto the stage, spelling the end of the line for Laplace's determinism. Poincaré published his discovery that the future was undetermined in a 270-page treatise titled *On the Three-Body Problem and the Equations of Dynamics.*

HOW COULD CHAOS HAVE ESCAPED NEWTON?

Uncertainty was lurking in the very midst of Newton's equations. Yet he failed to see it because he was looking for the cosmic order and harmony he believed in so deeply. His solution to the problem of the movement of planets appeared in his masterpiece, *Mathematical Principles of Natural Philosophy*, published in 1687. The answer was so total, complete, and definitive that the universe, which until then had seemed mysterious and uncertain, was suddenly transformed, as if by the stroke of a magic wand, into a well-oiled machine from which any indeterminism was banished and any whim excluded. The impact of the treatise was so momentous that it never occurred to any of Newton's contemporaries or successors to question the implicit message conveyed by his work. That message had already been proclaimed loud and clear as early as the fourth century B.C. by the Greeks: Nature was governed by universal laws that could be apprehended by human reasoning. This exuberant and boundless confidence in a rational world spilled beyond the domain of physics and pervaded all spheres of human endeavor in the ensuing centuries. Thanks to his power of logic, man could perfect social and political institutions as well. This unlimited trust in human reasoning culminated in the industrial revolution, not to mention the American and French Revolutions at the end of the eighteenth century.

That Newton missed the uncertainty hiding in the dark in no way diminishes his genius. On the contrary, it took all his intellectual talent to choose and pick from among the many challenges presented by Nature those problems that were most likely to be tamed by human reasoning and had a well-defined solution. Newton was well aware that the awesome theoretical edifice he had built did not account for everything. Much to his chagrin, he failed, for example, to sort out the movements of the defiant Moon, torn between the gravitational pulls of Earth and of the Sun. It fell to Poincaré's genius to discover the two faces of Newton's mechanical laws. The ostensible side of these laws is harmony, but their hidden side is chaos. Whereas Newton can rightfully be hailed as the prophet of order, Poincaré is the undisputed prophet of chaos.

Newton and Kepler were able to perceive the harmony of the world because we live in a solar system dominated by the mass of the Sun. The latter almost single-handedly presides over the gravitational show and dictates the dance of the planets. That is what enabled Newton to treat the orbit of each planet around the Sun as if it were a simple two-body problem—the Sun and each individual planet. Had our solar system included two suns, the problem would have involved three bodies (the two suns and each planet), and chaos would have been immediately obvious (Figure 31). Planets would have had erratic and unpredictable orbits, and creatures living on one of these planets would never have been able to perceive the slightest harmony. Nor would it have occurred to them that the universe might be ruled by laws and that it is up to man's intellect to discover them. Besides, it is not at all obvious that life and conscience could even emerge in such a chaotic system.

Figure 31. *Chaos in a three-body system*. The two diagrams illustrate the complexity of the orbit of a planet in a solar system dominated not by a single Sun, as we are familiar with, but by two stars of equal masses. In such a case, the planet follows an orbit that is extremely complex and unpredictable—in other words, chaotic.

NUMEROLOGY AND ASTEROIDS

Chaos pervades the solar system. Its insidious presence will help us under-stand the origin of those large rocky objects we call asteroids or comets, which, as we have seen, are all-important agents of contingency in the mak-ing of reality. Where they come from had long been a mystery. Two centuries ago, a map of the solar system would have included only the Sun, six planets (nobody knew about Uranus, Neptune, and Pluto back then), and a few ephemeral visitors—comets. Asteroids had yet to make their appearance in the world of astronomy. They would eventually gain recognition thanks to a bit of numerology sneaking into the study of the heavens.

Kepler was not alone in believing that numbers govern the heavens and the behavior of the planets. The German astronomer Johann Titius (1729–1796) shared that view. In 1766, he proposed a very simple recipe to account for the distances between the planets and the Sun in units of the Earth–Sun distance. Starting with the numbers 0 and 3, you can build a sequence by doubling the last number, which gives 0, 3, 6, 12, 24, et cetera. Next, add 4 to each number and divide the result by 10. The sequence becomes 0.4, 0.7, 1.0, 1.6, 2.8, 5.2, and 10. Now compare that sequence to the actual distances between the planets and the Sun, which are: 0.39 (Mer-cury), 0.72 (Venus), 1.0 (Earth), 1.52 (Mars), ?, 5.20 (Jupiter), and 9.54 (Sat-urn). Except for the question mark, the agreement is quite remarkable. Titius's recipe reproduces amazingly well the distances from the Sun to the six planets known at the time. Yet it failed to garner much attention at the time and promptly fell into oblivion. It experienced a second life when Johann Bode (1747–1826), another German astronomer, resurrected it and spread it among the general public. Indeed, the recipe is now known as the Titius-Bode formula. It acquired even more credence when in 1781 the British astronomer William Herschel (1738–1822) discovered the planet Uranus at a distance of 19.2 times the Earth–Sun distance, whereas the for-mula predicted 19.6.

Still, the mystery persisted about the missing member of the sequence. The Titius-Bode formula predicted the existence of a celestial body between the orbits of Mars and Jupiter, at 2.8 times Earth–Sun distance. Such a body remained conspicuously absent. An intensive search for the missing planet was initiated. In 1801, the Sicilian astronomer and monk Giuseppe Piazzi (1746–1826), working at the observatory of Palermo, discovered a celestial object of very low brightness at the distance predicted by the formula. The object was too dim to be a planet, but it did move against the background of stars. Piazzi named it Ceres, after both the patron saint of Sicily and the god-dess of agriculture in Roman mythology. This initial discovery was followed by an avalanche of others in the wake of the development of the photo-graphic plate in the latter part of the nineteenth century. Faint objects mov-ing relative to the stars were found in record numbers in the ensuing years. We know today that Ceres is in fact one of the largest members of a swarm of one million or so rocky masses with a jagged surface, which came to be called asteroids. They orbit between Mars and Jupiter, forming a kind of rocky belt (Figure 32). An intriguing question is whether those innumerable rocky objects might be the remnants of a planet blown to pieces in some cataclysmic event. The hypothesis appears rather implausible, because if all the asteroids were put back together, one would get a sphere barely 1,500 km in diameter, less than half the size of the Moon, clearly a far cry from anything resembling a planet. Besides, we have not the slightest idea what might have caused a planet to shatter into a million pieces. It seems more reasonable to assume that asteroids are left-over debris that failed to coalesce during the accretion process of planetesimals taking place 4.6 billion years ago, when the solar system was forming.

What is the significance of the Titius-Bode law? It did pass the tests of Uranus and the asteroid belt with flying colors. Does this imply that numbers preside over the arrangement of planets? Does it mean that the agreement between the formula and the actual distances between the Sun and the planets is not pure coincidence but is somehow preordained by the laws governing the formation of solar systems? Should one conclude that in similar systems around other stars, in some other Milky Way, the same numerical sequence should hold? The answer to all these questions is decidedly no. The reason is that the Titius-Bode law miserably failed subsequent tests.

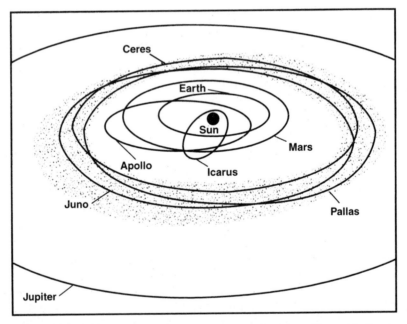

Figure 32. *The asteroid belt.* Most asteroids orbit around the Sun between Mars and Jupiter, at distances between 2.2 and 3.3 times the Earth–Sun distance. About 100,000 of them are sufficiently bright to be photographed from Earth. Ceres is the largest, with a diameter of 900 km and a mass equal to 30 percent of the mass of all the other asteroids put together. Combining all the asteroids in the belt would produce an object barely 1,500 km in diameter, far smaller than a planet. We may thus conclude that asteroids are not the remnants of a planet shattered to pieces, but are debris left over from the primordial solar system.

Neptune, discovered simultaneously in 1846 by the Frenchman Urbain Le Verrier (1811–1877) and the Englishman John Couch Adams (1819–1892), turned up at a distance of 30.1 times Earth–Sun distance, rather than at the 38.8 predicted by the formula. Pluto, the farthest planet out, was discovered in 1930 by the American Clyde Tombaugh. It lies at only 39.3 times Earth–Sun distance, not at the expected 77.2. Of course, Pluto's discrepancy could be brushed aside, since many astronomers believe that it is not a usual planet but is probably the most prominent member of a family of thousands of ice-covered asteroids in orbit near the periphery of the solar system. Indeed, Pluto has a very small mass (only $2/1000$ of Earth's mass) and a very modest size (66 percent the diameter of the Moon). The inclination of Pluto's orbit also happens to be highly abnormal, making an angle of 17.1 degrees with the ecliptic, whereas all other planets orbit the Sun in or very nearly in that plane. Moreover, Pluto has the most eccentric orbit of all the planets, so much so that at its closest point to the Sun, Pluto actually lies inside Neptune's orbit by some 100 million kilometers.

The failure of the Titius-Bode law in Pluto's case may have plenty of extenuating circumstances. But there is simply no excuse in Neptune's case, and that makes it doubtful that it is a fundamental law of Nature, for it would have to apply to all planets without exception. Must we conclude that it constitutes nothing more than simple numerical coincidence? It is difficult to give a categorical answer, as no one can deny that it accounts surprisingly well for seven planets and the asteroid belt. Stay tuned for further developments.

ASTEROIDS AND THE MISSING ORBITS

At the present time, astronomers have catalogued the orbits of about 5,000 asteroids among the million rocky objects populating the asteroid belt. Since this is such a crowded region, one might think that it would be the scene of innumerable impacts and collisions between these chunks of space rock speeding along their elliptical orbits around the Sun, between Mars and Jupiter. In fact, such events are quite rare. Space is so vast that, on average, any given asteroid is separated by several million kilometers from its nearest neighbors. Under these conditions, head-on collisions are the exception rather than the rule.

Yet there remained a mystery surrounding the world of asteroids. While studying the orbits of these rocky objects in 1857, the American astronomer Daniel Kirkwood noticed that some of these orbits were missing. Asteroids systematically avoided orbits that were "in resonance" with that of Jupiter. These so-called "resonant" orbits have a period (the time required to complete one revolution around the Sun) that is an exact fraction (like $1/2$, $1/3$, and so on) of that of Jupiter. What this means is that when Jupiter completes one orbit around the Sun every 12 years, an asteroid with a four-year period will have completed exactly three of its own orbits, at which time it will find itself in the same position relative to Jupiter and the Sun, and thus experience the same gravitational forces. This particular orbit is said to be in a 3-to-1 resonance with Jupiter. All of us have experienced the very same type of resonance as children, when we used to play on swings. Whoever was pushing us had to synchronize his or her actions with the oscillations of the swing to get us to go ever higher. To be effective, the pushing had to be done at the precise instant when the swing reached its highest point, before starting back down in the other direction. In other words, to amplify the back-and-forth motion, the pushes had to be timed to occur "in resonance" with the oscillation of the swing. Short of that, the swing would never gain any height. Likewise, an asteroid on the right orbit is pushed "in resonance" by the gravitational forces of Jupiter.

Kirkwood concluded that these repeated impulses from Jupiter must have gradually nudged asteroids out of resonant orbits, which would explain why they were missing. The explanation was not fully satisfactory, though, because it said nothing about what happened to the asteroids after they had left their resonant orbits. Did they remain placidly in some other nearby orbits in the asteroid belt, or did they leave their place of origin and venture out, crossing the orbits of Earth and Mars, perhaps even wiping out dinosaurs? Astronomers just don't live long enough to monitor firsthand how asteroids change orbit. This takes place on time scales of 100,000 years or more. Once again, it is necessary to resort to Newton's differential equations to peer into the future. But here again, we are dealing with a multibody problem. We have to take into account the gravitational influence not only of the Sun but of Jupiter and Mars as well. As we have explained earlier, such a problem does not have an exact solution. Astronomers tried perturbation methods, treating the influences of Mars and Jupiter as small corrections to the action of the Sun, which is really not quite justified. As in the case of the Moon, the approach led to a dead end. Astronomers gave up, and the problem of the missing orbits was left dormant.

A FEW VISITS TO EARTH'S NEIGHBORHOOD

The mystery of the missing orbits became a hot topic again in the 1970s. Granted, something new and very important had happened in the meantime: The computer had been invented. Its computational power had improved so much that extrapolating asteroid orbits hundreds of thousands of years out into the future was no longer wishful thinking. The American Jack Wisdom was working at the famous California Institute of Technology (Caltech) on developing a computer model of the solar system. He placed 300 asteroids on a specific missing orbit—namely, the one that is in a 3-to-1 resonance with Jupiter. This implied a period of four years, exactly one-third that of Jupiter. So as to obtain a bird's-eye view of all the orbits at once, Wisdom used the technique developed by Poincaré. He analyzed the movement of the asteroids not in real three-dimensional space but in a multidimensional mathematical phase space, in which the changes in position and velocity of each asteroid are represented by the quirks of meandering trajectories. Like Poincaré before him, Wisdom got a simultaneous view of the dynamics of all 300 asteroids by studying the intersections of their associated curves with a vertical plane. And also like Poincaré, Wisdom immediately recognized chaos.

In phase space, chaotic regions where the movements of asteroids were haphazard and unpredictable, and the future not well-determined, were intermixed with orderly regions where movements were regular and predictable, and the future well-determined. In some cases, asteroids uneventfully retraced their steps from one orbit to the next for hundreds of thousands of years. In some other cases, though, after some 100,000 years of orderly behavior, an asteroid would suddenly leave its normal elliptical orbit between Mars and Jupiter and go flying toward Mars and Earth. Such excursions were not without risks, and once in a while a wayward asteroid would collide with one of the inner planets. The computer simulation told Wisdom that, for every five asteroids placed on an orbit in a 3-to-1 resonance with Jupiter, one of them could suddenly escape its assigned trajectory and switch to a completely chaotic orbit in about half a million years on average. Thus would the mystery of the missing orbits be resolved. They are orbits confined to chaotic zones where minute changes in position or velocity can open the door to completely different paths. These chaotic conditions would send asteroids on a path toward Mars or Earth, eventually emptying out resonant orbits in the asteroid belt. In this model, Jupiter would not be nudging asteroids out of their regular orbits gradually with a series of small gravitational impulses, contrary to what Kirkwood had believed. Rather, its gravity would create chaotic conditions, resulting in sudden and dramatic changes of orbit.

Once again, we see that the perfectly deterministic laws of Newton's universal gravitation harbor the seeds of chaos. Barring this grain of madness, Newtonian gravity would never have sufficed to send us visitors from the asteroid belt. Without chaos lurking in the shadows, perhaps the dinosaurs might still be ruling Earth, and perhaps the Moon would not exist to cast her soft radiance on the nocturnal embraces of lovers.

IS THE WORM GOING TO CONSUME THE ENTIRE APPLE?

If Newton's equations contain in their very heart the germ of chaos, what does it imply about the movements of planets in the solar system? Since chaos has invaded the realm of asteroids, why should it not do the same with the planets? If the worm is already in the apple, what is to stop it from devouring it whole? Is it conceivable that the unending movements of the planets, which humanity has always looked up to as a symbol of perenniality, permanence, and constancy in the face of the changeability, impermanence, and inconstancy of everyday life, hide chaos and unpredictability? Does the celestial machinery, which appears so well-oiled, contain the seeds of its own destruction? Is the solar system, which seems so stable on a time scale of a few thousand years, in fact inherently unstable over much longer intervals?

The issue of the stability of the solar system is not new. It was already a topic of debate more than 200 years ago. The French astronomer and mathematician Pierre Simon de Laplace (Figure 33), was one of the first to propose tentative answers.

A YOUNG MAN IN A HURRY

Born in 1749 on a farm in Normandy, Laplace showed great aptitude for mathematics at an early age. After brilliant studies at the University of Caen, he went to Paris at age twenty to rub elbows with the greatest minds of the period and forge his own reputation. Armed with glowing letters of recommendation from his professors, he went knocking at the door of the philosopher and mathematician Jean d'Alembert (1717–1783), one of the most influential personalities in French university life at the time. D'Alembert, too, had pondered over problems of celestial mechanics, including the mysteries of the Moon's movements. He had a brilliant mind, and had a reputation for often coming to the aid of promising young intellectuals. At first he refused to see Laplace, sending word that he had no interest in people showing up with letters of recommendation. Not discouraged, Laplace went home and wrote a long dissertation on the general principles of mechanics, which he submitted to the master. The originality of thought and obvious mastery of the subject Laplace demonstrated were enough to win d'Alembert's good graces. He opened his door wide to the young genius and offered him all his support. Things went so smoothly from there on that only a few days later, Laplace was appointed mathematics professor at the prestigious Military Academy.

D'Alembert never had to regret his decision. From that moment on until his death in 1827, Laplace did more than justify the trust his mentor had placed in him. He not only excelled in astronomy and mathematics, but he also demonstrated a great capacity for political survival as he went through the turmoil of French history, from the Revolution to the Empire, and then on to the Restoration, without ever putting himself in danger of losing the favor of the men in power. Napoleon Bonaparte even appointed him Interior Minister in 1799. But the talents that served him so well in the sciences (precision, attention to detail, constant inquisitiveness) did not help him much as an administrator. "He was unable to grasp the whole picture, seeing subtleties everywhere," Napoleon would later write about him from his cell at St. Helena. To use a modern term, Laplace was prone to "micromanagement," to the point that decisions were never made. His tenure as minister lasted only six weeks. But Napoleon continued to hold him in high esteem, and later granted him the title of marquis. The emperor was astute enough to recognize the prestige that prominent scientists could bring to his regime. In the context of the endless rivalry between France and England, he was not at all displeased to have the "French Newton" at his side.

Figure 33. *Pierre Simon de Laplace, champion of determinism*. He is shown here in a nineteenth-century lithograph, reading his work *Celestial Mechanics*. Laplace believed that the universe was strictly deterministic, that it worked like a clock mechanism and was governed by precise mathematical laws. One of the preeminent mathematicians of his day, Laplace not only contributed to our understanding of the movement of planets but was also one of the first to theorize about black holes, which he called "occluded celestial bodies."

THE HYPOTHESIS OF GOD IS NO LONGER NECESSARY

While this brief foray into politics hardly qualifies as a resounding success, Laplace's scientific career, on the other hand, was exceptional. As early as 1773, he tackled the problem of the solar system's stability. And a formidable problem it was, as Laplace's goal was nothing less than to take into account the mutual gravitational interactions of the Sun and the six planets known at the time. The three-body problem was tough enough as it was. But here was Laplace diving headlong into a seven-body problem!

People already knew at the time that several planets refused to comply with Newton's equations and exhibited irregularities in their motions. Newton believed that these deviations could lead to the breakup of the solar system unless the celestial machinery was reset from time to time as if by magic—Newton went so far as to postulate some divine intervention.

The cases of Saturn and Jupiter proved particularly hard to solve. In 1,000 years, Jupiter deviated from its calculated position by nearly one degree (twice the angular size of the full Moon), while Saturn was more than two degrees off where it was supposed to be. The young Laplace attacked the problem and showed that there was a resonance between the two planets. Saturn completed two orbits around the Sun in approximately the same time it took Jupiter to complete five. That brought them in roughly the same position in the ecliptic plane every 59 years or so. Just as a swing soars higher and higher when it is being pushed in synchronism with its oscillations, these periodic rendezvous between Jupiter and Saturn create repeated gravitational impulses that accumulate over time and eventually cause both planets to deviate from their normal orbits. However—and this is a crucial point—Laplace demonstrated that these deviations could not grow indefinitely. They reach a maximum and subsequently decrease back toward a minimum, only to grow again toward a new maximum, in a 900-year cycle. Irregularities are held back and wild swings are kept in check. They cannot grow out of control and become so large as to cause the solar system to break apart. Laplace convinced himself that the planets will pursue their appointed rounds about the Sun until the end of time. As far as he was concerned, the solar system was a marvelous cosmic clock, a well-oiled piece of machinery, moving solely through the force of universal gravitation. In his major work, *Traité de mécanique céleste* (Treatise on Celestial Mechanics), he set out to prove how mathematics rules the world, and to show that the great law of Nature discovered by Newton can explain with unmatched precision the most diverse motions, from the fall of an apple in an orchard to the movement of the Moon around Earth. It was in fact that work that inspired him to make his remarkably optimistic statement that an intelligence knowing at any given instant all the forces acting on the things in Nature and their positions could apprehend at once the past, present, and future of the universe. Nothing would remain uncertain. The cosmic clock is so well-lubricated that it keeps going on its own. Divine intervention is no longer required. As he was explaining his *Mécanique céleste* to Napoleon, the emperor asked him why he had made no mention of the Lord Architect. Laplace replied "Sire, I have no need for this hypothesis!"

THE DIGITAL ORRERY

Laplace's prestige and reputation, as well as his works of popularization (he gets credit for introducing the term "celestial mechanics" in the language of physics), contributed greatly to the diffusion of his ideas among the general public. His deterministic views were debated in many Parisian "salons" by mathematicians, philosophers, and other brilliant minds of the time. Doubts were often expressed that the solar system was really as stable and eternal as Laplace would have it. To be sure, he had demonstrated that the deviations of planets from their ideal orbits were small and would remain so. Yet examples abounded in ordinary life of small causes that could lead to large effects. What mountain climber has never watched the fall of an innocuous stone trigger a massive rock slide? We are all familiar with the precarious equilibrium of a sand pile. Add just a few grains and the whole pile collapses. Even with minute deviations, is it not possible for planetary orbits to become unstable? Is it preposterous to think that they might collide and end up ejected from the solar system? Despite Laplace's great stature, doubts lingered on. Laplace himself had to concede that Newton's laws did not explain everything. After all, everyone knew that the Moon stubbornly refused to conform to Newton's equations. Could there be aspects of the theory that had escaped this great physicist, illustrious as he may be?

Suspicions persisted through the next century. They were finally borne out when Poincaré discovered chaos at the end of the nineteenth century. As we have seen, he was confronted by the coexistence of order and disorder when he took on the three-body problem. The discovery reinforced the plausibility that the solar system might be unstable after all, and that it could carry the seeds of its own destruction and switch without warning from order to disorder.

To know for sure, one might argue, why not just extrapolate Newton's equations for the planetary orbits far enough out into the future? After all, Laplace had studied Saturn and Jupiter over just a few 900-year cycles, a mere eyeblink in the history of the solar system, which is 4.6 billion years old and only halfway through its life expectancy. The only anticipated catastrophe will occur 4.5 billion years from now, when the Sun will become a "red giant." At that time, it will have consumed all its hydrogen fuel and will start dipping into its helium reserves. This new energy source will inflate the Sun to one hundred times its present size. Its fiery envelope will engulf Mercury and Venus. Living beings on Earth will see the sky become red hot, and the inferno will extinguish all life. In order to survive, our descendants will be forced to migrate to the edge of the solar system, perhaps to Pluto, to escape the burning tentacles of the red giant.

That said, what is the best way to extrapolate equations into the future?

We already know that a three-body problem does not possess an exact alge-braical solution. Things are even messier for a problem involving ten bod-ies—the Sun and nine planets. Algebra being powerless, astronomers have turned to the most precious ally of modern astrophysics—the computer. But even computer calculations pose formidable challenges. There are many reasons for this. First of all, planets move around the Sun at very different rates. The farther out the planet, the longer it takes to complete one orbit around the Sun (this is one of Kepler's laws of planetary motion, which states that the period of revolution is proportional to the distance between the Sun and the planet of interest raised to the power $3/2$). Thus Mercury, the closest planet to the Sun, hurriedly completes a revolution in 88 days, fully justifying its designation of "prompt messenger of the gods." Pluto, on the other hand, the most distant planet, takes its sweet time and moves at a snail's pace, requiring nearly two and a half centuries to go once around the Sun. Now, to calculate the evolution of the solar system as a whole, it is always the fastest planet—in this case, Mercury—that dictates the tempo. Since Mercury needs to be tracked on a daily basis as it speeds around the Sun, there is no choice but to do the same for Pluto as well. Needless to say, this entails an enormous waste of calculations for the outer planets, since their tempos are hundreds, if not thousands, of times slower than that of Mercury. Pluto needs to be checked only once every few years, not days. It is as if you were to photograph a race between a hare and a turtle. Because of the swiftness of the hare, you would be forced to take a picture every few minutes to keep track of the relative progress of both runners. The position of the hare would change quite a bit from one snapshot to the next, whereas the turtle would hardly move at all. A lot of film is going to be wasted as far as the turtle is concerned.

In addition, chaos may remain in hiding for hundreds of thousands of years, perhaps millions, before something interesting happens. Therefore, computers with phenomenal computational power are required to handle the job. Until the 1970s, even the most advanced machines did not have the necessary power. Researchers were forced to play with what might be called "toy solar systems," cutting out the inner planets, to retain only the outer ones (Jupiter, Saturn, Uranus, Neptune, and Pluto). That strategy allowed them to get rid of the hare and work only with the turtle, which made it pos-sible to fast-forward the calculation of orbits. Instead of following the evolu-tion of the system day after day, they could safely leapfrog across several years at a time. But even with a truncated and simplified solar system, by the end of 1983, no one had ever pushed a calculation farther than about 5 mil-lion years into the future. In that time span, no chaotic phenomenon had turned up. Could Laplace have been right all along? Could the solar system really be stable? Will the dance of planets go on forever?

The plot thickened in 1984 when two American researchers at MIT, Jack Wisdom (the very same one who uncovered chaos in the asteroid belt, and who in the meantime had left Caltech) and Gerald Sussman, developed a computer specially designed to calculate planetary orbits, which they nicknamed a "digital orrery." Just like the mechanical marvels of the eighteenth and nineteenth centuries, the digital orrery simulated the movements of planets. But instead of a mechanical assemblage of spheres of various sizes mounted on metal arms rotating at different velocities around a central point by means of a system of cogwheels as elaborate and precise as a watch, the digital orrery was made of electronic circuits, and planetary orbits were represented by numbers. However, before using this clever device to embark on a trip into the future of the solar system, it was of paramount importance to make sure that any orbital irregularity or deviation exhibited by a planet was real, and not due to the limited memory of the computer. As we know, computers can store in their memory only numbers of finite length. They must of necessity truncate them. For instance, if a computer can store only six decimal places, the number 1.4576952 will be rounded off to 1.457695. Such errors are extremely small for individual numbers, but they unfortunately add up and can grow quite large after millions of operations. This limited numerical precision can cause artificial orbital irregularities that have nothing to do whatsoever with reality. Another source of errors is the fact that the movement of a planet is obviously continuous, whereas integrating an equation numerically is an inherently discrete process. Time is divided into small intervals and planets are moved in finite steps from one interval to the next. The trick is to make sure that all these errors remain negligible and do not lead to erroneous conclusions.

This can be verified by subjecting the planet to a complete round-trip in time. For instance, the positions of Pluto can be calculated up to 200 million years into the future, at which point time is reversed and the evolution of the planet is followed all the way back to time zero. A perfect, errorless process would bring the planet back exactly to its starting point. If not, the initial and final positions will show a discrepancy. By minimizing all possible sources of errors, Wisdom and Sussman managed the extraordinary feat of transporting Pluto 845 million years out into the future and back, with the starting and ending positions differing by no more than one-sixth of a degree, which amounts to only one-third the angular size of the full moon. With such a small error, the stage was set to start hunting for chaos in the solar system.

UNPREDICTABLE PLUTO

The two scientists began an odyssey through time, traveling 845 million years into the future, or roughly one-fifth the age of the solar system. Here again, to get around the problem of the hare and the turtle, they decided to dispense with the four inner planets (Mercury, Venus, Earth, and Mars). Despite this simplification, it took the digital orrery five months of uninterrupted calculations to complete this journey through time. At the conclusion of this long trip, did chaos manage to rear its head? Did planetary orbits allow themselves to be taken over by disorder? The answer was negative for most planets. Chaos was conspicuous by its absence in the case of the four giant planets—Jupiter, Saturn, Uranus, and Neptune. To be sure, they did show some minor variations in their inclinations and in the shape of their orbits, but nothing to rave about. Only Pluto, the most distant planet, named after the god of hell, did not exhibit quite the same exemplary behavior.

Pluto has always been somewhat of a misfit among the outer planets. As we have seen, its orbit is so abnormally elongated that it can come closer to the Sun than Neptune, as it did last in 1989. Since their orbits intersect, one might worry that Pluto and Neptune could collide at some point. Yet that has never happened. The two planets skillfully avoid each other in a carefully choreographed ballet. That is due to yet another resonance. It turns out that Neptune goes around the Sun about three times during the time it takes Pluto to execute two revolutions. As a result, whenever Pluto crosses Neptune's track, Neptune is on the opposite side of its orbit, and the two planets are as far away as possible from each other.

In their journey forward in time, Wisdom and Sussman uncovered other resonances of Pluto. They found that its orbital inclination oscillates with periods of 3.8 and 34 million years, and that its closest approach to the Sun (called perihelion) occurs at the same place in its orbit every 3.7, 27, and 137 million years. Yet, in spite of all these periodic phenomena, Pluto adamantly refused to go hog-wild and just went on to pursue its ho-hum trek around the sun.

A nagging suspicion persisted, though, that these periodicities might be telling us that Pluto is just on the verge of chaos, and that it would take very little to push it over the edge from order to disorder, from tranquillity to turmoil. Perhaps Pluto is like a pile of snow in precarious equilibrium on a ledge, poised to turn into an avalanche at the slightest disturbance from a small rock, or like a sandpile about to collapse the moment a single grain is added.

To get to the bottom of the matter, Wisdom and Sussman decided to conduct an experiment. They placed two fictitious Plutos side by side, with nearly identical positions and velocities, and followed their respective orbits in phase space with their digital orrery. If the trajectories diverged, chaos is present; if they tracked each other, it is not. As it turned out, the trajectories did diverge exponentially. The distance separating the two Plutos doubled every 2 million years and the doubling continued unabated, so that after 25 million years, the initial separation had been amplified by a factor of some 5,800. The inescapable conclusion is that Pluto's motion is fundamentally no more predictable than next month's weather. A tiny change in velocity or position leads to completely different trajectories. Chaos is an integral part of Pluto's future.

But that does not mean Pluto will show up tomorrow in Earth's backyard, crash into Mars, or that it is about to leave the solar system for good. After all, it has spent more than 4 billion years revolving uneventfully around the Sun. Chaos is indeed embedded in the motion of Pluto, but it is controlled and kept in check. Perhaps it is responsible for the unusual orbit of the planet. Some people think Pluto was born in the ecliptic plane, just like all its cousins, but that a series of chaotic events left it in its present orbit.

The discovery of chaos in Pluto raises some intriguing questions about the other planets. Perhaps chaos has not been detected in them simply because the digital orrery has not pushed calculations far enough into the future. We should bear in mind that chaos just loves company. It is contagious! It seems the time has come to use even more powerful computers, able to handle a complete solar system, including all nine planets, and perform calculations farther out in time.

THE UNCERTAIN FUTURE OF THE SOLAR SYSTEM

The torch was picked up by the French astronomer Jacques Laskar, who works at the Bureau des Longitudes in Paris. This venerable institution was created in 1795 by the National Convention, at the urging of Abbé Grégoire, for the purpose of reclaiming mastery of the seas from the British. That could only be done if one knew how to determine accurately the position of ships at sea, which required calculating longitudes from the position of stars in the sky. The charter of the Bureau des Longitudes was to calculate as precisely as possible the trajectories of celestial bodies. Once a year, it publishes an ephemeris—that is, a set of tables listing the positions of the Sun, Moon, and all the planets during the coming year. Not only do these tables make it possible to calculate the positions of ships, they are also useful for a number of other purposes. Astronomers check them to plan their observations, and NASA uses them to send space probes toward celestial objects in the solar system. Calendars and public holidays are based on these ephemerides. For example, the Paris mosque traditionally consults with the Bureau des Longitudes to set the official date of Ramadan, which is tied to the appearance of the New Moon. The Bureau sometimes entertains some rather unexpected "clients." An insurer may want to know if a driver had the sun in his eyes at the time of his accident. A police investigator may inquire whether there was a full Moon or it was pitch black when a crime was committed. An architect may need help in properly orienting a house he is building in relation to the Sun's position. A film director may want to find out if the Moon will be out when he plans to shoot a scene at night. On a completely different time scale, a paleontologist may request details on the amount of sunlight falling on Earth 50 million years ago. Meanwhile, astronomers want to learn the fate of the solar system in a few hundred million years from now: Will it remain stable, or will it break apart, each planet going off in a different direction?

We have seen that Laplace, one of the founding members of the Bureau des Longitudes, had opted in favor of stability. Following in the footsteps of his illustrious predecessor, Jacques Laskar encountered chaos instead. He entered into his computer a mathematical expression that included some 150,000 algebraic terms describing the average behavior of planets in their orbits around the sun. Laskar was not interested in short-term variations, small deviations here and there, over thousands or even tens of thousands of years, after which time the planet returns to the fold. Rather, he was looking for long-term changes, much as economists, in order to study the stock market, pay little attention to the daily fluctuations of individual shares but concentrate on the trend of a pool of shares (for instance, the Dow Jones industrial index) averaged over a period of thirty days.

Laskar was able to transport the solar system 200 million years into the future. To look for chaos, he had to repeat the calculations with slightly different initial conditions. As usual, the key test is whether the two trajectories in abstract phase space follow each other closely, with deviations increasing *proportionally* with time, or whether they diverge *exponentially*, doubling and doubling again in a time interval that is increasingly shorter as the system is more chaotic.

The answer provided by the computer was unambiguous. It turns out that the entire solar system, including the inner planets, is subject to chaos. Pluto's unpredictability is the rule rather than the exception. Tiny differences in initial position get amplified to such an extent that the resulting trajectories diverge wildly. This extremely sensitive dependence on initial conditions implies that the present is disconnected from both the future and the past. The future cannot be predicted, and the past is lost forever.

Laskar noticed that the separation between two trajectories of any one of the planets with very slightly different initial conditions would double every 3.5 million years. Two imaginary Earths separated by only 100 meters at the start would find themselves 40 million kilometers apart, or one-third of Earth–Sun distance, 100 million years later. This can only mean that planetary orbits have an undetermined past and an uncertain future, since the position of a planet is never known with unlimited precision. Luckily, as we found out with Pluto, chance cannot always do as it pleases. It is subject to restrictions. The uncertainty concerning the future evolution of orbits does not automatically imply that Earth is about to wander toward Venus or that it is ready to escape from the solar system altogether. In all likelihood, Earth will continue to revolve around the Sun in her present orbit. But no one can be absolutely sure. There will always remain a slight chance that it will not. Laplace's dream of absolute determinism has been shattered into pieces.

GOD PLAYS DICE WITH THE PLANETS

Our ability to predict the future of the inner planets (Mercury, Venus, Earth, and Mars) diminishes considerably beyond a few tens of millions of years. The demise of Newtonian and Laplacian determinism allowed chance to take over the macroscopic world. It had already invaded the realm of atoms with the advent of quantum mechanics in the early part of the twentieth century. Bohr, Heisenberg, and Pauli taught us that we should not think of the trajectory of an electron the way we think of the orbit of Jupiter around the Sun. We can never specify both the position and the velocity of an electron. Like Janus, it has two faces: It is at once particle and wave. It behaves like a particle whenever you try to observe it. But when you look the other way, it takes on the attributes of a wave. Instead of dutifully following a single orbit around an atomic nucleus, it jumps around, twirls and pirouettes, and occupies all of an atom's empty space, a bit like the ripples caused by a stone tossed in a pond spread over the entire surface. The electron is not constrained by the shackles of determinism. It is impossible to predict where it is at any given instant. The best we can do is estimate the probability of finding it somewhere. Like waves in the ocean, the wave associated with an electron has crests and troughs. An electron is more likely to turn up at a crest, but there is no guarantee that it will actually show up there. The probability that it will is high, but is never equal to one.

Thus it is that the world of atoms fell under the sway of chance. People used to think that at least the macroscopic world—particularly celestial mechanics—would never fall to the assaults of indeterminism. Henri Poincaré discovered breaches in the fortification as early as the end of the nineteenth century. Chaos came rushing in and the stronghold of planets—until then the epitome of determinism and stability—fell apart before the twentieth century was out. "God does not play dice," Einstein assured us. He was an inveterate determinist when it came to quantum mechanics. He could not bear the thought that chance should rule the world of atoms. What would he have said had he found out that God in fact plays dice even with planets?

The notion that planets are an impregnable symbol of constancy was so universally accepted that Laskar's results were met with considerable skepticism at first. Criticisms came from all sides. Some argued that the average behavior of planets was not the right thing to focus on. Others suggested that the divergence of planetary orbits may have been the result not of chaos but simply of round-off numerical errors. Laskar tried to answer these charges by identifying plausible reasons that may favor chaos in the solar system. That led him straight to the issue of resonance. Subtle gravitational interactions exist between Mars and Earth, on the one hand, and between Mercury, Venus, and Jupiter, on the other. These resonances undermine predictability. They are the enemy of determinism. Just as pushing a swing in resonance with its oscillations propels it ever higher, to the squealing delight of little children, repeated gravitational nudges amplify what starts out as small differences until unpredictability takes command.

In science, the best way to validate a result is to do the calculation over again or repeat the measurement. Better yet, let other people do it, preferably with completely different methods or instruments. It is in that spirit that other astronomers began to look for chaos in the solar system. Some would not be content to just follow the average behavior of planets, as Laskar had done. Instead, they scrutinized the future evolution of each planet in considerable detail. Others built specially designed computers to move the solar system forward in time. In the end, the verdict was unanimous. Regardless of the calculation technique or the computing hardware, the fictitious solar systems studied invariably exhibited the telltale signs of chaos. Tiny changes in the initial conditions would consistently make orbits diverge exponentially after periods ranging from 3 to 30 million years. What's more, chaos was not restricted to the inner planets; it affected the outer ones with equal vigor. There are many orbits between Uranus and Neptune that turn out to be unstable. Place some celestial bodies there, and half of them will find themselves thrown out of the solar system after about 5 billion years. One can reasonably infer from this that a few billion years ago the solar system may well have included a number of additional smaller planets (in addition to the nine we know today), roughly the size of the Moon, which subsequently were ejected. Only the more massive ones, less vulnerable to expulsion, remained.

The role chaos played in the early stages of the formation of the solar system remains shrouded in mystery. Has the solar system formed rapidly, evolving in a few million years from a primitive nebula to its present state, with most of the matter accreting into planets and leaving behind very little debris? If so, chaos could not have been a leading player. Or was the process more gradual, taking billions of years, with lots of debris swarming between embryonic planets, in which case chaos would have acted as "street sweeper"? No one knows for sure. Chaos may even have influenced the fate of a putative tenth planet in the solar system. The additional mass might have been enough to upset the precarious equilibrium of the system, causing the extra planet to be ejected.

NONCATASTROPHIC CHAOS

Computer models of solar systems all display the symptoms of chaos. But if the disease is so endemic, an intriguing question comes up. The solar system didn't pop up just yesterday; it formed 4.6 billion years ago. Yet the planets have stuck to their stately orbits around the Sun. How is it that during all that time, none has gone berserk, crossing paths with their neighbors, and been reduced to smithereens in catastrophic collisions? The answer is that chaos is not necessarily synonymous with disorder, madness, and instability. Pluto has been a model of restraint and moderation for billions of years, even though we cannot predict its orbit more than a few tens of millions of years out. We are faced once again with what might be called "controlled chaos." There is always a small risk that the solar system will disintegrate at some future time and that the planets will go off in random directions in the next few million years, but the probability of that happening is extremely small. If a computer were to calculate the evolution of a billion "toy solar systems," nothing would happen in the 4.5 billion years to come for the vast majority of them. Earth would carry on its annual trek around the Sun, spring would keep on enchanting us with its exuberance of colors, and winter would continue to torment us with its biting cold. Out of a billion toy solar systems, perhaps one might suddenly go mad and break apart.

When gravitational perturbations remain weak enough and are not amplified by resonance phenomena, as seems to be the case in the solar sys-

Figure 34. *The chaos of star orbits*. The disk of the Milky Way, viewed face-on, would look very similar to the galaxy M 51 (M stands for "Messier") seen here. It is a spiral galaxy—so-called because the young stars arrange themselves in the form of spiral—located some 15 million light-years away from the Earth. The diffuse luminous blob at the end of one of the spiral arms is a dwarf galaxy (NGC 5195, a satellite of M 51). The stars in the Milky Way (the Sun being one of them) orbit tirelessly around a central mass called "bulge." In particular, the Sun completes one revolution around the Milky Way every 250 million years. Its orbit is stable, although the orbits of stars with a higher kinetic energy can become chaotic. (Photo courtesy of Anglo-Australian Observatory)

tem, randomness is said to be "controlled." It is precisely because chaos cannot give free rein to all its eccentricities that Newton and Laplace thought of the solar system as a perfectly oiled piece of machinery whose past, present, and future could be determined with absolute certainty. The chaos lurking in the corners of the universe thus constitutes a bridge between the abstract and idealized laws of physics and the complexity and disorder of the concrete world in which we live.

CHAOTIC STAR ORBITS

During clear summer nights, a swath of pale white arches across the vault of heaven. Its whitish tinge inspired the ancients to name it the "Milky Way." We know today, of course, that it is simply the disk of our galaxy, a collection of some 100 billion suns herded together by gravity. Because we are part of that disk, we only get to see it edgewise—hence this luminous band slicing through the sky. Stars are not standing still in this flattened disk. They revolve around the galactic center (Figure 34) in a huge cosmic carousel. Our Sun is only a suburban star toward the periphery, some 30,000 light-years from the center of the Milky Way and two-thirds of the way out toward the edge of the galaxy. The Sun carries the entire solar system with it through space at 250

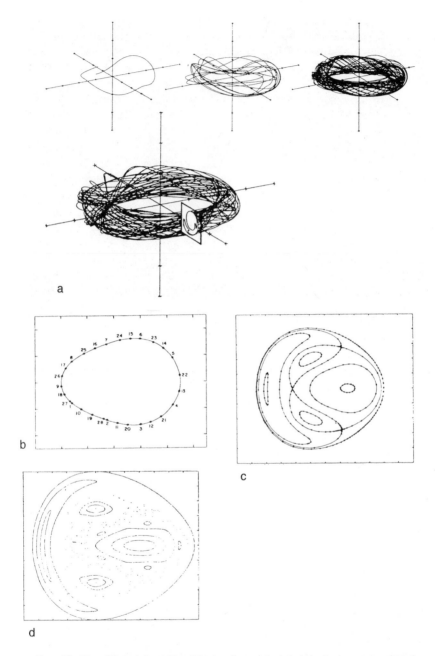

Figure 35. *Star orbits and chaos*. Star orbits describe complex trajectories in phase space, shaped somewhat like a torus *(a)*. To visualize these orbits, Poincaré used a vertical plane slicing through phase space. The patterns traced by the intersection points of the trajectories with Poincaré's plane turn out to be continuous and closed curves as long as the kinetic energy of the stars remains below a certain critical value: Star orbits are then stable *(b)* and *(c)*. As soon as the kinetic energy exceeds a threshold, the orbits become chaotic and the trajectories form patterns in Poincaré's plane characterized by zones of stability intermixed with zones of chaos *(d)*.

kilometers per second, completing a revolution in 250 million years. Since it was born 4.6 billion years ago, the Sun has gone 18 times around.

Since Newton's universal gravitation dictates the movements of the stars in the disk, a question arises: If chaos lurks in the heart of Newton's equations, why should it not also infect the stars?

The French astronomer Michel Hénon, of the observatory of Nice, decided to settle the question. To visualize the movement of stars in the galactic disk, he, too, resorted to Poincaré's vertical plane method (Figure 35a). Each point on the surface would represent a crossing of the plane by a star's trajectory in phase space. If the star's orbit is constant, the point is fixed. If not, the orbit does not close on itself in phase space, and the associated point starts wandering in Poincaré's plane. Hénon watched the points move about and saw chaos sneak into the world of stars as well. In fairness, chaos did not manifest itself right away. The first orbits Hénon calculated for stars with a moderate kinetic energy behaved quite well. Even though they were not completely regular and did not repeat themselves exactly, they were by no means unpredictable. The points did not move haphazardly throughout Poincaré's plane but traced curves with well-defined shapes, somewhat resembling an egg (Figure 35b). That meant that stars moved in phase space in a volume shaped like a bicycle tire, called a *torus* (Figure 35a). Hénon became curious about what would happen if he increased the stars' kinetic energy. When he tried it, the egglike curve got distorted into more complicated shapes, such as a figure eight, or even split into separate loops (Figure 35c). But the orbits remained stable and there was still no sign of chaos. Hénon increased the energy some more, and suddenly, as if by magic, chaos made its appearance. The points started moving erratically in Poincaré's plane. In some regions, it was still possible to join them by a continuous curve. But in others, that became virtually impossible. Islands of order popped up here and there, lost in a vast ocean of disorder (Figure 35d). Star orbits had become unstable, and chaos triumphed. Two nearly contiguous points in Poincaré's plane could belong to completely different orbits. That is the signature of chaos: Change the energy of a star ever so slightly, and its orbit becomes unpredictable.

STRANGE ATTRACTORS

Just as we learned in the case of planets, chaos does not mean total disarray. Chaos is deterministic and disorder controlled. In abstract phase space, the points where orbits intersect with Poincaré's vertical plane trace well-defined figures. Instead of being a shapeless and jumbled scribble, they follow lovely patterns that are extremely pleasing to the eye. The points do not scatter randomly, but seem "drawn" to curves with unusual and wonderful shapes. Physicists have dubbed them "strange attractors."

The word "strange" emphasizes that these lovely figures result from two antagonistic trends. On the one hand is a tendency to converge, as all orbits are irresistibly drawn to particular patterns, just as a nail is attracted to a magnet. On the other is a tendency toward divergence, since orbits that are initially almost indistinguishable drift apart exponentially after a sufficiently long time. These patterns are also strange because they are self-similar, meaning that they look alike on any spatial scale. Admire the delicate lines of a strange attractor; then take a magnifying glass and zero in on a small part of it: You will find that the pattern is the same, only smaller. Magnify it some more, and once again you'll run into the same pattern. The same leitmotiv keeps recurring no matter what the magnification.

Hénon discovered that the strange attractor of star orbits is shaped like a banana. The lines tracing this curved figure turn out to have a finite thickness, and they have a most peculiar property: They keep on dividing endlessly (Figure 36). If you were to use ever more powerful magnifying glasses, you would find that two lines that have just subdivided from a single one in turn split into

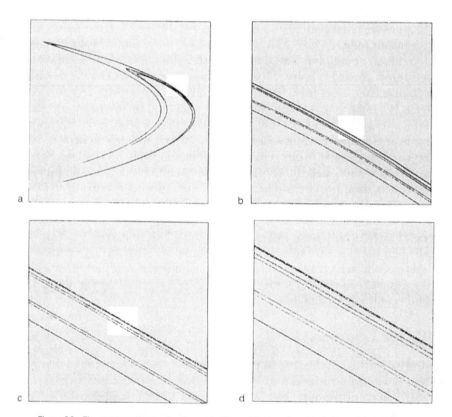

Figure 36. *The strange attractor for star orbits*. Star orbits describe a kind of torus in phase space (see Figure 35a). The intersection of this torus with Poincaré's vertical plane produces a figure shaped like a banana *(a)*. Magnifying the small square in *(a)* yields *(b)*. Each line that seemed to be single in *(a)* turns out to split in two in *(b)*. Likewise, magnifying the small square in *(b)* yields *(c)*, where again each line splits in two. Likewise, magnifying the small square in *(c)* yields *(d)*, and so on. The splitting of lines continues *ad infinitum* with each successive magnification. Orbits whose strange attractor is this banana-shaped figure are chaotic; it is impossible to predict which particular line and which location the point corresponding to the next star orbit will happen to fall on.

two more each, only to do it again farther on down, and so on *ad infinitum*, like an endless set of Russian Matryoshka dolls nested in one another. A point representing one star orbit will always lie on one of those lines, but it is impossible to guess exactly where the point corresponding to the next orbit will turn up, except that it will be somewhere on the strange attractor.

Strange attractors do not apply exclusively to the orbits of stars. Edward Lorenz, the man who exposed chaos in meteorology, as mentioned earlier, discovered a strange attractor there too. The complex movements of air masses that bring us rain and sunshine can be represented by a point that darts and flitters about in abstract phase space, tracing beautiful patterns. Lorenz has shown that the points always collect in a strange attractor shaped like a pair of butterfly wings, with loops and spirals crowding endlessly but never repeating themselves (Figure 37). Here again, things look like an endless set of Matryoshka dolls. A magnifying glass reveals curves splitting into two branches, then four, eight, and so on to infinity.

This endless division raises a troubling question. How can a finite space contain something that is infinite? The answer is as astonishing as it is profound: What seems impossible can happen because strange attractors have a fractional dimension.

THE COAST OF BRITTANY IS A FRACTAL OBJECT

Back in our school days, we were all taught that familiar objects have a dimension that can be expressed as an integer. A straight line has a dimension of 1, a planar surface 2, while the space in which we go about our normal business has a dimension of 3 (up-down, left-right, and forward-backward). In the early 1970s, however, we became aware of a new category of objects whose dimension can be expressed only in terms of fractions, such as $9/5$ or $5/2$. Such objects were dubbed "fractals" by Benoit Mandelbrot (1924–), the French-American mathematician who discovered them.

You might think that objects whose dimension is not an integer must be pretty bizarre, with convoluted and phantasmagorical shapes existing only in the fanciful imagination of mathematicians. You would be quite wrong. On the contrary, objects with fractional dimensions are so familiar that we come in contact with them every day. Some can even elicit in us a feeling of indescribable beauty, such as a snowflake on a window in the wintertime, the delicate contour of a cloud drifting in the blue sky, or the intricate pattern of a leaf growing on a tree. How do we recognize such objects with a nonintegral dimension? The answer is quite simple: They have irregular shapes that do not fit into the framework of Euclidean geometry. Their irregularities also repeat themselves on any magnification scale.

Consider the example of the coast of Brittany in France (Figure 38). If you look at it through the window of an airplane cruising at 35,000 feet above the ground, you will recognize its overall irregular shape, but you won't see any of its finer details, like its beautiful beaches or its glorious bays filled with crystal clear water. Now, if you drive along the coast, looking at

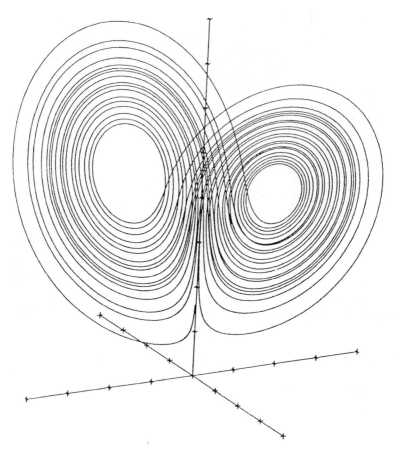

Figure 37. *The Lorenz attractor*. Air is subject to convection movements owing to the fact that Earth's atmosphere is cooler at higher altitudes. Warm air rises, which causes it to cool down and come back down. As it comes down, it becomes warmer, which sends it back up. While studying these air convection movements in phase space, the meteorologist Edward Lorenz discovered that the representative point traced a strange and attractive figure in Poincaré's plane, shaped like a pair of butterfly wings. The point never repeated the same movement twice, but kept on tracing new loops. It was irresistibly drawn toward those loops—hence the name "strange attractor" given to the pattern traced in Poincaré's plane. The point might describe one loop in the left wing, followed by two loops in the right one, only to switch back to the one at left. The movement of the point was chaotic in the sense that it was impossible to predict where on a loop and on which loop it would show up next.

the landscape from a few tens of meters away, you will be able to appreciate those beaches and bays in all their beauty. But you would still miss the smaller nooks and crannies. Those you can see only if you walk along the beach. But even then, you would fail to catch more subtle details, such as tiny grains of sand or minuscule fragments of shells or rocks. Imagine now that you become an ant crawling along the shore a millimeter at a time. You would then discover a whole new world in which the tiniest grain of sand looks like a mountain and the smallest crevice like a deep canyon. Thus it is

Figure 38. *The coast of France is a fractal object.* This image, taken by the satellite Météosat, shows the coast of France with a resolution of a few tens of kilometers. Even at this relatively coarse resolution, the coast is clearly not smooth but jagged. If the satellite were at a lower altitude and looking at the coast in greater detail, pictures would show even more irregularities. In fact, the irregularities of the coast manifest themselves at any magnification, with the same patterns repeating themselves from one scale to the next. The degree of irregularity of the coast can be quantified by a dimension, although that dimension is not necessarily an integer, as is the case with a Euclidean geometrical figure; rather, it is a fractional number. The dimension of the coast turns out to be between 1 (the dimension of a line) and 2 (the dimension of an area). The coast of France is an example of a fractal structure in Nature. (Photo courtesy of the European Space Agency)

that irregularities manifest themselves at any magnification, and patterns repeat themselves from one scale to the next. Bays, coves, and beaches reveal sub-bays, sub-coves, and sub-beaches that, in turn, unveil sub-sub-bays, sub-sub-coves, and sub-sub-beaches, with no end in sight. This replication goes on all the way to the level of atoms, on a scale of one hundredth of a millionth of a centimeter.

You may ask yourself how long the coast of Brittany is. You will find out that there is no single answer to the question, because it depends on the scale on which you set out to do your surveying. From an airplane, many twists and turns smaller than one kilometer will be neglected. The answer you will get will for sure be smaller than the true length of the coast. From a car, you will get a better sense of the jagged lines of the bays and coves, but features smaller than a few meters will still escape you. The length you will measure then will be greater than before, because more details will have been taken into account, but it will still fall short of the true value. Suppose now you walk along the coastline, armed with a yardstick. This time, your answer will be closer to the truth, but it is still bound to be smaller than the real number, because nooks and crannies less than a meter will have been ignored. And so on *ad infinitum.* The ant, which has to go around the tiniest grains of sand, is sure to come up with a greater length than you will with your yardstick. In short, the answer depends on how the observer looks at the object being measured (in this case, the coast of Brittany). How large or small the estimate comes out is directly related to whether the measuring is done from far away or up close. This interrelation between observer and observed is reminiscent of what happens in the world of atoms, where the mere act of observing an atom perturbs and alters its properties.

THE GEOMETRY OF THE IRREGULAR

Common sense would tell you that the ever-increasing estimates of the length of the coast of Brittany have to converge toward a final value that is the true length. And you would be right if the coast of Brittany were describable by a Euclidean geometrical figure. A circle is a good example of such a figure. To measure its length, the method of summation of increasingly shorter inscribed segments does indeed converge toward the true perimeter of the circle. You start by approximating the circle by an inscribed triangle, then a square, and on to a pentagon, a hexagon, and so on. The perimeters of these figures contained within the circle will approach ever more closely that of the circle itself. Unfortunately, that turns out not to be true of any irregular figure, such as the coast of Brittany. Mandelbrot discovered that as the scale of the measurement decreases, the measured length of the coast keeps on increasing without limit until it becomes infinite, which seems to fly in the face of common sense.

In fact, Euclidean geometry breaks down the moment one tries to describe any object as irregular as the coast of Brittany, and more generally anything that is convoluted, dislocated, and rough. It cannot describe what is not smooth and rounded. It is helpless when faced with objects that are irregular, jumbled, and intertwined. As it happens, the vast majority of objects around us are characterized not by regularity, but by irregularity. As Mandelbrot is quick to point out, "Clouds are not spheres, mountains are not cones, coastlines are not circles, and thunderbolts are not straight lines." The shapes dealt with by ordinary Euclidean geometry (spheres, circles, cones, and other regular figures) constitute a powerful abstraction of reality. This type of geometry has been enormously useful and has influenced our view of the world for over two millennia. It is what most children still learn in school to this day. It was a guiding principle for Plato, who believed the world to be but an imperfect reflection of a perfect world with perfect Euclidean shapes. Because the circle was viewed as the most perfect Euclidean shape, Ptolemy believed that planets moved on circles whose centers rotated themselves on celestial spheres centered on Earth. This erroneous description of the world survived for more than twenty centuries. Yet, as useful and venerable as they were, Euclidean concepts had run their course. In order to describe the complexity of the world, it became necessary to invent a new language—the language of the irregular. Since the traditional Euclidean concepts of length, width, and thickness gave answers that did not mesh with reality, it was time to discard them. Mandelbrot came up with a different notion to describe the realm of the irregular—the concept of dimension.

FRACTAL OBJECTS, OR REGULARITY IN THE MIDST OF IRREGULARITY

The concept of dimension is not new. It exists as well in the geometry invented by Euclid (ca. 365–ca. 300 B.C.). However, Mandelbrot seized on the idea, proposed in 1919 by the German mathematician Felix Hausdorff, of fractional dimensions—that is to say, dimensions expressed by fractions rather than integers. A fractional dimension provides a means to express the roughness and irregularity of an object. The length of the rugged coast of Brittany, as we have explained, is not very a well-defined quantity. On the other hand, its jaggedness, or degree of irregularity, can be quantified by a fractional dimension, as the following argument will make clear.

Since the coast of Brittany does not follow a smooth curve, its dimension must be greater than 1. By the same token, it has to be smaller than 2, since the coastline does not spread so much as to occupy a whole surface. Thus, it is somewhere between 1 and 2. The number will be closer to 1 for a coast that is relatively smooth and regular, and closer to 2 for one that is jagged and zigzags to enclose a larger surface.

The fractional dimension is thus a measure of the unevenness of an object—in other words, of how efficiently it occupies space. Mandelbrot coined the word *fractal* to describe such irregularly shaped objects, both because the Latin word *fractus* means broken, irregular, and because *fractal* evokes the word *fraction*.

Fractal objects have a rather peculiar property: Their lack of regularity is not random. Indeed, there is a measure of regularity in the midst of their irregularity. The degree of irregularity remains the same on different scales. What is more, a fractal object has the same appearance whether it is looked at from a distance or from up close. As you get closer to it, small details come into focus, but the object looks just as irregular as it did when viewed in its entirety from afar. Fractal objects possess what is called "scale invariance." They appear the same regardless of the magnification. They are characterized by patterns within patterns within patterns, all looking the same, like an endless series of Matryoshka dolls nested in one another, or like a fish swallowing a smaller one, itself swallowing a smaller one still, and so on ad infinitum. The same irregularity shows up on every scale (Figure 39).

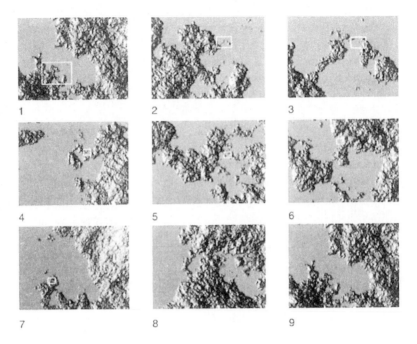

Figure 39. *A computer-generated fractal coast.* This series of nine pictures shows with increasing magnifications the topography of a coast with constant fractal dimension. Each picture is a magnified version of the area indicated by a white rectangle in the previous picture; the magnification increases from left to right and from top to bottom. The degree of irregularity is seen to remain the same regardless of the magnification. This computer-generated coast bears a striking resemblance to real-life coasts: Nature resorts to fractal structures to construct reality.

NATURE LOVES FRACTAL STRUCTURES

Fractals have found applications in the most diverse and surprising areas, from geology to physiology to botany. Nature seems to have a great affinity for fractal structures. In the field of geology, fractal geometry proved to be a powerful tool to describe the topography of Earth's surface. A case in point is a pile of rocks lying at the foot of a mountain. Viewed from afar, the rock slide outlined against the horizon seems to have no depth whatsoever. It seems to have a dimension of 2. But as you get closer, the rock slide reveals a multitude of individual stones, some of which are as large as an automobile. These rocks almost fill a three-dimensional volume, although not completely, as they do not fit perfectly and leave a network of interstices between them. Therefore, the fractal dimension of the slide is less than 3—its exact value turns out to be 2.7.

Fractal structures can also be found in the human body. The network of blood vessels, from the main aorta to small capillaries, has a fractal nature. The large arteries branch off into medium-size arteries, which in turn branch off into arterioles. That is the most efficient solution Nature could find to cram the enormous surface area of blood vessels within the limited volume of a human body. Nature did a marvelous job at handling this dimensional trick. Even though the circulatory system occupies only about 5 percent of a human body's volume, its fractal structure ensures that, for most tissues, no cell is ever very far away from a blood vessel.

There are other organs in the human body that possess a fractal structure. The goal is always to pack a large surface area into as small a volume as possible. The surface area of human lungs is larger than that of a tennis court! Yet, all this surface fits miraculously within the thoracic cage. Once again, Nature has resorted to a strategy of ramifications: Bronchia split off into sub-bronchia, which in turn split off into bronchioles. The urinary system, the bile duct in the liver, the fibers that transmit electrical signals to the heart, all of these turn out to have a fractal organization. All have structures that branch off into substructures themselves branching off into sub-substructures, in a pattern that reappears from one scale to the next. It is but a small step from the pulmonary system to botany when it comes to branching patterns. Indeed, the ramifications of the branches of a tree, the delicate motif of a leaf, or even the complex patterns of bark constitute as many fractal structures (Figure 40).

a

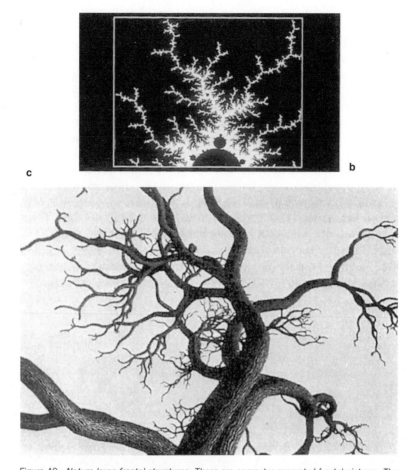

Figure 40. *Nature loves fractal structures*. These are computer-generated fractal pictures. The information needed to produce such pictures is programmed with just a few very simple rules of the type $z(final) = z^2(initial) + C$, as described in the legend of Figure 42. Yet these pictures are remarkably similar to real-life objects: ferns and leaves growing on trees *(a)*, the zigzags of storm lightnings *(b)*, and the ramifications of tree branches *(c)*. Nature seems to apply the rules of fractal mathematics in building complexity.

But fractals are more than just Nature's language. They provide the indispensable conceptual framework for the study of chaos. As it happens, strange attractors associated with chaotic systems (those geometrical figures in abstract phase space that the movements of a system are irresistibly drawn to) also exhibit this repetition of a common pattern recurring on any scale—in other words, a fractal structure. Fractal objects and chaos are intimately and inextricably intertwined.

TURBULENCE ON JUPITER

Chaos is not restricted to the movements of air masses that bring us summer storms or spring showers. It shows up in every fluid in the world that surrounds us. Take a stroll along the river Seine and follow with your eyes the flow of water under the magnificent bridges of Paris. You will see eddies forming near every pier. As if moved by a will of its own, the water breaks into complicated, irregular, and seemingly disordered movements. The layman would describe them as turbulent; scientists prefer to say chaotic. Chaos is also present in the air flowing through organ pipes, as they fill the church with their exalting music. The sprays of incandescent lava that Earth, in an awesome display of wrath, spits out toward the sky through the giant mouth of volcanoes are also chaotic. The curlicued spirals a smoker exhales voluptuously through his lips, water flowing out of a faucet or gushing into a limpid fountain, the cataracts of white water at Niagara Falls—all of these fall within the purview of chaos.

These phenomena of chaotic turbulence are not the exclusive province of our beloved Earth. They also show up in the atmosphere of Jupiter, the giant planet of our solar system, where a mysterious phenomenon has long puzzled scientists. There is a large oval-shaped feature in the Southern

Figure 41. *The Great Red Spot on Jupiter.* In this photograph, taken by Voyager 1 in February 1979, the Great Red Spot in Jupiter's atmosphere measures roughly 20,000 km in length by 10,000 km in width. These dimensions have varied slightly over the last three centuries, the time span over which the Great Red Spot has been observed. At maximum, the spot can reach 40,000 km by 14,000 km, large enough to swallow three entire Earths. It spins in a counter-clockwise direction once in approximately 6 days. This spot corresponds to a place in the Southern hemisphere of Jupiter where the movements of gases in the Jovian atmosphere are relatively ordered, as opposed to the disorder and turbulence of the gas surrounding it. Strong winds, which blow from east to west in the region north of the spot and from west to east in the region south of the spot, are responsible for the colored bands parallel to the equator. The spot is a self-organized system generated and regulated by the chaos responsible for the turbulent agitation of the surroundings. The photo also shows two of Jupiter's Galilean satellites—Io, directly above the Great Red Spot, and Europa to its right. (Photo courtesy of NASA)

Hemisphere, resembling a giant eye staring at the cosmos (Figure 41). It is known as the Great Red Spot. This huge, turbulent mass of swirling gas has existed at the top of Jupiter's atmosphere for at least four centuries. Galileo (1564–1642) saw it in 1609 when he trained his telescope for the first time toward the heavens. How is it that this swirl of gas never dies down? What keeps its fury going? The most outlandish and implausible hypotheses have been proposed. In the nineteenth century, some believed that the Great Red Spot was a giant flow of lava escaping from a volcano on the surface of the planet. We know today that this explanation is incorrect since Jupiter has no solid surface. In the 1940s, the hypothesis of a huge bubble of hydrogen and helium floating in the atmosphere, like an egg floating on water, gained popularity. There the matter stood until the space probes Voyagers 1 and 2, amazing technological marvels, visited the outer planets and collected a harvest of extraordinary discoveries that would forever change our view of the solar system. Voyager 2 flew by Jupiter and took a close-up look at the Great Red Spot. What it saw was a gigantic swirl of gas with brown and orange hues as dazzling as an impressionistic painting. The size of the spot has varied during the course of the centuries. When Voyagers 1 and 2 flew by, it was only slightly larger than Earth. But at its maximum size, the spot gets so huge (40,000 km by 14,000 km) that it could swallow three entire earths. This gargantuan maelstrom is confined within enormous horizontal bands of clouds, whose colors alternate between light (high-pressure zones) and dark (low-pressure zones). These bands are parallel to the equator because of Jupiter's extremely fast rotation.[2] The spectacle looked much like a huge hurricane here on Earth. The analogy should not be taken too literally, though, since the hurricanes we are familiar with derive their destructive power from the heat released when masses of humid air condense into rain. As far as we know, there are no downpours of rain unleashed in Jupiter's atmosphere. On Earth, a hurricane would swirl clockwise in the Southern Hemisphere (and counterclockwise in the Northern). Yet, even though the Great Red Spot lies south of the equator, it stubbornly rotates counterclockwise. Finally—and that is the primary difference—hurricanes here on Earth dissipate in a few days, luckily for earthlings who find themselves in their path, whereas the Great Red Spot has persisted for centuries.

Thus, a different explanation is called for, and chaos is able to provide it. This centuries-old eddy can be understood as a self-organizing system, a stable region created and maintained by the chaos responsible for all the surrounding upheaval and turbulence. It is an island of stability in an ocean of instability, a haven of organization in the midst of a storm of anarchy. Once again, we are faced with the coexistence of order and disorder that Poincaré had foreseen while studying the three-body problem. The Great Red Spot on Jupiter tells us that both determinism and chaos are embedded deep in the equations describing fluid turbulence. Disorder, which had been ignored and repressed until the end of last century, manifests its presence and demands recognition on a par with order.

CHAOS AND THE THINGS OF LIFE

Chaos does not reside only in the weather here on Earth or on Jupiter, or in the orbits of the planets of the solar system or of the stars in the Milky Way. It can manifest itself just about anywhere, in the most ordinary circumstances. We have all gone through events that seemed completely innocuous at the time but ended up radically altering the course of our lives. A man gets up a little late because his alarm clock did not go off; he misses his appointment, loses the job he was hoping to get, and ends up with a completely different career from what he had planned. A woman is a few seconds late because the elevator in her building is out of order; thanks to this short delay, the flower pot from the tenth floor, which would have struck her on the head, misses her by inches. These situations characterize chaos: Minute changes in our lives can have dramatic consequences. Events that are a priori unimportant can determine the course of an entire life. Change the initial conditions ever so slightly, and your whole destiny is completely altered.

Chaos has become a fashionable subject. It often makes the front page of newspapers, and international conferences devoted entirely to it proliferate around the world. Books explaining the concepts of "strange attractors" and "fractal objects" (Figure 42) filled with fascinating pictures whose beauty rivals that of art books, crowd bookstore shelves. Chaos has been elevated to the dignified status of object of science, and research institutes are being created worldwide to study it. It has spilled beyond the domain of the natural sciences and found its way into disciplines and specialties as diverse and varied as anthropology, biology, ecology, geology, economics, history, Islamic architecture, Japanese calligraphy, linguistics, music, the stock market, radiology, telecommunications, urban planning, and zoology, to cite but a few.

THE EBB AND FLOW OF LIFE

Chaos has even invaded the ebb and flow of life. The world is a caldron harboring a disparate mix of millions of living species. The Amazonian forest teems with all manner of reptiles, exotic birds, and buzzing insects. Rivers and oceans are filled with fish and crustaceans of all ilk, while jungles abound in still-preserved rare specimens. How can such diverse populations interact with one another? Chaos has something to say about the evolution of species as well. Let us see how.

Consider a population of hares living in the woods. What would happen if the food supply in the forest suddenly diminished or disappeared altogether? What would the result be if a wolf happened to come along and took to happily consuming one hare for every meal? What would the fate of the human race be if an epidemic were to spread or a new virus appeared? It is to answer these types of questions that a new scientific discipline has emerged recently—ecology. Some biologists with a good proficiency in mathematics began to create models to study how populations evolve. The simplest model is a standard Malthusian scheme, in which the demography

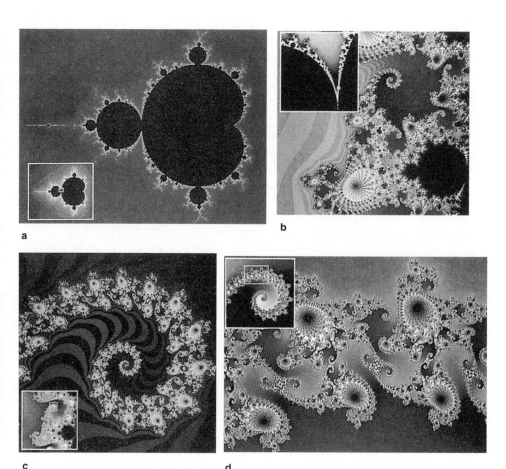

a

b

c

d

Figure 42. *The Mandelbrot set.* These pictures, which possess a strange and complex beauty, with patterns shaped like sea horse tails, spirals, curlicues, and thorn-studded disks, represent increasing magnifications of a mathematical object called the "Mandelbrot set." The magnified region is indicated by the inset. The Mandelbrot set is a fractal structure whose pattern repeats itself *ad infinitum* at any magnification scale [the magnification changes by roughly 1 million from (a) to (d)]. Any drawing on a planar surface has a fractal dimension between 1 and 2, depending on its intricacies. The Mandelbrot set turns out to have a fractal dimension of 2 (the maximum possible), which qualifies it as "the most complicated object in fractal mathematics." In spite of its highly complicated shape, it can be described completely in very simple mathematical terms. All it takes to generate it is to choose a complex number C (a complex number is made of two parts—a real part, which is an ordinary number, and an "imaginary" part, proportional to a quantity i defined as the square root of the negative number -1; by way of example, the quantity $2 + 3i$ is a complex number), and add to it the square of another complex number z that can vary from 0 to infinity. The recipe for the operation is thus defined by a recursive relation of the form $z(final) = z^2(initial) + C$. The Mandelbrot set is generated by repeating the above operation many times over for a given complex number C. (Photos reproduced from James Gleick, *Chaos* [New York: Viking, 1987], all rights reserved)

keeps expanding indefinitely, without any restriction, be it moral, territorial, or food-based. Such a model, originally proposed by the English economist Thomas Malthus (1766–1834), is characterized by a growth rate determined by fertility. For instance, if the initial population starts at 100,000 and the growth rate is 1.1 (meaning a 10 percent increase annually), the population will be 110,000 the following year. The new population is obtained simply by multiplying the current population by the growth rate. The population thus increases year after year, just as a savings account keeps growing by accruing interest. Such a model is, however, not very realistic because it does not take into account the harsh realities of life. Food shortages can trigger famines, and fierce competition for what little supplies there are provoke wars, diseases, and epidemics that decimate populations. These factors and others all combine to slow growth whenever the population explodes.[3]

By incorporating into the model some factors limiting growth, you might think that, after an initial expansion phase leading to overpopulation, followed by periods of shrinkage, the population ought to rapidly reach some kind of equilibrium, at which point it will either remain relatively constant or fluctuate in a steady cycle. And indeed you would be right. Computer calculations confirm that, as long as the growth rate (reflecting the tendency for demographic explosion and overpopulation) remains moderate—with values between 1 and 3 (values smaller than 1 lead to extinction), the population does remain stable. Nothing mysterious so far. But things get out of control as soon as the growth rate exceeds the value 3 (corresponding to a population tripling every year), in which case equilibrium is destroyed and the population starts oscillating between two distinct values from one year to the next. If the growth rate increases even further, a second bifurcation takes place. The preceding two levels become four. The population takes on successively four values, each of which recurs every fourth year. For greater growth rates still, a new bifurcation occurs, and the population now takes on eight values recurring on an eight-year cycle. Further doublings (16, 32, 64, and so forth) keep occurring more and more frequently, and the repeat cycles become increasingly longer as the growth rate continues to go up.

Amazingly, despite the complexity of this behavior, the same values keep recurring on a periodic timetable. Regularity still prevails in complexity. However, even that regularity disappears altogether when the growth rate reaches the critical value of 3.57. At that point, chaos takes over. Periodicity is then swept aside in favor of randomness. The variations in population become completely haphazard. Yet, in the midst of such chaos, order is not totally banished. It still manifests its presence from time to time. Some stable cycles reappear for certain values of the growth rate, only to be once again overwhelmed by chaos for others. Regular patterns exist side by side with irregular ones, and order is mixed with chaos. Completely chaotic stretches are invariably accompanied by others that seem perfectly regular (Figure 43).

Before the advent of the science of chaos, ecologists were divided into two camps. There were those who believed that populations generally evolve

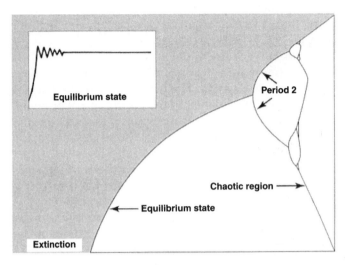

a

Equilibrium state

Period 2

Chaotic region

Equilibrium state

Extinction

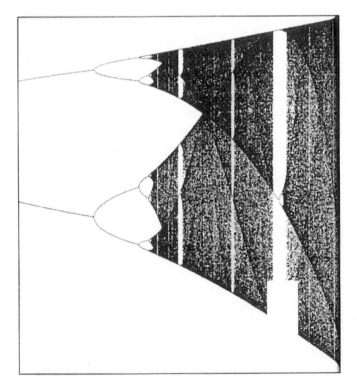

b

Period 2

Period 4

Chaos

c

d

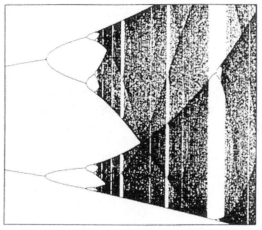

e

Figure 43. *Chaos and the evolution of populations*. Diagram *(a)* shows the evolution of a population (plotted vertically, increasing from bottom to top) as a function of its rate of increase (plotted
horizontally, increasingly from left to right). Growth rates that are too low lead to extinction. Past a
threshold, higher growth rates produce larger populations. The population remains in stable equilibrium as long as the growth rate remains below a critical value. When that value is exceeded,
equilibrium is broken and the population oscillates between two distinct values, then four, then
eight, and so on. The doublings (schematically shown in *b*, where the vertical axis corresponds to
the population size, and time is plotted horizontally) continue with increasing growth rates, until the
evolution turns completely chaotic, going through an infinity of values in a completely unpredictable
manner. Diagram *(c)* is a magnified view of the chaotic region of diagram *(a)*. Complex structures
appear, with the same pattern repeating itself at all magnification scales. For instance, when the
small rectangle at the lower right corner of *(c)* is magnified, one gets *(d)*. Likewise, magnifying the
small rectangle in *(d)* produces *(e)*. The structures seen in *(d)* and *(e)* are very similar indeed to
those in *(c)*.

in a regular and ordered manner and that they are subject to completely deterministic laws. Others believed that most of these variations were totally random, following no particular rule, being subject to the influence of environmental factors so unpredictable and powerful as to wipe out any deterministic trend. Chaos theory reconciles both points of view. Just as chaos lurks in wait in Newton's deterministic equations and invaded the seemingly orderly world of planets, simple deterministic models describing the evolution of a population can lead to randomness. Studies of how the population of fish or insects evolves in well-controlled environments (such as a pond in the case of fish) have demonstrated that they do indeed go through chaotic periods alternating with periods of stability. The simple model described above does not apply directly to human populations; their case is more complicated because of their propensity to migrate and interact with other populations.

CHAOS AND EPIDEMICS

While chaos may not be directly observable in the evolution of human populations, its effects can nevertheless be seen in epidemics that from time to time rage across the globe. Epidemiologists have noticed that they strike in cycles, that diseases come and go periodically. What is the outcome when health officials set a vaccination campaign in motion? If your intuition tells you that the epidemic will eventually lose momentum and die down, you would be right. The long-term trend will certainly be to get things under control. But it is not uncommon to witness surprisingly chaotic swings. After the start of a vaccination campaign, there will typically be a period when an epidemic, far from diminishing, will actually pick up steam and reach all-time highs. You can now understand that such wild oscillations are due to chaotic phenomena triggered by the perturbation effect of the vaccination campaign itself. With that in mind, if there is an initial recrudescence of measles or tuberculosis, doctors should not lose heart and conclude prematurely that the vaccination campaign has been a failure. In spite of these chaotic excursions, the epidemic will ultimately be brought under control.

THE RHYTHMS OF THE HEART

The science of chaos has even helped doctors to better understand how the human body functions. Our body is perhaps the most complex dynamical system we can study. Being the seat of various movements and currents— movements of muscles, fibers, and cells, circulation of ions in muscle tissue, electrical currents in neuron circuits, and so on—the human body has many subtle and diverse rhythms that interact with one another. It has often been described in reductionistic terms, as a collection of organs, each of which has its own structure, function, and chemistry. Medical students are familiar with such a reductionistic approach. They spend years memorizing the names of the different parts of every organ and their functions. This strategy has proven quite useful in the past, but it has limitations. The science of chaos offers a unique opportunity to further our understanding. It provides a more holistic view, considering the human body as an integrated whole where movements and oscillations act in concert. By exploiting the techniques characteristic of this new science, researchers have been able to study numerous physiological dysfunctions, such as respiratory disorders, the periodic or haphazard nature of the growth of cancerous cells, the fluctuations in the number of white and red cells in blood diseases, and more. As such, chaos has given birth to a new physiology, whose greatest success has been some surprising discoveries concerning a vital organ—the heart.

A normal heart has a regular beat. But if, by a stroke of bad luck, it gets into a pathological state in which the beats become erratic, the result can be death. Heart irregularities have been inventoried, catalogued, and studied. A physician with a trained ear can recognize dozens of abnormal patterns when he places his stethoscope on a patient's chest. There is one particularly dangerous condition, called "ventricular fibrillation." It can be deadly because, instead of contracting and expanding rhythmically to pump the blood through the vascular system, the tissues of the heart muscle flutter erratically. As a result, the heart is never quite contracted or expanded, and blood circulation becomes spasmodic. One of the most intriguing aspects of fibrillation is that individual parts of the heart seem to function quite normally, whereas the whole goes hopelessly awry. Each signal generator in the heart continues to send normal electrical impulses and each heart cell continues to respond as it is supposed to, contracting upon receiving the proper stimulus, relaying it to a neighboring cell, and relaxing until the next stimulus comes along. But taken collectively, the heart no longer functions as intended. That is why the science of chaos, which deals with systems as a whole, can shed light on this condition.

CHAOS IN THE HEART

How can chaos overrun an organ like the heart? Why does a rhythm that has been steady for an entire lifetime, with over 2 billion uninterrupted cycles, suddenly lose its bearings and go into uncontrolled, and often fatal, convulsions? A team of researchers at McGill University in Montreal, Canada, led by Leon Glass, set out to understand this conundrum. The scientists removed tiny ($^1/_{80}$ of a centimeter) aggregates of heart cells from one-week-old chicken embryos. When they placed them together and shook them, they observed an astounding phenomenon under the microscope: Without any cardiac pacemaker, the little pieces of chicken heart began to spontaneously beat in unison, following the same natural and regular rhythm. The researchers went on to study the behavior of a single small lump of chicken heart under the influence of an externally applied periodic stimulus, in this case a small electrical shock. Under these conditions, the chicken heart was subjected to two completely distinct stimuli: the natural heart rhythm and the artificial rhythm of the electrical shocks. The researchers found that the heartbeats depended on the relationships between the two rhythms. They stepped straight into the exotic and varied dynamical behavior exhibited by any chaotic system. When the period of the natural rhythm was a multiple of that of the electrical perturbation, the chicken heart would beat once every one, two, three, or more shocks. Varying the frequency of the electrical stimulus could cause period doubling: The heart would beat at a certain pace some of the time, and then switch to a completely different pace at others. This period doubling phenomenon is entirely similar to the phenomenon discussed earlier concerning the evolution of populations. In some cases, the heartbeats would follow no particular pattern at all, becoming completely random and irregular. The chicken heart was then taken over by chaos resulting from interactions between the natural physiological rhythm and the frequency of the electrical stimulus.

These experiments suggest that fibrillation in humans can be triggered by abnormal secondary sources internal to the heart, sending signals that conflict with the natural rhythm of the heart muscle. Interactions between these secondary impulses and the primary rhythm sends the heart into a chaotic state and triggers the onset of fibrillation. Viewed from this angle, this medical condition is of a "dynamical" nature. It comes about because the heart is a system that, starting from a normal beat, can stop beating entirely, or start beating in new and unpredictable ways. Fibrillation is a form of enduring chaos that does not disappear by itself. Only an electrical discharge from a defibrillator, applied through a patient's chest, can jolt the heart back into a normal state. For the time being, without the benefit of a detailed understanding of the dynamics of the human heart, the intensity and duration of the electrical jolt to be applied are determined by trial and error. The study of cardiac chaos is only in its infancy. By attacking the problem from a holistic angle, cardiologists are hoping to be able to identify subjects at risk, to design better defibrillators, and to prescribe more effective medication.

LET CHAOS BE, AND HEALTH WILL FOLLOW!

Some physiologists have even come to the belief that a certain amount of chaos is desirable for the well-being of the human body. Indeed, researchers are trying to develop a "chaos-based" approach to bringing relief to epileptics. Epileptic seizures appear to be related to large "peaks" in the electrical activity of the brain, as if large numbers of neurons were all firing at the same time. By avoiding such cooperative peaks—that is to say, by forcing neurons to behave more chaotically and randomly—it may be possible to alleviate and perhaps even prevent seizures. The idea is to "tickle" the brain by applying to it small electrical impulses that would trigger a more chaotic behavior of neurons. If it works, chaos would then paradoxically be called upon to regulate and control erratic behavior!

Irregularity is thus a fundamental component of life. Living beings draw order from an ocean of disorder. "It is by concentrating this flux of order in himself that man escapes disintegration and atomic chaos." So said the Austrian physicist Erwin Schrödinger (1887–1961), one of the founding fathers of quantum mechanics, who also reflected on the problem of the origin of life.

CHAOS AND THE STOCK MARKET

Economics and the world of finance are also domains in which chaos theory has something to say. These are clearly areas where "nonlinearities" reign supreme, where a small cause can result in incommensurably large effects. Take the example of a simple and idealized game often given as a training exercise to students at MIT's business school. It involves the distribution of beer. The game is intended to expose the students to chaotic situations apt to develop in certain ostensibly quite simple economic systems. Each participant assumes one of the following roles: retailer, wholesaler, distributor, or brewery plant manager—in other words, one of the links in the distribution chain connecting the beer consumer to the brewery that manufactures the product. The objective is for each member of the chain to stock just enough beer cases to satisfy the demand of his or her client, and no more. Players are penalized if they have either too much or not enough beer on hand. The retailer must order just enough beer cases from the wholesaler to meet the demand of consumers during the following week, at least as he anticipates it. Likewise, the wholesaler projects the needs of the retailer and places his order with the distributor accordingly. In turn, the distributor forecasts how many beer cases the wholesaler will need before placing his own order with the brewery manager. Finally, the plant manager produces just enough beer to fill the anticipated orders of the distributor.

The game starts with a steady demand from the retailer's customers—let's say ten cases of beer a week. As long as that demand remains constant, an equilibrium sets in. Everybody orders ten cases of beer each week from his supplier, and all the brewery has to do is produce the same number to keep everybody happy. But assume now some perturbation in the equilibrium of the system. For some unknown reason, the retailer's customers all of a sudden double their consumption and start ordering twenty cases of beer per week. Your intuition might tell you that all members of the distribution chain will quickly and painlessly adapt to the new situation by doubling their orders (or doubling the production in the case of the brewery). That is hardly what happens! Instead of a smooth transition, the weekly orders start fluctuating wildly. By the thirtieth week, it is not uncommon to see the distributor order forty cases of beer from the brewery, when clients consume only eight. Evidently, a small perturbation in the beer distribution chain is more than enough to completely upset the orderly workings of the system and lead to bizarre and unexpected behavior. If chaos is able to take over as simple a system as a beer distribution chain, it should not come as a shock to anyone that the stock market is vulnerable too. After all, financial markets are extremely complex dynamical systems, reacting to the slightest bit of rumor. A fall in the price of gold, a hike in interest rates, a presidential election, even a strike at a particular plant—all of these can have dramatic repercussions. This is another situation where a small perturbation can lead to consequences that seem out of proportion. If you don't believe it, you need only observe the financial centers in New York, London, Tokyo, or Paris, abuzz with thousands of operators, all shouting at the top of their voice, outgesticulating one another to place the magic order at just the right time, all the while watching the competition and keeping an eye on an avalanche of fresh information flooding in from around the world by telephone, TV screens, and telecopiers. The rule of the game is to maximize profits while minimizing risks. Each operator has to assimilate new developments, anticipate which way the market will turn, jump in the water, and place an order.

Despite this frantic activity bordering sometimes on madness, could there be a rational description of these temples of capitalism? As everyone knows, the dream of any economist and speculator is to be able to predict the highs and lows of the market and derive the greatest amount of profit. Can chaos theory help? Even though the workings of financial markets are exceedingly complex, there are self-regulating mechanisms that come about, based on a mix of human psychology, social behavior, and rational thinking. For instance, if the price of a product becomes too high, demand will subside, which will drive the price back down. This feedback has important and unexpected implications for the behavior of markets, prices, and entire economies. Indeed, they can lead straight to chaos. The price of gold behaves very similarly to the fluctuations of populations described earlier. In order to maximize profits, gold merchants push the price as high as the market will bear. Assume they jack up the price every week, the way a Malthusian population increases yearly. When the price reaches excessive levels, fewer buyers are willing to step forward and the price drops, very much as a runaway population growth is brought in check by food shortages, diseases, and wars. In fact, in this simplified model, the equations describing how the price of gold varies look exactly the same as those that are used to model the evolution of a population.[4] As we saw before in the case of a population, the price of gold can reach a steady state when its yearly increase is not too high. Jack up the price increase a little, and the price of gold starts oscillating between 2, 4, 8, 16, and so on, possible values, successive doublings that remind us of the way populations fluctuate. With steeper increases still, the price of gold becomes chaotic. We are a long way from predicting the highs and lows of gold prices, but the considerations we have just discussed suggest that chaos theory has all the ingredients necessary to help us better understand how markets and economies behave.

Chaos has thus entered physiology, biology, ecology, and economics. But quantitative studies of chaos in these so-called "soft" sciences are far from having reached the degree of sophistication they enjoy in the "hard" sciences, such as astrophysics of the solar system, hydrodynamics, or even meteorology. Two main reasons explain this disparity. First, biological and economic systems are so complex that it is difficult to zero in on an equation that describes accurately how they evolve in time. Approximations are often too simplistic. Second, we are dealing here with what is referred to as "complex adaptive systems," meaning systems that have the ability to learn, remember, and adapt, thereby changing the very nature of the initial system. Because of that, even if we knew the equations describing the time evolution of biological and economic systems, these equations would slowly change with time. The path to the ultimate goal remains long and arduous.

Despite these difficulties, chaos theory has already made us aware that some dynamical situations in biology, economics, or politics, far from producing equilibrium, can lead to chaotic fluctuations impossible to control. Politicians, legislators, and other officials would be well advised to consider the possibility that their decisions, despite their best intentions to produce equilibrium, can actually cause wild and unpredictable oscillations, with potentially disastrous consequences. Chaos theory can at least help us develop a better understanding of these fluctuations and perhaps even suggest ways to more effectively control them.

That said, chaos has been instrumental in liberating Nature. It has given her the freedom to exercise her creativity. She does so through subtle principles of symmetry that manifest themselves in a profound unity of the physical world. That is our next topic.

The Austere Beauty
of Symmetry

THE BEAUTY OF A CIRCLE

Given a choice between a circle, a square, and a rectangle, which one would you find the most attractive? Which is most likely to stir your sense of aesthetics? If a group of people were to be polled, that question would elicit different answers. But by and large, most people would tend to put the circle first. In doing so, they echo the ancient Greeks, who also believed that the circle is the most perfect geometric shape. That is what prompted Aristotle to proclaim that the planets move along circles, since the heavens were seen as the realm of perfection. This obsession with circular planetary orbits persisted until the beginning of the seventeenth century, when Johannes Kepler had to admit almost reluctantly that the planets follow elliptical, rather than circular, trajectories.

You may well feel intuitively that the circle is the most beautiful shape, but you would probably be hard-pressed to articulate why. Physicists, on the other hand, have a ready-made answer: The circle is the most beautiful because it possesses the highest symmetry. What exactly is symmetry? The dictionary tells us that the word comes from the Latin *symmetria*, an architectural term itself derived from the Greek *summetria*, which means "proper measure, proportion." Thus, the word *symmetry* was first used in architecture to designate the "consonance and concordance of the different parts of a building among themselves and with the whole, combining to produce beauty." By extension, the word eventually came to include "the regularity and harmony of the parts of any object." It spread to other fields besides architecture, particularly during the classical period when it was used to describe a work of art (1660). It was subsequently applied to literature in the eighteenth century (around 1770), and to music (1847). In his *Encyclopedia*, Diderot defined *symmetry* more specifically as a "regular distribution of parts and similar objects about an axis passing through the center." The term later found its way into botany (1866), zoology, and, more significantly, into geometry (1872). Figuratively, it came to be used to denote similar ideas or arguments, and situations that show a correspondence. Thus, the notion of symmetry evokes congruence, proportion, regularity, harmony, and beauty.

Whenever physicists want to rely on beauty as a guide in searching for the physical laws describing Nature, they must begin by uncovering the symmetry of the systems they are studying. To do so, they need a precise operational definition of the concept: An object is declared symmetrical if it does not change appearance when it is subjected to certain operations (it is then said to be invariant). We have already seen that fractal objects possess scale invariance, since they keep displaying the same patterns regardless of the scale on which they are examined. There are other objects that are invariant with respect to a reflection or a rotation. For instance, some objects look the same if you cut them in two, keep one half, and reflect it in a mirror. They are said to possess right-left symmetry.

RIGHT-LEFT SYMMETRY

It is not surprising that the word *symmetry* is rooted in architecture. The greatest architectural masterpieces humankind treasures almost always have bilateral symmetry, which is another way to say right-left symmetry. Take, for instance, the Taj Mahal in India. You can insert a giant vertical mirror right through the middle of the structure. The reflected image of the left half will match the right half precisely. Every pattern of the right side corresponds to a similar one on the left side, as though it had been reflected through a mirror (Figure 44, top). That is why right-left symmetry is also called reflection symmetry. Through the ages, the greatest architects in every culture have capitalized on this right-left symmetry to convey a feeling of order, beauty, and perfection, and to impress onto the mind an ineffable sense of harmony. To appreciate it, you need only feast your eyes on the Arc de Triomphe in Paris, or Le Nôtre's gardens in Versailles, or even the marvelous Gothic architecture of Notre Dame. The spires of the cathedral of Chartres rising toward the sky high above the rich plain of the Beauce province do not obey this right-left symmetry. This is because the Roman spire was built in the twelfth century, while its Gothic counterpart was not completed until the sixteenth. But the dissimilarity of the two spires does not mean total lack of symmetry. As a matter of fact, everything else in the cathedral's architecture constitutes a glorious illustration of the principle of right-left symmetry (Figure 44, bottom).

Figure 44. *Symmetry and architecture*. Bilateral (or right-left) symmetry has been a crucial component of architecture in all cultures at all times. The ineffable beauty of the Taj Mahal, in India (top), is due in part to its perfect bilateral symmetry. By contrast, the two spires of the cathedral of Chartres, in France (bottom), are not symmetrical because they were built in different periods: The Roman spire was completed in the twelfth century, while its Gothic counterpart was built in the sixteenth century. However, the lower portion of the cathedral obeys bilateral symmetry, which gives it a beautiful quality of harmony and proportion.

Another example of bilateral symmetry is the human body. A human face is right-left symmetrical. You can take a photograph of yourself, cut it down the middle right through your forehead and the middle of your chin, place one half edgewise against a mirror, and you will see a perfectly normal portrait of yourself. To be sure, there may be subtle differences, because the symmetry of a face is not absolutely perfect. Nevertheless, this bilateral similarity is what enables you to recognize your friends from either side. It probably results from biological evolution. The symmetrical arrangement of the eyes and ears enables us to see landscapes in relief and determine which direction a sound comes from, which was crucial for our ancestors to survive in the face of predators some 2 million years ago. Virtually all animal species possess this right-left symmetry, so much so that we are usually astounded when we run into exceptions, such as the peculiar flatfish called a flounder, which has both eyes on the same side.

THE SYMMETRY OF A SNOWFLAKE

There are other symmetries besides right-left. A circle does not change appearance when it is rotated about its center. It has what is called rotational symmetry. In fact, a circle has the highest possible degree of rotational symmetry, because you can turn it by any angle at all, be it 5 or 70 or 180 degrees, and it always remains equal to itself. That is why you may have intuitively declared the circle to be the most symmetrical shape. A square also has a rotational symmetry, but to a lesser degree than a circle, because if you want to keep it looking the same you can only rotate it by four possible angles—90, 180, 270, and 360 degrees. A rectangle is even less symmetrical, since only rotations by 180 and 360 degrees leave it unchanged. So you can congratulate yourself if you voted the circle the most beautiful geometrical shape. It is more attractive than a square or a rectangle because it exhibits a higher degree of symmetry.

Snowflakes glittering on the branches of trees stripped bare by the rigor of winter, or those appearing on windows after a storm has spread an immaculate white blanket over the serene countryside, have a great beauty of their own. They are a delight to our eyes because the laws of crystallization craft ordinary water drops into an almost infinite variety of crystalline arrangements with stunningly beautiful symmetries. They appeal to our sense of aesthetics because they combine both types of symmetries discussed before—reflection and rotation (Figure 45).

THE SYMMETRY OF PHYSICAL LAWS AND EXTRATERRESTRIALS

Up to this point, we have talked about the symmetry of objects. We have marveled at the harmony of the gardens of Versailles and of the Taj Mahal. We have admired the beauty of a human face or of a snowflake. In fact, Nature is far more subtle. In order to weave the rich and complex fabric of reality, she resorts to the symmetry not just of things but of laws as well. Sym-

Figure 45. *The beauty of snowflakes*. Nature loves symmetry. She builds snowflakes by taking advantage of both reflection and rotation symmetries.

metry implies constancy and invariance. A law of Nature is said to be symmetrical if it remains the same regardless of the observer or conditions of observation. For instance, the laws of Nature are the same independently of the orientation of the laboratory in which they are being studied. What now seems to go without saying was not always that obvious. Aristotle believed that natural movements in space were vertical, that everything went preferentially from up to down, and that Nature abhorred anything horizontal. It was not until the seventeenth century that Newton, building on Galileo's work, proved that all directions in space are equal. An apple falls vertically to the ground not because that is a privileged direction in space, but simply because we happen to live on a spherical body that, by virtue of its large mass, attracts the apple gravitationally toward its center.

Nor do the laws of Nature depend on where you study them. Whether I observe a galaxy from the Kitt Peak Observatory, in the arid splendor of the Arizona desert, or from the Hawaii Observatory, atop the moonlike landscape of the extinct Mauna Kea volcano, its properties are the same. An extraterrestrial observing the same galaxy from the other end of the Milky Way will measure exactly the same properties.

The laws of physics are also invariant with respect to time. It does not matter that Johannes Kepler discovered the laws governing the movements of planets in 1609 or in 1997, or that Edwin Hubble discovered that the universe is expanding in 1929 rather than at the end of the twentieth century. The laws of Nature have always been, are now, and always will be the same. This independence of laws with time is particularly useful to astronomers. It enables them to use telescopes as time machines. Much like explorers have gone back to the source of the Nile, astronomers can travel back to the origin of the universe, both in time and in space. Even though the information carried by light travels across space at the greatest speed possible (300,000 km/s), it still crawls along at a snail's pace on the scale of the universe. Accordingly, we always see things with a certain time lag. The Moon appears to us as it was a little more than a second earlier, and the Sun as it was eight minutes earlier. Likewise, light from the nearest star reaches us with a four-year delay. For Andromeda, the nearest galaxy of size comparable to the Milky Way, the delay is 2 million years, or back to when man was just appearing on Earth; for the nearest galaxy cluster, it is 40 million years; the record, held by the farthest galaxy to have been detected thus far, is about 12 billion years.

Because they are invariant with respect to time and space, the laws of physics discovered here on Earth, our little corner in the vast universe, enable us to understand the properties of stars and galaxies in the far reaches of the cosmos. These objects are so distant that the light they emitted to reach our telescopes today began its long intergalactic and interstellar journey a very long time ago indeed, long before the atoms making up our bodies were manufactured by the nuclear alchemy of a massive star and were blown out into interstellar space by the star's violent explosion as it was coming to the end of its life. Yet, despite distances that confound the imagination, the properties of these distant worlds can be apprehended by human reason. If that feat is at all possible, it is because the laws of physics are not capricious. They do not vary from galaxy to galaxy, or from one corner of the universe to another. The mass of an oxygen atom or the charge of an electron in these faraway galaxies is exactly the same as it is here on Earth. This independence of physical laws in time and space may come in very handy the day we eventually come in contact with an extraterrestrial civilization, since it will provide us with a common language—the laws of Nature—to begin a dialogue.

THE WORLD SEEN THROUGH A MIRROR

If the laws of Nature are invariant in space and time, do they also enjoy reflection symmetry, the right-left equivalence that is so pleasing to the human mind, which has inspired the greatest architects and has produced the most beautiful monuments and gardens to grace the face of Earth?

More specifically, does Nature have a preference for right or left? Does she act like a hostess fussy about the rules of etiquette, insisting that the guest of honor always be placed to her right at the banquet table? To answer the question, we have to step into the world of mirrors.

Take any physical phenomenon, such as the collision between two billiard balls or the movement of the hands of a clock, and reflect it in a mirror. The rules of the game are as follows: If the sequence of events reflected in the mirror conforms precisely to the laws of physics as we know them in the real world, these laws are said to possess symmetry by reflection. In the mirror-image world, left and right are reversed. You will see billiard balls come from and recoil in inverted directions, and the hands of a clock turn backward. But none of the known physical laws would be violated, from which we may conclude that these laws obey right-left symmetry. As in *Alice's Adventures in Wonderland* by Lewis Carroll, you have to step into the mirror-image world to verify symmetry by reflection. Of course, this world is not identical with the one we operate in. When you look at yourself in a mirror, the person you see does have your face and smile, but a particular lock of hair hangs down on the left rather than the right side, your heart is now slightly to the right instead of the left, and the twisted double helix of your DNA molecules coils in a direction opposite to normal. But, as far as we can tell, none of this contravenes any known physical law. It is probably an accident of biological evolution that our heart is slightly to the left rather than to the right. Chemists have managed to synthesize in the laboratory some molecules that are truly mirror images of those found in Nature. They have exactly the same properties as the original ones. And so if, like Alice, you were to step into the mirror-image world and meet there a physicist with his heart slightly to the right, and if you were to ask him to describe the laws of physics in his world, they would agree so completely with the physics in our own real world that you could only conclude that they must obey symmetry by reflection.

PARENT AND DAUGHTER NUCLEI

Or is that really true? Is Nature completely impartial as far as right and left are concerned? Until 1956, every physicist thought so. The idea that Nature could favor one side over the other, that she could behave like a hostess, conscientiously reserving the seat to her right for the guest of honor, was entirely too distasteful. As you might have guessed, Nature promptly showed that she has a mind of her own and reminded physicists of the hazard of being too sure of themselves.

In the mid-1950s, new particle accelerators were being designed and built. They were capable of sending beams of elementary particles smashing into each other at great speeds. Matter shattered to smithereens gave birth to a plethora of new elementary particles. There was one group of particles in particular that no one had expected. Their properties were so bizarre that, for lack of better information, physicists dubbed them "strange particles." Such particles did not last forever. They would decay spontaneously after a while, and their behavior as they decayed just did not conform to the known laws of physics. Two Chinese-American physicists, Chen-Ning Yang (1922–) and Tsung-Dao Lee (1926–) were working on the problem. They were soon forced to conclude that the only way to understand the behavior of these "strange particles" was to abandon one of the sancrosanct postulates of physics—namely, that Nature is right-left symmetrical.

How does one go about verifying such a radical proposition? Lee and Yang suggested analyzing the behavior of a radioactive nucleus spinning about itself. Before being introduced to this radioactive nucleus, though, we must first explore the heart of an atom. The first thing we find there is the atomic nucleus, made of minuscule building blocks of matter—protons, which carry a positive electrical charge, and neutrons, which do not, as their very name implies. Protons and neutrons are bound together by a strong glue called the "strong nuclear force." Next, we encounter electrons, the other building blocks of matter, each of which carries the same electrical charge as a proton, but of opposite sign. There are as many electrons as protons in an atom. As a result, their charges cancel out and an atom is electrically neutral. The size of an atom is only one hunded of a millionth of a centimeter, microscopic compared to ordinary objects in life. Yet its volume is gigantic when compared to that of a nucleus, the latter occupying only one part in a million billionth (10^{-15}) of the volume of an atom. The nucleus at the heart of the atom can be compared to a grain of sand lost in the immensity of a football stadium. Electrons have a grand time twirling around this huge ballroom that is an atom. Because the volume of the nucleus is so incredibly tiny compared to that of the entire atom, all the matter that surrounds us, the chair you are sitting on, the vase of flowers brightening up your room, the book you are holding in your hands, all these objects are made almost entirely of vacuum on a submicroscopic scale.

Like the vast majority of objects in the universe, atomic nuclei are not immobile but spin about themselves. The sense of rotation is well determined and defines a specific direction in space. For instance, if a nucleus spins from west to east, like Earth, the north pole will be up and the south pole down. If another spins in the opposite direction, its poles would be switched around. Imagine for a moment that Earth reverses its sense of rotation. Not only will Japan lose its designation of "Land of the Rising Sun" and acquire that of the Setting Sun, but Australia would be in the Northern Hemisphere and Alaska in the Southern.

We are now ready to get acquainted with the nucleus of a so-called "radioactive" element. Radioactivity refers to that property of certain atomic nuclei whereby they spontaneously lose some of their mass by emitting particles or electromagnetic radiation. The original nucleus, called "parent nucleus," transmutes into a "daughter nucleus." The force responsible for this decay is called the "weak nuclear force," because it is much less intense than the strong nuclear force that, as we have seen, binds protons and neutrons in a nucleus. Decay is accompanied by the ejection of an electron escaping at high speed. This electron is created from scratch; it is not part of the original crowd of electrons dancing about in the vast enclosure of the atom. Those are so far away from the nucleus that they are not affected at all by the decay reaction. The ejected electron proceeds along a well-defined path—north, south, or some intermediate direction.

NATURE VIOLATES RIGHT-LEFT SYMMETRY

What we are trying to find out is whether the decay of a radioactive nucleus obeys reflection symmetry. To get to the answer, we will follow Alice in Wonderland and step into the mirror-image world. The question is whether the laws of physics that govern the decay of a radioactive nucleus are the same in the mirror-image world as in the real world. Suppose that the electron ejected in the real world proceeds north, north being defined by the spin direction of the nucleus, from west to east. What happens in the mirror-image world? The reflected nucleus will spin in the opposite direction, from east to west, just as the hands of a clock would go counterclockwise. But since the spin direction defines a direction in space, reversing it flips the poles upside down. In the mirror-image world, the north pole is down and the south pole up. An electron flying north in the real world will be heading south in the mirror-image world. A physicist living in the mirror-image world, with his heart slightly to the right, would arrive at conclusions diametrically opposite those of his counterpart in the real world, with his heart slightly to the left. At least that would be true of the behavior of the electron ejected during the decay of a radioactive nucleus.

In practice, the experiment involves not just one but a great many radioactive nuclei. Physicists collect data coming from many decay processes to determine if they show a preference for any particular direction. If Nature respects symmetry by reflection, as many electrons should head north as south in the real world. A physicist in the mirror-image world would come to the same conclusion. On the other hand, if it turned out that in the real world more electrons go south than north, it would mean that more electrons go north than south in the mirror-image world. In other words, the two conclusions would contradict each other and symmetry by reflection would be broken. If so, Nature would not be impartial toward right or left, and she would be like a hostess consistently placing the guest of honor to her right.

Lee and Yang convinced a colleague of theirs, Madame Chien-Shiung Wu, to try the experiment. Madame Wu, also a Chinese-American, had a reputation for bringing elegant simplicity and meticulous care to her experimental work. She faced a difficult challenge, because at ordinary room temperature nuclei are constantly jostled around by thermal agitation, and their spin axes point in all possible directions. Madame Wu had to chill them to a very low temperature to quiet them down and force their spins to all line up along the same north-south direction. Having done that, she tracked the direction in which electrons were ejected during the decay of the radioactive nuclei. The verdict was not long in coming in: Electrons turned out to have a marked preference for the south and generally avoided the north. This implied that the mirror-image world did not follow the same laws of physics as the real world, at least as far as the decay of radioactive nuclei was concerned. By stepping into the mirror-image world, Alice discovered different rules! The news that Nature transgressed the tenet of reflection symmetry sent shock waves through the physics community. Another bastion that had been assumed impregnable had just fallen. It was as though the most respected authority in matters of etiquette had committed an unforgivable faux pas.

Nature does not violate reflection symmetry indiscriminately and in any circumstance. She does so very selectively and with considerable circumspection. Only when the weak nuclear force is involved does she take a malicious pleasure in breaking reflection symmetry. Whenever one of the other three forces calls the shots, she scrupulously respects right-left symmetry. The time has come to meet these other three forces.

First is the force of gravity, the one that keeps us glued to the Earth and prevents us from floating helplessly in the air, the one that brings us down when we trip. Its range of action encompasses the entire universe. It is responsible for the amazing architecture of the cosmos, from simple planets all the way to the largest structures of matter—superclusters of galaxies—not to leave out stars, star clusters, and galaxies.

Next is the electromagnetic force that holds atoms together, as well as molecules and the twisted double helix of DNA. Its range is that of ordinary things in life. It is responsible for all the shapes that brighten up our existence and make life worth living—the delicate pattern of a rose petal, or the elegant lines of a statue by Rodin. It gives objects their cohesiveness. It is what prevents you from walking across a wall or passing your hand through the pages of this book.

Finally comes the strong nuclear force, which we have already encountered. It holds together the building blocks of matter—protons and neutrons—to form atomic nuclei. Its range is quite short, being limited to the size of an atomic nucleus, or one-tenth of a thousandth-billionth (10^{-13}) of a centimeter.

Whenever one of these three forces is present, Nature recovers her impartiality, and both the real world and its mirror image observe the same physical laws.

ELUSIVE NEUTRINOS

Why does Nature not respect reflection symmetry when the weak nuclear force is in charge? The key to the answer lies in the existence of a second character that comes into play during the decay of a radioactive nucleus. As we have seen, the transformation from parent to daughter nucleus is accompanied by the emission of an electron. At least that is what physicists believed at first. But Nature can be mischievous, and she had a monumental surprise in store for those who set out to verify that the decay process obeyed another fundamental tenet of physics—the principle of conservation of energy. That principle states that in an isolated system the total energy can neither increase nor decrease. Now energy can take one of two possible forms. First is the kinetic energy associated with a mass in motion. An athlete has to expend energy to run, and a car has to burn gasoline to move. Einstein taught us that there is another form of energy—rest mass energy. Anything endowed with a mass possesses this type of energy, which is equal to the mass of the object multiplied by the square of the speed of light. Taking both forms of energy into account, physicists tallied up the total energy involved in the decay of a radioactive nucleus. The initial energy is the mass energy of the parent nucleus. The final energy is the sum of the mass energy of the daughter nucleus, the mass energy of the ejected electron, and its kinetic energy. The principle of energy conservation demands that the final energy be precisely equal to its initial value. Much to their consternation, physicists discovered that things were not so. The final energy turned out to be consistently lower than it should.

Where was the missing energy? The Austrian physicist Wolfgang Pauli (1900–1958), one of the founding fathers of quantum physics, had a brilliant inspiration in 1931. He postulated the existence of a new particle whose kinetic energy was precisely equal to the energy shortfall. To explain the fact that no one had ever detected it, Pauli gave his mystery particle some rather unusual properties. It had to have neither mass (therefore no rest mass energy) nor electrical charge, it would travel at the speed of light, and it would interact only with the weak nuclear force. The forces of gravitation and electromagnetism, and the strong nuclear force, would have no effect on it. It was precisely this lack of interaction that had made the particle so elusive, because elementary particle detectors, being made of ordinary matter, rely on electromagnetic and strong nuclear forces. Much like an evanescent ghost gliding through the thick of the night, the new particle was impalpable.

The Italian physicist Enrico Fermi (1901–1954) immediately made himself a champion of the particle postulated by Pauli and dubbed it "neutrino"—an Italian word for "small neutron." He picked the name to reflect the fact that the particle in question had no electrical charge, much like the neutron that would be discovered the following year (1932) by the Englishman James Chadwick (1891–1974), and also to differentiate it from the neutron, which is far more massive, having a mass slightly larger than that of a proton.

Unlike some contemporary physicists who have no compunction inventing the most outlandish and hard-to-detect particles to explain the slightest unusual phenomenon, Pauli feared he had committed the worst possible sin in the world of physics—to postulate the existence of a particle that may never submit to detection. His angst turned out to be unjustified. The probability of a neutrino interacting with an atomic nucleus is certainly exceedingly small, but it is not zero. The chances of a successful detection can be magnified by placing the greatest possible number of atoms in the path of a neutrino—for instance, by filling an enormous tank with a liquid. All that is needed afterward is a lot of patience, since waiting for a positive event can take months, if not years. Through sheer determination, the American physicists Frederick Reines (1918–1998) and Clyde Cowan (1919–1974) finally "saw" a neutrino in 1955, thereby confirming Pauli's brilliant intuition. Today, entire beams of neutrinos are routinely produced every day in particle accelerators like the one at CERN (European Center for Nuclear Research) in Geneva, Switzerland. Neutrinos have taken their rightful place in the zoo of elementary particles, and no one has the slightest doubt about their existence any longer.

As it happens, neutrinos also played a fundamental role in the history and evolution of the universe. The big bang theory tells us that a huge population of neutrinos burst onto the scene in the first fractions of a second following the primeval explosion. These primordial neutrinos continue to roam throughout the universe to this day and, by virtue of their sheer number, constitute the second most abundant population of particles, immediately behind photons, which constitute the so-called fossil radiation (the remnants of the fire of the creation). As you read these lines, hundreds of billions of neutrinos rush through your body every second.

Because they interact so little with other particles making up ordinary matter, the primordial neutrinos, as plentiful as they may be, have never been captured by our telescopes and detectors, since those happen to be made of just that plain, ordinary matter. As such, neutrinos remain fleeting entities.

Wolfgang Pauli believed that neutrinos have no mass at all. This hypothesis has been questioned recently in light of the so-called "Grand Unification" theories, which attempt to meld all the fundamental forces—with the exception of gravity—into one. These theories do attribute a finite mass to neutrinos. Even if that mass were as small as one ten-thousandth that of the electron, there are so many neutrinos around that they would dominate the mass of the universe. If that proved true, the future of the universe would be altered, because the extra gravitational attraction due to neutrinos would be enough to stop its expansion, reverse the flight of galaxies, and ultimately cause a big crunch (a big bang in reverse). But there is no cause for immediate alarm: In spite of sustained efforts, physicists have thus far failed to detect a significant mass for the neutrino. In 1998, a team of Japanese researchers, using a huge tank of 50,000 m^3 of purified water as detector, reported measuring a finite mass for the neutrino, although far too small (it is only one millionth that of the electron) to stop and reverse the expansion of the universe. Until further notice, the universe will go on expanding forever.

ANTIMATTER PAYS US VISITS FROM OUTER SPACE

Nature throws right-left symmetry overboard when the weak nuclear force comes into play. This shocking symmetry violation was a clue for physicists to reexamine another widely accepted tenet of modern physics—the perfect symmetry between matter and antimatter. Before 1929, no one knew about antimatter. That particular year, the English physicist Paul Dirac (1902–1984), another founding father of quantum mechanics, argued forcefully that the very existence of matter implied that of antimatter as well. Dirac had noticed that the solutions to his equations always came in pairs. Every solution corresponding to an electron with a negative charge had a matching one corresponding to a particle with exactly the same properties (mass, spin, and so forth) as the electron, except that it had a positive charge. At first Dirac thought that the antiparticle of the electron was simply the proton, since its charge was precisely equal and opposite to that of the electron. But that could not be right, because the proton does not have the same mass as the electron. In fact, it is 1,836 times heavier. Subsequent events confirmed that Dirac's equations did not lie. In 1932, the study of cosmic rays, those streams of energetic particles expelled by massive stars in their death throes and raining down on Earth from deep space, revealed corpuscular entities that had exactly the same mass as electrons but an opposite charge. The antielectron had just been detected and was named "positron." Not long thereafter, the antiproton and antineutron—antiparticles corresponding to the proton and the neutron—were also discovered. Antimatter is the image of ordinary matter, but viewed through a mirror that reverses the sign of electrical charges.

NATURE PREFERS MATTER TO ANTIMATTER

A question that comes immediately to mind is whether nature is symmetrical with respect to matter and antimatter. To come up with the answer, the ground rules are just the same as in the case of right-left symmetry. You have to step into the antiworld and ask an anti-physicist to describe for you the physical laws that govern his own universe. If they are the same as those governing the real world, then Nature abides by matter-antimatter symmetry. Until 1956, physicists believed that to be the case, for a number of good reasons. With the exception of the electrical charge, matter and antimatter appeared to have exactly the same properties. An antiproton could combine with an antielectron to form a hydrogen antiatom. Antiatoms could bind together to form antimolecules. These antimolecules could in turn assemble to produce amino antiacids and antiproteins, which could arrange themselves in long chains to create anti-DNA molecules with twisted double helices. Those would generate life somewhere in a remote corner of an antigalaxy, on an antiplanet orbiting around an antisun. And on this antiplanet, there might be an antiyou reading an antibook! Your life and that of the antiyou would evolve similarly. As long as each of you remains in your own corner, your lives will unfold along parallel paths. But should you and your antiyou ever come to meet and shake hands, the outcome would be disastrous—you would both instantly turn into light, for the matter in your body would annihilate the antimatter in the antiyou's body, and all would become pure radiation.

Antimatter seemed to behave in every respect like matter, except that the sign of electrical charges is inverted. The breaking of right-left symmetry in situations where the weak nuclear force rules had burned physicists once before. That made them anxious to check the matter-antimatter symmetry in cases where the weak force dominates. They focused their attention on the neutrino and its matching antiparticle—the antineutrino. Both elementary particles are sensitive to the weak force only. Once again, they zeroed in on the spin direction. A neutrino rotates about itself from west to east. If matter-antimatter symmetry is obeyed, an antineutrino should spin in the same direction, also from west to east, since nothing should change except the sign of electrical charges. And once again, physicists were in for a jolt: Antineutrinos turned out to stubbornly spin in the opposite direction. That meant that matter-antimatter symmetry was being violated.

In retrospect, this result was perhaps not entirely unexpected. We live, after all, in a universe made of matter. There are no pockets of antimatter lurking out in the shadows here and there, ready to leap out and annihilate matter. You do not instantly turn to light while shaking hands with a friend or sitting down on a chair, which would happen if your friend or the chair were made of antimatter. The probability of finding an antiyou somewhere in the universe is extremely minute. We know this because cosmic rays, those streams of charged particles reaching us from the edges of our galaxy, some tens of thousands of light-years away, tell us that it is so. They contain almost exclusively matter—mostly protons. Because we live in a universe made chiefly of matter, Nature cannot be completely impartial toward matter and antimatter. If she were, there would have been equal quantities of matter and antimatter during the very first instants of the universe, when it was so extremely hot, dense, and small (hundreds of billion billion times smaller than an atomic nucleus). Matter and antimatter would have been annihilated, and all that would be left would be a universe filled with light and nothing else. Elementary particles, stars, galaxies, planets, human beings, you, me, none of this would exist. The universe would be barren and we would not be around to debate the issue. As it turns out, astronomical observations tell us that Nature prefers matter over antimatter by a tiny one part in a billion. For each billion of antiparticles that popped into existence out of the vacuum during the first fractions of a second of the universe, a billion *plus one* particles appeared at the same time. The billion particles and the billion antiparticles destroyed each other in an orgy of annihilation, leaving behind a billion photons and a single particle of matter, rescued from the massacre because it had no matching antiparticle to annihilate itself with. Thus it is that all antimatter went out of existence in the early universe. The ratio of a billion to one between the numbers of light particles and matter particles is exactly what is observed in the universe today.

NATURE'S CAPRICIOUS BEHAVIOR

At this point, we know that Nature respects neither right-left symmetry nor matter-antimatter symmetry when she deals with the weak nuclear force. She seems unwilling to conform to what would be most pleasing and agreeable to the human mind—perfect and complete symmetry. But physicists would not give up. They surmised that two wrongs might make one right. Perhaps by inverting left and right as well as the signs of electrical charges at the same time, symmetry might be restored.

The rules of the game remain the same: Take a magic mirror that inverts not only left and right but also electrical charges. Are the physical laws that govern this magic world the same as those that prevail in the real world? The hopes of physicists were promptly dashed. In 1964, the American physicists Valentin L. Fitch (1923–) and James W. Cronin (1931–) demonstrated that Nature has no more respect for symmetry when left-right and charges are inverted simultaneously than when they are considered separately. Fitch and Cronin looked at the decay of a particle of matter called a "K-meson." If Nature respects left-right and matter-antimatter symmetries combined, a K-meson should always decay into two other particles called "pi-mesons." And indeed, that is what happens most of the time. But now and then, among thousands of normal decay events, Nature shows her capricious side. A K-meson suddenly decides to decay not into the normal two but into three pi-mesons! This is a clear case of symmetry violation. Such intermittent transgressions are most upsetting. It is as though Nature behaved like an inconsistent and capricious child, with sudden and unpredictable mood changes. Things were different when left-right and matter-antimatter symmetries were violated separately. As shocking as Nature's disregard for symmetry may have been then, at least she had the good taste to be systematic about it. She was consistent in her lack of respect. At least physicists knew what to expect. But with this business of intermittent violation of combined left-right and matter-antimatter symmetries, it is as though from time to time Nature thumbed her nose at physicists too curious about her secrets. On top of this infuriating inconsistency, Nature violates symmetry very selectively—only in the decay of K-mesons! After more than twenty years of hard work, physicists have never been able to observe similar symmetry violations in any other situation.

At this point in our initial exploration of symmetries in nature, what do we know? In the majority of cases, when gravitation, electromagnetism, or the strong nuclear force are the principal actors, Nature scrupulously respects both left-right and matter-antimatter symmetries. She violates them only when the weak nuclear force is involved, reserving the right to contravene each separately or taken together. In the latter case, she can be impulsive and fickle by breaking the rules only in fits and spurts during the decay of K-mesons. But Nature has more than one trick up her sleeve, and she can show herself infinitely more subtle. She can exhibit far more refined and clever symmetries, which link and unify phenomena and physical concepts a

priori completely different and disconnected—phenomena such as electricity and magnetism, concepts such as time and space. Let us explore this in
more depth.

AN ISLAND CALLED MAGNESIA

Electrical and magnetic phenomena have long exerted a profound fascination on man's imagination. The ancient Greeks were already aware that a
piece of amber rubbed against wool had the ability to attract small flakes of
papyrus. Without knowing it, children today often repeat the same experiment by rubbing a comb made of ebonite. People in Greece also knew that
there existed on the island of Magnesia a particular type of rock with the
strange power to attract iron. The Chinese used this rock, called magnetite,
to make compasses that showed explorers the north direction.

During the course of time, many bits of empirical facts accumulated here
and there, but all this knowledge remained disorganized and unstructured.
Discoveries were made haphazardly, and progress was slow and sporadic.
Electricity and magnetism were viewed as completely distinct and unrelated
phenomena.

The French physicist Charles de Coulomb (1736–1806) was the first
scholar to bring some semblance of order in this heap of disorder. He established that the force acting on two "electrified" objects varies as the inverse
of the square of the distance between them. If you place them 10 times farther apart, the electrical force becomes 100 times weaker. This dependence
on distance happens to be precisely the same as for the force of gravity; it,
too, decreases as the inverse of the distance squared. But unlike gravitation,
which is always attractive, electrical forces can either attract or repel. It
attracts if the electrical charges are of opposite signs, and repels when they
are of like sign. We all carry in our bodies a mixture of negative charges
(from electrons) and positive charges (from protons in atomic nuclei). If so,
why are we not electrically attracted or repelled by other people? The
answer is that Nature always sees to it that there are equal numbers of negative and positive charges in the things of life, whether it be a human body, a
tree, or a book. The balance is so perfect that the net charge, and any resulting electrical force, is always zero. When a man is attracted by a woman, it is
most assuredly not electrical forces that are responsible. If the man and the
woman had only 1 percent more electrons than protons in their bodies, they
would repel each other electrically with a force powerful enough to lift the
entire Earth!

GALVANI'S FROG

The next great step forward occurred thanks to a serendipitous discovery in
a rather unexpected area of research—the anatomy of frogs. In 1789, while
revolutionary fervor was raging in the streets of Paris, the Italian physicist

and physician Luigi Galvani (1737–1798) was busy dissecting frogs. He noticed a strange fact: On contact with a metal scalpel, the frog's muscles would contract as if jolted by an electrical current. That initial interpretation later proved to be correct. We know today that electrical impulses travel constantly through nerves and muscles in our bodies. Another Italian, Alessandro Volta (1745–1827), a physicist by training, and count and senator of the kingdom of Italy through the grace of Napoleon Bonaparte, made a bold conceptual leap: He realized that Galvani's frog had itself very little to do with the production of an electrical current, that it was all due to the tissues and fluids in the frog's body. From that insight he deduced that he could generate an electrical current just as easily if he replaced the frog by an appropriate chemical fluid, and the metal scalpel by metal plates immersed in that fluid. The electrical battery had just been invented.

The next decisive breakthrough came from the Danish physicist Hans Christian Oersted (1777–1851). He demonstrated for the first time a direct connection between electrical and magnetic phenomena. He noticed that an electrical current can cause the needle of a compass to deflect. Since a compass responds to magnetism only, the experiment meant that an electrical current can generate a magnetic field. Electromagnetism, the synthesis of electricity and magnetism, was beginning to take shape.

LINES OF FORCE

A new actor stepped onto the stage. He was the English physicist Michael Faraday (1791–1867). Recognized as one of the greatest experimenters of his day, Faraday had a most extraordinary destiny. Born in a disadvantaged family—his father was a blacksmith—he, unlike most of his peers, lacked the means to go to school and attend a university. He never acquired any formal training in physics. His fascination for electricity dated back to his adolescent days when, while employed at a local bookstore, he accidentally came across an article on electricity in the *Encyclopaedia Britannica* that captivated him. He began to regularly attend public lectures on electricity given by the then very famous chemist Sir Humphry Davy (1778–1829) at the Royal Academy in London. Chance smiled on Faraday when Davy, impressed by the young man's enthusiasm and intelligence, hired him as his assistant in his London laboratory. Faraday quickly learned in the trenches how to conduct physics experiments, and the pupil soon surpassed the master. It caused a certain amount of jealousy in Davy, who tried in vain to put up obstacles in the path of his assistant in the hope of slowing down the rise of his career. Eventually, Davy had to grudgingly admit that, of all his discoveries, perhaps the most important one was to have recognized Faraday's talent. Among the many magnificent accomplishments of Faraday, there is one that deserves to be singled out. He succeeded in 1831 in answering a question that had long stumped physicists. That was whether Nature was symmetrical with respect to electricity and magnetism. If, as Oersted had observed, an electrical current can generate a magnetic field, is the reverse also true? Can a magnetic field generate an electrical current? Faraday succeeded in proving that it was indeed the case. But on one condition: The magnetic field had to vary. A fixed magnet, producing a static magnetic field, is incapable of creating an electrical current. Only when the magnet is moved, which amounts to changing its field, is an electrical impulse set up in a nearby wire. The same result is produced by moving an electrical wire across the static magnetic field of a fixed magnet.

Guided by his experimental results, Faraday began to think about the nature of electrical forces. Before Faraday's time, it was well-known that there were two types of electricity, one positive and the other negative. But no one knew if they were truly distinct or the same, the positive kind corresponding to the presence of an electrical fluid, and the negative kind its absence. The fact that an electrical force could act at a distance, that an electrical current could cause the needle of a compass to deviate without any direct contact, was quite troubling. Some people believed that electricity must have a dual nature. On the one hand, it involved static charges that attracted bits of paper and flowed in a wire like a fluid. On the other, because of its ability to act at a distance, it had to also possess some immaterial quality that enabled it to pass unimpeded through the walls of metal objects and to propagate across space.

Faraday got rid of all these notions. Precisely because of his unorthodox academic development, he was unencumbered by the usual baggage of preconceived ideas. He was free to go wherever his intuition would lead him. To account for the ability of electrical and magnetic forces to act at a distance, he imagined lines of force emanating from an electrical charge or a pole of a magnet and spreading across space to form an extended electric or magnetic field. Near the electrical charge or magnetic pole, the lines of force would be densely clustered and the corresponding field and force would be strong. Farther away, the lines of force would spread apart and become less dense, resulting in a weaker field and force. The electrical force would decrease as the inverse of the square of the distance, exactly as Coulomb had predicted. A second electrical charge or magnet placed in the field would experience this force. Electrical charges or magnets would not act on one another directly, but through the bias of a force field produced by each. Faraday's ideas immediately had enormous practical repercussions. Thanks to his conceptual picture of lines of force, it became obvious how to convert a mechanical movement into electricity (for instance, by displacing an electrical wire in the field of magnet), or electricity into movement. The era of dynamos and electrical motors was ushered in.

Despite their extensive practical ramifications, Faraday's ideas were greeted by his peers with a certain skepticism. Because he was self-taught, Faraday lacked the mathematical tools to prove his ideas rigorously. That task fell on one of his compatriots, the Scottish physicist James Clerk Maxwell (1831–1879).

THE MARRIAGE OF ELECTRICITY AND MAGNETISM

Unlike Faraday, Maxwell was born into a wealthy Scottish family and received the best possible scientific education at Cambridge University. Impressed by Faraday's bubbling imagination and penetrating intellect, Maxwell was convinced that his colleague was on the right track and that the lines of force he had envisioned were quite real. He had no doubt that electrical charges did interact via a force field. He used his considerable mathematical talents to quantify Faraday's insight. In 1873, he managed to distill the entire body of knowledge about electricity and magnetism accumulated over the previous century. He captured it all in four masterly mathematical formulas, known ever since as "Maxwell's equations." They sealed the marriage of electricity and magnetism, celebrating the symmetry of Nature. With one skillful stroke of the pen, Maxwell united electricity and magnetism, which until then had lived separate existences, into a single "electromagnetic field." Maxwell's equations describe how the electromagnetic field evolves in time and space. One equation tells how an electric field changes in space in the presence of a varying magnetic field. Another tells how the electric field forms around a charge, and so forth.

But Maxwell's equations also revealed to him something completely unexpected—the existence of electromagnetic waves. That is the hallmark of great equations or theories: They take on a life of their own and surprise even their authors by giving back far more than what was originally invested in them.

Maxwell's equations told the following scenario. A time-varying electric field generates a magnetic field. By the very fact that it now exists while there was nothing before, the magnetic field itself varies in time and, in turn, generates an electric field, which generates a magnetic field, and so on. The electromagnetic field propagates in space much as a wave propagates on the surface of a lake in which you have just tossed a stone. Maxwell was able to calculate precisely the propagation speed of this electromagnetic wave. The result was stunning and completely unexpected. It was exactly the same as the speed of light (300,000 km/s)! Electromagnetic waves turned out to be nothing else but light! That was like magic. In one fell swoop, not only were electricity and magnetism united, but to that another whole field of physics was joined as well—optics—which up until that point had been believed to be completely distinct. The science of light suddenly found itself under the banner of electromagnetism. The laws of optics, worked out painstakingly by the Englishman Isaac Newton and the Dutchman Christian Huygens (1629–1695), proved to be embedded in Maxwell's equations. Maxwell would have been entitled to celebrate his triumph in quasi-biblical terms: "Let electricity and magnetism be, and there was light!"

NEW TYPES OF LIGHT

This discovery considerably broadened humanity's view of the world. From that day on, light was no longer limited to visible waves that our eyes, thanks to Darwinian evolution, are sensitive to. Nature's palette of light suddenly found itself extraordinarily enriched, and man now uses various types of light from the entire electromagnetic spectrum (Figure 46) to communicate with Nature and wrest her secrets. Electromagnetic radiation is characterized by an energy and a specific length (called "wavelength"), which is the distance between two successive crests of the associated wave. The more energetic the radiation, the shorter its wavelength. Gamma and X rays are the most energetic. Their wavelength is only about one-tenth of a billionth of a centimeter. They go through the human body with such ease that X rays are routinely used to image lungs and detect diseases like tuberculosis. Ultraviolet light, with a wavelength of a few millionths of a centimeter, is a little less energetic, but still enough to burn your skin and cause cancers in case of excessive exposure to the Sun. Next in order of decreasing energy comes our beloved visible light, with a wavelength of a few hundredths of a thousandth of a centimeter; infrared light, with a wavelength of roughly a micron (a millionth of a meter); microwave radiation (the same kind that cooks food in our microwave ovens), with a few millimeters; and finally radio waves, the least energetic of all, with a wavelength ranging from a few centimeters to several kilometers. It is the latter that carries your favorite radio or television program from the transmitter station to your TV set or stereo receiver. As you read this, you are surrounded by electromagnetic waves. They fill your house, and all it takes is a simple flick of a switch in order for radio waves to be converted to sound or images on radio or TV. Maxwell surely had no idea that his brilliant discovery would someday result in this small screen flooding mankind with images and stealing so much of its time. But that is not all. Mixed in with radio waves are visible light waves that come from sunlight filtering in through your window, bouncing off the surface of every object and redirected toward your eye, enabling you to see everything around you. Ultraviolet rays from the Sun and X rays from deep space come in and join the party. In short, a genuine kaleidoscope of electromagnetic waves pervades our existence at all times.

Maxwell's discovery fundamentally changed the way human beings communicate with one another. It gave birth to the entire telecommunications industry. The theory predicted that if one produced an electromagnetic perturbation at a given location—for instance, by jiggling an electrical charge— that perturbation would propagate across space at the speed of light and could be detected somewhere else. But it took time for technology to catch up with theory. Fourteen years would elapse following Maxwell's great synthesis before the German physicist Heinrich Hertz (1857–1894) succeeded, in 1887, in transmitting a signal from a signal generator to a detector less than a meter away. His achievement opened the door to wireless telegraphy via electromagnetic waves, also called "Hertzian waves" to honor his mem-

WAVELENGTH   SPECTRUM

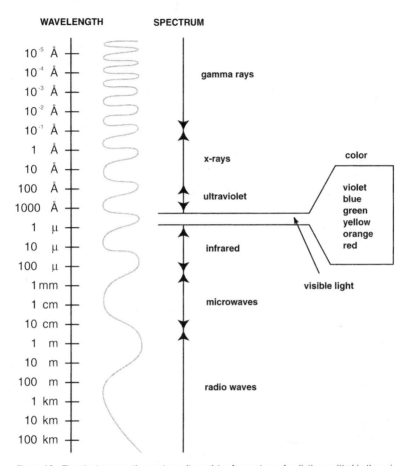

Figure 46. *The electromagnetic spectrum.* It consists of every type of radiation emitted in the universe, from gamma rays, the radiation with the highest energy and shortest wavelength (the wavelength is the distance between two crests of a radiation wave), to radio waves, which have the lowest energy and longest wavelength (the wavelength is inversely proportional to the energy). Visible light, to which our eyes are sensitive, constitutes only a very small part of the electromagnetic spectrum. The fact that our eyes detect visible light is a consequence of Darwinian evolution. Because the Sun emits most of its energy in the form of visible light, Nature has endowed us with eyes sensitive to that portion of the spectrum, giving us the ability to adapt to and survive in our environment. To explore the universe in all its glorious totality, astronomers have built not only optical telescopes capturing visible light, but also gamma-ray, X-ray, ultraviolet, infrared, and radio telescopes. Only optical and radio telescopes can be deployed on Earth, because the corresponding radiations are the only ones that can make it through Earth's atmosphere. The other types of instruments must be lofted into orbit above the atmosphere. The units used in the above figure are 1 μm = 1 micron = 10^{-4} cm, and 1 Å = 1 angstrom = 10^{-8} cm.

ory. The Italian Guglielmo Marconi (1874–1937) significantly increased the transmission range and implemented the first transatlantic link by Hertzian waves, from Cornwall, in England's extreme southwestern corner, to the island of Nova Scotia, off the coast of Quebec, thereby opening the era of intercontinental radio communications.

Barely 150 years after Galvani's frog gave out its last breath, motors, electric generators, and telegraphs had already come into existence, paving the way for an avalanche of inventions about to flood the modern world. Radars, radio and television sets, solar cells, transistors, stereos, telephones, answering machines, lasers, electronic chips, computers, optical fibers, fax machines, the Internet, and many more increasingly advanced devices appeared in rapid succession, devices that only yesterday were thought to exist only in man's dreams. They have so profoundly changed our way of life that we can hardly imagine them not being part of our daily environment.

But the benefits are not just material—they have a spiritual component as well. Because electromagnetic waves are the stuff of light, Maxwell's momentous synthesis literally sheds light on our own existence. It enables us to connect with the rest of the universe. The movements of electrons and protons in the atmosphere of a star lost in the far reaches of the Milky Way, the electrical impulses in the midst of some faraway galaxy, generate light destined to travel across space in an intergalactic and interstellar journey lasting millions, if not billions, of years, finally to be captured by our telescopes and come tickle the electrons in the retina of our eyes. By linking us with stars and galaxies, Maxwell's laws connect us to the entire cosmos.

THE IMPONDERABLE ETHER

In spite of Hertz's and Marconi's successes in transmitting electromagnetic waves across space at the speed of light, one question continued to trouble Maxwell's contemporaries. If radio signals move through space the way waves do on the surface of the ocean, what played the role of the ocean for Maxwell's light waves? What material support enabled these waves to travel? Maxwell believed that the electromagnetic field existed on its own, independently of any material substratum, but his colleagues were not convinced. They concocted a kind of hypothetical fluid, imponderable and elastic, in which light waves were supposed to propagate. They named it "ether," after the term the Greeks had used previously to designate the subtle fluid that, according to them, filled the space beyond Earth's atmosphere.

There was a simple way to verify the existence of this ether. Earth does not stand still. Rather, it drags us along in its yearly orbit around the Sun at a speed of about 30 km/s, roughly 0.01 percent of the speed of light. Because of this movement, the speed of light measured by physicists on Earth should in principle depend on the direction the light comes from. When light approaches Earth head-on, its apparent speed should be larger than its real value. On the contrary, in a direction opposite to Earth's movement, light has to catch up to Earth and its speed should appear lower. Albert Michelson (1852–1931) and Edward Morley (1838–1923), two American physicists, tried the experiment in 1887. The results they got were totally unexpected. The apparent speed of light did not seem to change one iota, no matter what direction it came from. It remained remarkably constant at 300,000 km/s. This was a classic case of an experiment supposed to confirm a predicted result, which turned out, instead, to completely contradict it.

The result of the Michelson-Morley experiment could be understood only if Earth was assumed not to move relative to the ether, which was patently absurd. Earth was known to move around the Sun, and it did not make any sense that the ether, supposedly filling the entire universe, should do the same and follow exactly the motion of Earth, that tiny speck of dust lost in the immensity of the cosmos. The model of an ether filling the whole of the universe, as a sort of theater stage on which the cosmic drama was unfolding, was in a state of crisis. The solution to the dilemma came from an obscure employee of the patent office in the city of Bern, Switzerland, who spent his free time pondering physics problems. His name was Albert Einstein (1879–1955), and he was about to come out of anonymity and become a true magician of modern physics.

THE RELATIVITY OF THE SPACE-TIME COUPLE

In the summer of 1900, Albert Einstein had just finished studying physics at the Polytechnic Institute in Zurich. An average student, he had applied for, but failed to get, a position of research assistant at his alma mater. The following two years was a difficult time, during which "at the threshold of life, [he] felt like a pariah kept on the sideline, disliked and abandoned by everybody." In order to survive, he was forced to fall back on temporary jobs as a teacher in mediocre Swiss high schools. Finally, in 1902, thanks to the intervention of the father of one of his friends, he managed to secure a position as "technical expert 3rd class" at the federal patent office in Bern. This job brought him relative security and made it possible for him to make plans to marry Mileva Maric, a classmate he had met at the university. It also provided him with enough free time and peace of mind to devote himself to his beloved science musings. In 1905, he published a paper that was to rock twentieth-century physics. In that paper, which laid the groundwork for his famous theory of relativity, Einstein resolved the problem of the constancy of the speed of light in an extremely original way. For him, "Nature may be subtle, but malicious she is not." She was bound to respect a symmetry principle dear to Einstein—namely, "the principle of relativity," which holds that the laws of physics must be the same anywhere in the universe, regardless of the speed of the observer who studies them. In particular, the speed of light must always remain invariant. However, in order to make that possible, Einstein was forced to discard two concepts that up to then had been central principles of Newtonian physics—the universality of time and of space. How he went about this bears examining.

One of the most gorgeous sights of an evening in Paris is the floodlit iron lace of the Eiffel Tower, whose slender silhouette stands tall above the city about to go to sleep. Every evening, following an immutable ritual, the floodlights illuminating the tower go out at the stroke of midnight, as if touched by a magic wand. One particular evening, just as the Eiffel Tower plunges back into darkness, two cars cross each other. Paul is at the wheel of his Renault, driving away from Trocadero Square, right across the river Seine from the Eiffel Tower; Ariane is approaching the same square in her Peugeot. Both have noticed the light go out on the venerable iron lady. They glance briefly at their watches; it is about midnight. Actually, it is a fraction of a second later, since it takes light a finite amount of time to travel from the now-darkened tower to the retinas of our two drivers. Knowing the distance between the tower and the cars, as well as the time interval needed for the last particle of light emitted by the tower to reach them, our two budding physicists can deduce the speed at which the information has been carried. They simply divide the distance by the time elapsed since midnight. For Ariane, who is traveling toward the Eiffel Tower and meets the oncoming light, the time elapsed since midnight is a little shorter, and hence the speed of light she calculates is slightly greater, since it is the true speed of light plus that of her Peugeot. For Paul, who is moving away, the situation is reversed. Light must catch up to him, and therefore the time elapsed since midnight is a little longer, and the apparent speed of light comes out slightly smaller, being the true speed minus the speed of his Renault. Nothing earthshaking in this. It all seems to hold together sensibly.

Or does it? Isn't there a slipup somewhere in the argument—and an important one at that? The Michelson-Morley experiment tells us that Ariane and Paul must come up with precisely the same speed of light, regardless of their own speed and direction of travel. Faced with this dilemma, Einstein proposed a revolutionary solution. Since speed is a distance divided by time, it can remain constant and observer-independent only if distances and time intervals vary together with the movement of the observer so as to keep their ratio invariant. With this bold proposal, Einstein was tossing aside the concepts of universal space and time, which had been the bedrock of Newton's theory. Time and space suddenly became malleable and flexible. Time stretches when you step on the accelerator to go faster. Your time and my time are slightly different. But they are both equally valid. One time is no better or more correct than any other. The demise of universal time meant that space, too, lost its universal character. From then on, space could shrink at will to accommodate the movement of an object. The concept of ether, which was supposed to act as an absolute frame of spatial reference, found itself swept by the wayside.

TIME STRETCHES

In his 1905 paper, Einstein told us precisely how time speeds up and slows down for an observer or an object in motion. The higher the speed, the more time slows down. One second can stretch into eternity if an object moves fast enough. If you go at half the speed of light, your second will be 15 percent longer than the second of a person standing still. Accelerate up to 99.99 percent of the speed of light, and a second becomes 1.18 minutes. Accelerate some more to 99.9999999 percent of the speed of light, and a second stretches to 6.2 normal hours. Time stretches more and more as you get ever closer to the speed of light.

You may tell yourself that this goes against common sense and that you do not understand it at all. But common sense is a very poor guide when it comes to high speeds. And, in truth, there is nothing to understand: That is simply the way Nature is.

Time dilation has been verified repeatedly—for instance, in the behavior of "cosmic rays." If you place a Geiger counter next to you, it will start clicking nonstop. This crackling chatter is caused by cosmic rays—streams of high-energy particles coming from space and constantly bombarding Earth. Cosmic rays are believed to originate in the violent death of massive stars (30 times more massive than the Sun) scattered throughout the Milky Way. During their final gigantic explosion, these supernovae, as they are called, eject large numbers of particles at ultrahigh speed; then these particles begin a long interstellar journey and some of them ultimately reach Earth. If it were not for the shielding effect of Earth's atmosphere, these cosmic rays would be so hazardous that life on Earth would be impossible. Some of these particles do make it all the way through the atmosphere and play a determining role in biological evolution on Earth by causing genetic mutations in living cells.

A fraction of these very-high-energy particles collide with atomic nuclei in the upper layers of Earth's atmosphere. The nuclei break up into myriad pieces, creating a shower of subatomic debris that rains down on Earth. The majority of the debris particles have a very short lifetime and decay before reaching the ground. But some particles, called "muons," survive sufficiently long to reach us and cause Geiger counters to "sing." Muons resemble electrons, but they are more massive. They interact very weakly with ordinary matter, and as they emerge from the atmosphere, most continue on into the depths of Earth's mantle. Their presence on the surface of Earth does, however, raise an intriguing question. Muons are notoriously unstable, existing for the very briefest instant before decaying. Their half-life is only two-millionths of a second, which means that from an initial population of 10,000, only 5,000 will be left after 2 microseconds, 2,500 after another 2 microseconds, 1,250 another 2 microseconds later, and so on. The population of muons decreases quite fast and is reduced to almost nothing in just an eyeblink. Now, in 2 microseconds, light covers only a distance of 600 meters. Muons, which do not travel quite as fast as light, cover even less distance. If that is the case, how can they last long enough to make it all the way through Earth's 20-kilometer-thick atmosphere and pay us a visit?

Enters Einstein with the explanation. Because muons move at nearly the speed of light, their lifetime measured with a terrestrial clock is stretched roughly 1,000-fold. A half-life of 2 microseconds translates into two-thousandths of our own seconds. And that is plenty of time for muons to make it to Earth's surface and cause Geiger counters to click furiously. Muons are not the only particles to have their lives stretched. So do the elementary particles racing in the CERN nuclear accelerator. Accelerate them to 99.7 percent of the speed of light, and their lifetime will increase 13-fold. At the 99.9 percent mark, they will live 45 times longer. In all cases, time dilation manifests itself exactly as Einstein had predicted.

You remain skeptical and object that all of this seems too abstract. It is easy for physicists to talk of time dilation for particles that are invisible, not directly accessible to our senses. The only way to convince you would be to show you firsthand how time stretches for a concrete object, such as a clock, when in motion. As it happens, you are in luck, because that is precisely the experiment done in 1971 by a couple of American physicists. They installed atomic clocks—devices that can track time with unparalleled accuracy— aboard jetliners.

Until quite recently, in this world of ours obsessed with time, a second was defined by means of the rotation of Earth. Unfortunately, the rotation rate is not constant. As we have seen earlier, Earth's spin slows down more and more because of the tidal forces exerted by the Moon. Calculations show that when Earth was born 4.6 billion years ago, it rotated roughly four times as fast as it does today. A day lasted only six hours and the Sun hurried across the sky in about three hours. Clearly, this slowing down is not noticeable on the scale of a human life, but it is large enough to affect the most accurate clocks. That is why atomic clocks were invented. Nowadays, a second is defined by the vibrations of a cesium atom. In one second, a cesium atom oscillates 9,192,631,770 times. That is now the standard for the time broadcast by the Bureau international des poids et mesures (International Bureau of Weights and Measures, located in Sèvres, a suburb of Paris), against which clocks and watches in the entire world are synchronized.

In the experiment conducted on commercial jets, four atomic clocks traveled east, and another four west. Upon completion of their respective trips, the times indicated by the clocks were compared, not only among themselves but with reference clocks on the ground. Just as Einstein had postulated, time differences showed up in exactly the proportions predicted by his relativity theory. Because jetliners generally fly at less than one-millionth of the speed of light, the observed differences were extremely small, on the order of one-millionth of a second per day of flight. Nevertheless—and that is the power of atomic clocks—such tiny changes could easily be measured. The clocks that had traveled east had lost 59 billionths of a second when compared to clocks on the ground, whereas those that had traveled west had gained 273 billionths of a second. The discrepancy between these two times was due to the west-to-east rotation of Earth. It, too, causes time dilation. When that additional effect is taken into account, one concludes that both sets of clocks have slowed down by precisely the amount predicted by Einstein.

In our everyday lives, we move far more slowly than a commercial jet. When we barrel down an expressway in a car at 130 miles per hour, we reach only two-tenths of a millionth of the speed of light. Even on board the TGV (the high-speed train that is the pride of France), hurtling along its tracks at 200 miles per hour, the time experienced by its passengers slows down by less than one-millionth of a second. Such a minute difference can be detected by an atomic clock, but it is completely unnoticeable with a common wristwatch. And a good thing it is for our mental health! Think of all the missed appointments and blown opportunities that would beset us if time changed appreciably at the whim of our every movement.

FASTER THAN LIGHT?

One important consequence of the fact that the speed of light is constant is that no material object can move faster than light. If any object could accelerate from a sedate ordinary pace to a breakneck speed greater than that of light, it could catch up to a light ray racing ahead of it, overtake it, and leave it in its wake. The apparent speed of light for anyone riding on such an object would then decrease, go through zero, and eventually increase in the negative direction; that would be a flagrant violation of the experimental results, which demand that any observer must always observe the same speed of light, no matter what. As such the theory of relativity forbids anything from crossing the light-speed barrier. Going from below to above 300,000 km/s is not permitted. Actually, contrary to what is often erroneously stated, relativity theory does not preclude the existence of particles or objects traveling faster than light. What it does forbid is *crossing* the barrier of the speed of light. The ban applies to both directions. It is no more allowed to go from a speed lower than that of light to one that is higher than the other way around. Physicists have in fact come up with a name to designate particles traveling faster than light; they dubbed them "tachyons," which means "speed" in Greek.

Up to now, tachyons exist only in the fertile imagination of researchers. And it is just as well, because the existence of such particles would cause many paradoxes in physics. Traveling faster than light could bring you back to the past, which would violate the principle of causality. We are accustomed to causes occurring before effects, actions preceding their results. An egg breaks because we tap against its shell. A leaf comes loose from a tree because the wind blows in gusts. A nail penetrates the wood of a chair because we pound on it with a hammer. A target shatters because we fire at it. I came into the world because my grandmother gave birth to my father, who in turn conceived me. If relativity tossed absolute time aside and gave each of us our own individual time, one question comes inevitably to mind: Can it also rearrange the order of events, make the effect come before the cause, force the nail to sink before it is struck by the hammer, or allow me to be born before my grandmother? Fortunately for our psychological well-being, relativity cannot alter the order of events when it deals with objects that travel below the speed of light. In our world, the target is always blown to pieces after the bullet has left the gun, never the other way around. To be sure, relativistic time is elastic, and the time interval separating the moment the bullet emerges from the gun and that when it hits the target varies with the speed of the observer and—as we will see shortly—with the intensity of the gravitational field in which he happens to be. But the cause will always precede the effect.

CAUSALITY TURNED UPSIDE DOWN

But with tachyons traveling faster than light, the order is no longer respected. Let us examine how that comes about.

Imagine a stationary gun firing not ordinary bullets but tachyons racing at twice the speed of light. Imagine also Jules, speeding along in his spacecraft in the same direction as the tachyons. Being made of ordinary matter, Jules can only move slower than light. Assume he is traveling at, say, 80 percent of the speed of light. Under such conditions, Jules would see the target disintegrate before the tachyon-bullet ever leaves the barrel of the gun. Causality would be reversed. In fact, he would actually see tachyons emerge from the shattered target and return to the gun's barrel, like a movie played backward.

The analogy is not entirely correct, though. In a movie, the sequence of events is reversed because the movie reel rotates backward—in other words, because you have inverted the arrow of time. There is no such trick with our tachyon gun. Time continues to move forward in a normal fashion. The past fades away to be replaced by the present, and the future is yet to come. If the effect comes before the cause in this particular case, it is only because tachyons travel faster than light. In fact, Jules would see the tachyons bullets slice through space from the target back to the gun twice as fast as light, as long as he himself is traveling at more than half the speed of light. The interval between the instant the target is blown apart and the time the tachyon-bullet reenters the gun barrel would diminish as Jules slows down closer to half the speed of light. At exactly the 50 percent mark, Jules would see the tachyon bullets go instantly from target to gun, as though they had infinite velocity. Only after he slows down below half the speed of light would he see causality restored. Only then would the order of events return to normal, and the target will once again be blown to pieces after the tachyon bullets leave the gun.

COMMUNICATING WITH THE PAST

A world containing tachyons would be a place where logic as we know it becomes meaningless. Causality would be topsy-turvy and the relationship between cause and effect turned on its head. Being made of ordinary matter, we cannot travel back to the past, even in a world containing tachyons. The time machine imagined by H. G. Wells remains in the realm of science fiction. But if a trip to the past is not in the cards for ourselves, we can in principle utilize tachyons to send signals back to the past. Let us see how that might be done.[1]

Jules, our adventuresome space traveler, is speeding along in his spacecraft at 80 percent of the speed of light, while his friend Jim, the sedentary one, enjoys peaceful days here on Earth. The two friends carefully synchronized their watches before going their separate ways. They promised to stay in touch by sending each other messages by way of tachyon transmitters emitting signals traveling at four times the speed of light. Let's say Jules leaves Earth at 10 o'clock in the morning. At noon, Earth time, Jim decides to use his transmitter to send a message to his friend Jules. Jules has just traveled for two hours at 80 percent of the speed of light; he has thus covered a distance of 2 x 0.8 = 1.6 light-hour (one light-hour is the distance traveled by light in one hour, which is equal to 1.08 billion kilometers, or 671 million miles). Since the tachyon message travels at 4 times the speed of light, it takes 0.4 hour (or 24 minutes) to cover that distance. But during that 0.4-hour period, the spacecraft continues on its way and travels another 0.4 x 0.8 = 0.32 light-hour, which means that the message needs another 0.08 hour to reach Jules. Of course, during this additional time, the spacecraft keeps on moving and covers an additional distance of 0.08 x 0.8 = 0.064 light-hour, which adds an extra 0.016 hour for the tachyon message to catch up to the spacecraft, and so on. We are dealing here with an endless series of additional times that become increasingly smaller. In the end, Jules will not receive his friend's message until half past noon, Earth time. But the time indicated by his own watch is altogether different. Since he is traveling at 80 percent of the speed of light, his time passes more slowly than Jim's. For each Earth hour, Jules's watch advances only 60 percent as fast, or 36 minutes.

It is easy to calculate by how much time slows down. You divide the spacecraft's speed by the speed of light, which in this case yields 0.8. Take the square (0.64), subtract it from 1, which gives 0.36, and take the square root of the result, which gives 0.6, or 60 percent. The 2.5-hour-long trip according to an Earth watch seems to Jules to have taken only 1.5 hour. Therefore, his watch says 11:30 when he receives Jim's message. As far as Jules is concerned, the distance between him and Earth is only 1.5 x 0.8 = 1.2 light-hour.

Jules, being a highly conscientious fellow, replies to his friend's message without delay. His own tachyon transmitter also sends signals traveling at four times the speed of light. For Jules, things are now reversed. His message must catch up to Earth, which is receding at 80 percent of the speed of light. To cover a distance of 1.2 light-hour, the tachyon message needs 0.3 hour. But in the meantime, Earth will have receded some more, and some extra time is required to make up the difference. From Jules's point of view, it will take three-eighths of an hour all told, or 22.5 minutes, for Jim to receive his message, at which point Jules's watch will indicate 11.5 hours + 22.5 minutes, or 11:52:30. That is 1 hour, 52 minutes, and 30 seconds (or 112.5 minutes) after Jules took off from Earth. From Jules's vantage point, his friend Jim is being carried by an Earth receding at 80 percent of the speed of light. It is therefore Jim's time that is being slowed down by 60 percent. According to Jim's Earth watch, the reply will thus have taken only 112.5 x 0.6 = 67.5 minutes to reach him. His watch will then say 11:07:30. What this means is that

Jim is going to receive the reply a full 52.5 minutes before he sent his original message at high noon, which blatantly violates causality. With the help of his tachyon transmitter, Jules is able to send signals back into Jim's past.

Hopefully, this example will have convinced you that a world containing objects or particles traveling faster than light would play havoc with logic and causality as we know them. Einstein was acutely aware of this. He declared categorically in his 1905 paper that velocities higher than the speed of light were forbidden. The impassable barrier that is the speed of light thankfully exists to prevent the notions of future and past from turning topsy-turvy and to help preserve our sanity. Having said that, there is absolutely no mathematical restriction in the special relativity theory unveiled in Einstein's 1905 paper that explicitly precludes the existence of tachyons. Physicists would dearly love to discover such a restriction, but if it exists, it has thus far remained elusive. In any event, physics, logic, and causality would be in for some rough times indeed if tachyons were ever discovered.

THE SECRET FOR ETERNAL YOUTH?

Here is an exciting piece of news: Einstein taught us that motion can slow time down to a trickle. Might he have discovered there a fountain of youth and the secret for eternal vigor? Is going fast all it takes to remain young? Nice try, but don't pop open the champagne just yet.

To begin with, you would have to get fairly close to the speed of light for time dilation to become really appreciable. If Jules travels at 80 percent of the speed of light, he will age by 60 years compared to Jim's 100. At 86 percent of the speed of light, he will age by only 50 years, and at 99.5 percent by no more than 10 years. The closer he gets to the speed of light, the more slowly he ages compared to Jim. But there is a catch. Approaching the speed of light entails a hefty price, because the faster Jules's rocket goes, the heavier it gets. As a matter of fact, the mass increases by exactly the same factor by which time slows down. At 99.5 percent of the speed of light, Jules may age 10 times more slowly than Jim, but his rocket also becomes 10 times more massive. And the more mass it gains, the more fuel it requires—which adds even more mass to the spacecraft, and thus increases the fuel requirements further. A vicious circle sets in, with no end in sight. To reach the speed of light, Jules would need an infinite mass of fuel, which falls in the realm of science fiction.

But even assuming for a moment that this fuel problem could be solved, Jules can slow the rate at which he ages only in relation to someone else's time, in this case Jim's. As it turns out, as long as Jules moves at a constant speed, without accelerating or decelerating, the situation is completely symmetrical between Jules and Jim. Jules, sitting in his rocket, thinks that Jim is carried away by Earth at 80 percent of the speed of light, which slows Jim's time to 60 percent of normal, and he himself is the one who ages at a normal rate. At the same time, Jim, firmly planted on Earth, thinks that Jules is moving away at 80 percent of the speed of light and, therefore, that it is his friend's time that is being slowed by 60 percent compared to Earth's time. How can Jules's time simultaneously go faster and more slowly than Jim's? Is there a paradox here? Could there be something terribly wrong with special relativity?

The answer is decidedly no. As long as Jules and Jim are apart, the situation is totally symmetrical and each man thinks the other's time is slowed down compared to his own. Each can even verify this by watching television images sent by cameras aimed at the clock in Jim's living room on the one hand, and at the one in Jules's spacecraft cabin on the other. These images are carried by radio waves, which, as Maxwell discovered, propagate at the speed of light. Let us first examine things from Jules' vantage point. Because Jim seems to be moving away from him at 80 percent of the speed of light, Jules's sees his friend's clock advance by only 36 minutes during a complete hour in his space capsule. Actually, Jules sees Jim's clock move even more slowly than that, because in addition to the slowing down predicted by relativity theory, the radio waves carrying the image of Jim's clock take a certain time to travel the distance between Earth and the rocket, which introduces an extra delay. This additional effect is known as the "Doppler effect," named after the Austrian physicist who discovered a similar phenomenon for a source of sound moving with respect to an observer.[2] Next we consider what happens from Jim's point of view. Things are completely symmetrical. Jim sees Jules move away at 80 percent of the speed of light and observes that Jules's clock is slower than his. The delay, caused by the combination of relativity and the Doppler effect, is precisely the same as that observed by Jules as he keeps track of Jim's clock.

Jules finally reaches his destination, a planet orbiting a star somewhere in the Milky Way. After completing his mission on the distant planet, Jules begins his return trip back home. The situation is now reversed. The images of Jim's living room now meet the rocket head-on, rather than having to catch up to it, and the Doppler effect makes Jim's clock seem to run faster. Of course, the slowing down predicted by relativity continues to apply, but that is smaller than the Doppler effect, and the net result is that Jules sees Jim's clock gain time compared to his. What about Jim? Here, too, the situation is symmetrical. Jim receives television images carried by radio waves that meet Earth straight-on. Once again, the Doppler effect dominates over relativistic effects, and Jim concludes that Jules's clock runs faster than his. In fact, each friend thinks the other's clock outpaces his own at precisely the same rate.

BROKEN SYMMETRY

The situation remains perfectly symmetrical as long as Jules and Jim do not meet face-to-face. What happens now when Jules makes it back to Earth and can finally compare his clock with Jim's without having to rely on video images carried by radio waves? Surely, at that point things can no longer remain symmetrical. His clock must be either ahead of or behind Jim's, but it cannot be both. Here is where Einstein steps forward to sort it all out: Time will have passed more slowly for Jules the adventurer than for Jim the sedentary one. If Jules has kept a constant speed of 80 percent of the speed of light during the entire round trip, his time will have flowed 60 percent slower than Jim's. Let's say his trip lasted 20 years as counted on Earth. He left in the year 2000 and came back in 2020. As far as Jules is concerned, though, his trip will have lasted only 20 x 0.6 = 12 years. The calendar on board his spacecraft tells him he is back in the year 2012. That difference is quite real. Jules's heart will have beaten fewer times, his lungs will have absorbed less air, his hair will not have turned nearly as white, and his face will not show as many wrinkles; he will not have brushed his teeth as often or eaten as many meals.

In short, symmetry will have been broken. Why? Because during the course of his trip, Jules has had to change speed; he has had to accelerate and decelerate. He had to accelerate to break free of Earth's gravity and reach his cruising speed of 80 percent of that of light, and then had to decelerate to nearly zero speed to come to rest on the surface of the distant planet. Then he had to accelerate again to start his return trip, and decelerate once more for a soft landing back on Earth. These accelerations and decelerations are very real. Astronauts experience them firsthand when they are pressed hard against their seats as their spacecraft lifts off. We experience it in a car that leaves a red light burning rubber or in an express elevator starting a run to the top floor. The situation is symmetrical between Jules and Jim as long as Jules's speed remains steady. Under these conditions, to say that Jules's rocket moves away from a fixed Earth or that Earth moves away from a stationary rocket amounts to the same thing. Symmetry prevails as long as relative velocities are constant. The moment there is a change in speed, symmetry is broken. Jules's time flows more slowly than Jim's because it is Jules who is subjected to accelerations and decelerations.

And so the fountain of youth discovered by Einstein takes on a very peculiar character. All you have to do to pile up years less rapidly is to step on the accelerator. But you cannot use speed to slow your rate of aging in your own time frame of reference, only relative to someone else's time. As long as he was traveling, Jules felt that he was aging quite normally. Only after his return to Earth did he realize that he had aged by only 12 years, rather than the 20 Jim had accumulated.

THE FIRE OF STARS

Not only does mass increase with velocity, but Einstein found out that it can also be converted to energy. This discovery was to completely change the course of mankind's history and give physicists a sense of guilt.

For Newton, all the energy of an object resided exclusively in its motion. The faster it went, the more energy it contained, the latter being proportional to the square of the velocity. A stationary object would have no energy.

Einstein completely changed all that. He demonstrated that an object possesses energy even when at rest, because of its mass. This mass energy is calculated by simply multiplying the mass by the square of the speed of light. That is the meaning of the equation $E = mc^2$, which has come to be perhaps the most recognizable physics equation of all times. Because the speed of light is so great (300,000 km/s), mass energies can be mind-boggling. For instance, if you weigh 150 pounds and know how to convert that mass into energy, the energy released would be 30 times larger than that of the most powerful nuclear bomb ever detonated by mankind.

Converting mass into energy is no easy task. During World War II, the best and brightest physicists of the Allied nations gathered at Los Alamos, in a remote corner of the New Mexico desert, to attempt to unlock the power confined within matter. The Allies believed that Hitler had begun a similar effort and that it was crucial to beat him to it. The Los Alamos physicists met the challenge brilliantly. The first atomic bomb test took place in the dawn hours of a July morning in 1945 at Alamogordo. The serenity of the desert was shattered by the explosion of an enormous fireball. Upon seeing the giant black mushroom rising up in the pale sky, two lines from *Bhagavad-Gita*, a Hindu poem, raced through the mind of the American physicist Robert Oppenheimer (1904–1967), the director of Los Alamos:

> *I am become Death,*
> *the shatterer of Worlds.*

Sadly, his premonition came true only a few weeks later when, on August 6 and 9, 1945, two atomic bombs were dropped on the Japanese cities of Hiroshima and Nagasaki, causing more than a hundred thousand immediate casualties, and leaving an indelible stain on humanity's soul. Will we find the wisdom never to forget the pain and grief and forever protect our beloved blue planet from nuclear peril?

Einstein had no inkling of atomic bombs when he discovered the equivalence between mass and energy. His chief concern was objects in motion. But discoveries in fundamental science often have unexpected ramifications and can take unpredictable turns. They do not harbor good or evil in themselves. It is how politicians and the military choose to use them that ultimately determines their moral value.

Einstein's discovery also made it possible to resolve several important scientific enigmas. Why stars shine was a complete mystery in the early twentieth century. No one had a clue about what fueled their fire—particularly over such a long time, since it had by then been established through radioactive dating of elements in Earth's crust that the Sun was a few billion years old. Thanks to the mass-energy equivalence, we now understand that it is by converting a very small portion of its hydrogen mass that the Sun shines, warms us, and is the source of all life on our planet. Granted, this conversion process is not very efficient—only 0.7 percent of the Sun's hydrogen mass is converted into radiation—but it is enough to turn our Earth into a cozy cocoon in the frigid cold of interstellar and intergalactic space, where the temperature is near absolute zero.

It also helped resolve the mystery of how the universe was born. It is currently believed that it came into existence by way of a fantastic explosion—the big bang—in a vacuum filled with energy. The mass-energy equivalence enables us to understand how the vacuum energy could have given birth to the entire material content of the universe, from galaxies to stars, from planets to men and microbes. Einstein's discovery also explains the behavior of elementary particles in the subatomic world. In that world, particles appear and disappear at the whim of collisions, as if by magic. When two protons hurtling at blinding speed collide, the result is not just the two original protons but a plethora of other particles. Part of the kinetic energy of the two protons is converted into new particles. Newtonian physics is unable to account for such a phenomenon. For Newton, two billiard balls collide and the same two balls rebound in other directions, but the appearance of new particles was totally inconceivable. We have also mentioned how the Austrian physicist Wolfgang Pauli used the equivalence of mass and energy and the principle of energy conservation to infer the existence of neutrinos.

GRAVITY AND ACCELERATION

The symmetry between Jules and Jim was broken because Jules was subjected to accelerations and decelerations, while Jim was not. In the wake his 1905 paper, Einstein started thinking about the nature of accelerated motion as well as that of gravity. He soon realized that the two phenomena were intimately related. The insight, which he would later describe as "the happiest thought in my life," came to him in 1907 as he was daydreaming at the Patent Office in Bern. Like most great ideas that have altered the course of history, this one, too, was of limpid simplicity. Einstein imagined himself standing in an elevator suddenly starting on its way up. We have all experienced this sensation. The acceleration pushes you against the floor as if gravity had abruptly increased and your weight had just shot up. Conversely, when the elevator accelerates down, you have the queasy feeling that your stomach cannot quite keep up with the rest of your body, that you are less firmly planted on the floor, and that you have suddenly shed many pounds. You can experience the same sensation in an airplane suddenly encountering air turbulence, when the pilot quickly descends to avoid the area of disturbance. Elevators made Einstein realize that the effect of accelerated motion is equivalent to that of a gravitational field.

Let us return to Jules flying aboard his spacecraft. At the moment, he is far away from any planet or star, which means that he is not subject to any gravitational force. Jules has shut down all engines on his rocket. Carried by its momentum, it keeps on gliding at constant speed in the quiet vastness of interstellar space. Inside his cabin, Jules has fallen asleep, secured by a belt to his seat. Around the cabin, an empty glass, a pencil, an eraser, and a writing pad are drifting about. Jules was writing a letter and drinking a glass of water when he dozed off. Had he not been restrained by his seat belt, he, too, would be floating along with the pencil and eraser. Jules wakes up and decides to restart his engines. Glass, pencil, and everything else continue to float freely, as if nothing had happened. But the floor of the cabin catches up to them at increasing speed, and the freely floating objects end up hitting it, all at the same time. A passenger unaware that Jules had just restarted his engines would have thought that the pencil and eraser had fallen to the floor because they were attracted by a gravitational field, just as a ripe apple falls to the ground because of Earth's gravity. The effects of the rocket's acceleration and those of a gravitational field are strictly identical.

What Einstein did was to elevate this profound similarity to the status of principle. The "equivalence principle," as it is known, constitutes an extremely powerful tool for exploring Nature's secrets. To understand, for instance, the behavior of light in a gravitational field, all one has to do is ask oneself how it would appear to an observer in accelerated motion, which is a considerable and very effective shortcut. Einstein, the magician, had thus pushed the notion of symmetry to its highest level. He was not content simply decreeing that two observers in constant relative motion must perceive exactly the same physical reality. He went a step further and asserted that any difference between the physical reality perceived by an observer in constant motion and another in accelerated motion could be interpreted in terms of gravity.

A FEATHER FALLS AS FAST AS A CANNONBALL

Einstein incorporated in his equivalence principle a result derived by one of his illustrious predecessors—the Italian physicist Galileo Galilei. Galileo had also spent much time pondering the nature of accelerated motion and had discovered that all objects fall at the same rate in Earth's gravity field, irrespective of their weights. Centuries earlier, Aristotle had proclaimed that heavy objects fall faster than lighter ones. By a clever argument, Galileo demonstrated that view to be wrong. Suppose, he argued, that we drop two objects, one heavy and the other light, from the top of a tower. The two objects are not free, but are tied together by a rope. The question is whether the light object is going to slow down or accelerate the fall of the heavy one toward the ground. If Aristotle was right, the lighter object would fall more slowly, thereby pulling on the rope and slowing the fall of the heavy one. But one could just as easily argue that the two objects tied together form a new system more massive than the heavy object alone, and that the two objects tied together ought, therefore, to fall faster than the heavy one taken in isolation. Thus, if Aristotle was right, one would come to the absurd conclusion that the lighter item should simultaneously accelerate and slow down the fall of the heavy object. The only way to avoid this absurdity is to conclude that the lighter object does not affect the fall of its heavier counterpart one way or the other, which means that both have to fall at the same speed.

History books hold that Galileo went on to drop objects of various weights from atop the Leaning Tower of Pisa to verify the result he had worked out through reasoning alone. Ideally, the experiment would have had to be conducted in a complete vacuum, so as to eliminate any resistance to movement caused by air. In such conditions, a bird feather and a cannon-ball would fall equally fast. Some 350 years later, the experiment was actually done on the airless surface of the moon by an American astronaut, using a golf ball and a steel hammer.

MATTER DISTORTS SPACE

With his equivalence principle, Einstein would go on to construct the magnificent edifice of general relativity, which describes the behavior of any object in a gravitational field and constitutes one of the most marvelous achievements of the human intellect.

An immediate consequence of the equivalence principle is that space is curved by gravity. To appreciate how this comes about, we once again visit our old friend Jules as he is in the process of accelerating aboard his space-craft. He happens to have a gun capable of firing a laser beam (a laser is a type of light source). Standing against one wall of the cabin, Jules aims at a target on the opposite wall and fires. The laser beam takes a small fraction of a second to get across the cabin. As it makes its way, the spacecraft continues accelerating, dragging with it floor, walls, and target. As a result, the laser beam hits not the intended target but a spot slightly below. If Jules were to retrace the trajectory of the laser beam, he would obtain not a straight line but a curve. The acceleration of the rocket caused the laser beam to bend. But since, as we have emphasized several times, an accelerated motion is equivalent to a gravitational field, it follows that light is bent by gravity as well, even though photons—the elementary particles of light—have no mass. To assert that gravity is caused by a material mass is tantamount to saying that matter bends light.

Light does not like to waste time. It always chooses the shortest path to go from one place to another. In a flat space, one without any curvature, that path is a straight line. But in a space that is curved, it itself becomes curved. The fact that matter bends light therefore means that matter distorts space. The Moon completes its monthly orbit around Earth by following an elliptical trajectory. For Newton, this elliptical orbit was determined by a gravitational force exerted by Earth on the Moon and transmitted via a mysterious medium called "ether."[3] Einstein did away with the notions of force and ether. From his point of view, Earth's mass distorts the space around it, and the shortest path for the Moon to orbit Earth in such a curved space happens to be an ellipse. That is why the orbit of the Moon is shaped like one.

Einstein even thought of ways to verify that matter does indeed warp space. He proposed to observe distant stars whose positions projected against the sky were close to the Sun's. If space is truly curved by the Sun's gravitational field, then the path followed by the light emitted by these distant stars should bend as it passes near the Sun. This bending should manifest itself as a slight angular deviation (by 1/2000 of a degree, or roughly the angle subtended by a quarter at a distance of 6 miles) compared to the apparent position of the same stars photographed six months later, when Earth is located diametrically opposite in its orbit around the Sun and the starlight no longer has to go through the Sun's gravitational field to reach us. To photograph stars whose angular coordinates are close to that of the Sun, one has to wait for a solar eclipse, when the Moon obscures the Sun's blinding glare. In 1919, two British expeditions were organized to go observe a solar eclipse, one in Brazil and the other in Spanish Guinea, to attempt to verify Einstein's predictions. The results were spectacular. Starlight did turn out to be bent by the solar gravitational field, and by precisely the angle calculated by Einstein. This decisive confirmation of relativity made the front page of newspapers around the globe. A world in disarray, freshly emerging from a devastating world war, tired of hearing of nothing but bolshevism, war reparations, and famine, became captivated by this fantastic piece of scientific news. Einstein found himself instantly propelled to the pinnacle of fame.

GRAVITY SLOWS TIME

But that was not all. The equivalence principle led Einstein to an even more provocative conclusion—namely, that time is slowed by gravity.

To convince ourselves of it, let us join Emily and Claire on a visit to the Eiffel Tower. Emily stays at ground level while Claire goes up to the very top of the 1,050-foot tower. Budding physicists, they want to compare their respective times. In this case again, light is to play the role of messenger between the two friends. They agreed ahead of time that once a second, according to her own watch, Emily would send a flash of light to her friend Claire. The question is whether Claire is going to see the signals arrive at one-second intervals as measured by her watch as well. Einstein provided the key to the answer with his equivalence principle. According to him, the situation experienced by Emily and Claire is exactly equivalent to transplanting the Eiffel Tower away from Earth's gravity and into a huge fictitious spacecraft in constant acceleration. Analyzing the new situation is now fairly straightforward. Because of the rocket's acceleration, Claire moves away at an increasingly faster speed from the light signals sent by Emily. These take more and more time to catch up with Claire atop the tower. Accordingly, Claire sees successive light flashes suffer increasing delays, which causes them to arrive separated by more than a second. The conclusion is that Claire, standing at the top of the tower, sees Emily's time, down at the bottom, stretch and slow down in comparison to her own. Earth's gravity has therefore slowed time. At the base of the tower, Emily is closer to the center of Earth and experiences a stronger gravity—remember that its intensity decreases as the square of the distance to Earth's center. Conversely, at the top, Claire is farther away from Earth's center; she experiences a diminished gravity, and her time runs faster.

This effect is quite real. People living on the top floor of apartment buildings do indeed age more rapidly than those who live on the ground floor. Likewise, time runs faster for Ecuadorians living close to the equator than for Eskimos near the North Pole. That is because, as Earth spins, the centrifugal force causes the planet to bulge slightly (by 0.3 percent) near the equator when compared to polar regions. As a result, Ecuadorians are a little farther away from Earth's center and experience slightly less of its gravity than Eskimos. Even though Earth's gravity prevents us from floating freely in the air and the space shuttle needs to burn tons of fuel to escape its grip, it is weak enough for these time differences to be insignificant on the scale of a human lifetime. A clock on Earth loses about one-billionth of a second in an hour compared to one that is in space, far away from any gravitational influence. If you lived on the ground floor of your building during your entire life, you would gain barely a microsecond over your neighbor up on the tenth floor. It is just as well that the difference is so minute. Otherwise the world might face a serious housing crisis, as nobody would want to live on the upper floors. And it would be a shame if everybody on Earth deserted the sunny equatorial regions and cooped themselves up in igloos near the North Pole just to gain a few additional breaths of life.

Our crude wristwatches are incapable of detecting such minute time differences. There are, however, sophisticated instruments than can do it. Two American physicists at Harvard University, Robert Pound and Glen Rebka, succeeded in measuring a fractional change of 2.5 millionths of a billionth between the top and bottom of a 75-foot tower on the Harvard campus. This change corresponded to a time lag of one second in 100 million years between the clocks at top and bottom.

Astronomers have also used the Sun's gravity to measure time slowing down. Because of its enormous mass, the Sun's gravity is far greater than Earth's. Therefore, time should run slower near the Sun than on Earth. The fractional delay is of the order of two parts in a million. Because Earth and all the other planets revolve around the Sun, their positions in the sky change. There are times when Earth, another planet, and the Sun are nearly aligned, with the Sun in between. One can send radar signals from Earth, bounce them off the planet on the far side of the Sun, and measure the time it takes the radar waves to complete the round trip. Because the radar beam passes near the Sun, where time slows down, the measured round-trip travel time is slightly longer than when the planet is in some other region of the sky and the radar beam is no longer affected by the Sun's gravity. In other words, radar echoes should arrive back later than expected when they have skimmed over the Sun. Literally hundreds of radar echoes have been measured to this day, and they have invariably vindicated Einstein.

TIME ACQUIRES A SPACE COMPONENT

Gravity slows time and bends space. Step on the gas pedal to pick up lots of speed, and you will see time stretching and space contracting. As Jules races at 80 percent of the speed of light, not only does Jim see Jules's time elapse at 60 percent of the rate of his own, but he also sees Jules's rocket shrink to 60 percent of its normal size. It is as though the stretching of time is exactly offset by the contraction of space. In a Newtonian world, time and space were independent actors free to do as they pleased on the grand cosmic stage. Einstein does not allow that independence any longer. From now on, the two would form an inseparable and indissoluble pair, always acting in concert. Disjointed time and space are but mere shadows of reality. The only thing that has a tangible existence is a united space-time. Space takes on attributes of time, and time of space.

Some 20,000 years ago, our Paleolithic ancestors were already intuitively giving time a spatial dimension when they carved notches into animal bones to record the passing days. By destroying the universality of time, Einstein provided a scientific justification for adding a space component to time. By becoming malleable and flexible to accommodate movement, this new type of time swept concepts of universal simultaneity, past, present, and future, overboard. What is "now" for me can well be perceived as the future by you or the past by my friend. It all depends on my motion relative to yours and the distance that separates us. For instance, with the simple act of putting this book down and walking across your living room, what you call "now" can change by a whole day from the vantage point of someone living in the Andromeda galaxy, 2 million light-years away. If the two of you are sitting and reading quietly, "now" is the same for both of you. But if you are stepping across your living room, a few meters closer to Andromeda, your "now" will move one day forward into the future of your counterpart in Andromeda, at least if you happen to get a few meters closer to the galaxy. On the other hand, if you move in the opposite direction, your "now" will slide back by one day into the past of the inhabitant of Andromeda.

Losing the concept of simultaneity does violence to our traditional notions of past, present, and future. We think of the past as done with, gone and lost in the folds of our memory, and nothing can be done to change it. Only the taste of a madeleine (a French pastry) and the magic of writing can enable someone like Marcel Proust to relive time gone by. On the contrary, we think of the future as something to build; it carries our dreams and hopes, and our present actions can influence and mold it. We feel time passing by, and our language reflects that fact: We speak of the "course" or "flow" of time. Yet this psychological time does not conform to Einstein's physical time. If, as he taught us, my future can be my friend's past, and if my neighbor's past can be my present, then the past cannot be completed and the future is not yet to be made. Past and future must coexist with the present. Like a landscape extending as far as the eye can see, physical time exists in its entirety at once. The canvas of time stretches from the horizon of the past to the horizon of the future. All distinction between past, present, and future is but an illusion. In contrast to psychological time, physical time neither flows nor passes. It exists as a single entity; it simply is. In that sense, it echoes the poetic intuition of William Blake in *Jerusalem*:

> *I see the Past, Present and Future*
> *Existing all at once before me. . .*

or T. S. Eliot in *Burnt Norton*:

> *Time present and time past*
> *Are both perhaps present in time future,*
> *And the time future contained in time past,*
> *If all time is eternally present.*

Physical time resolves many paradoxes posed by psychological time. If time really flows by, at what speed does it do so? Obviously, this is an absurd question. But if time simply is and does not flow at all, there is no longer any reason to pose the question in the first place. Yet everything in the human experience points to an arrow of time, carrying us from our cradle to our grave. Just what creates this sensation of motion? Is this perception of flow merely a mental construction? Is it an illusion concocted by the brain, with no real existence of its own? Could appearances be nothing more than shadows of reality, as Plato maintained? Or is there an aspect of physical time that we have yet to apprehend? Einstein revolutionized our concept of time at the beginning of the twentieth century by marrying it with space. Will there be another revolution in the twenty-first century that will resolve the profound dichotomy between the two kinds of time—physical and psychological? One thing is clear, though: Such a solution will remain beyond our reach until we understand how the brain works and perceives time, and how all of this is related to our concept of free will.

EINSTEIN AND "BLACK HOLES"

Gravity slows time and bends space. Such notions fly in the face of common sense. Nevertheless, they have been confirmed by clocks placed atop towers running faster than at the bottom, by delays of radar signals, and by light rays bent when they pass through the Sun's gravitational field. But these effects are so minute that they can be detected only by the most sophisticated instruments. If those had been the only consequences of the theory of relativity, Einstein's ideas might have lost much of their vitality. Instead, the second half of the twentieth century witnessed such a strong revival of Einstein's ideas on such diverse fronts that this era came to be seen as relativity's "golden age." That was due in large part to a string of discoveries of astrophysical objects, one more extraordinary than the last. Without a doubt, the most peculiar creatures in this fantastic astrophysical zoo are "black holes."

The concept of a black hole had been put forth long before the advent of relativity. The English philosopher John Michell (1724–1793) had proposed it as early as 1783. The French mathematician Pierre Simon de Laplace (the same one who had sung the praises of determinism and proudly replied to Napoleon that he had no need to postulate a Lord Architect in order to explain the world) had mentioned it in his masterwork *Le Système du monde* (System of the World), published in 1796. Their reasoning was quite simple. A minimum velocity is required to break free of a gravitational field. For instance, a speed of at least 11 km/s is needed to escape Earth's gravity, and 617 km/s to overcome the Sun's. Assume now a massive object with such strong gravity that the escape velocity exceeds the speed of light. In such a case, light cannot escape that mass, and the object will appear black. Of course, the name "black hole" had not appeared yet at the time—it was coined by the American physicist John Wheeler (1911–) in 1967. Laplace called such objects "occluded celestial bodies." There matters languished for more than a century, for lack of a more advanced theory and more accurate observations. The idea was too advanced for the times.

When Einstein set forth the equations of general relativity in 1915, he immediately realized that they predicted the existence of objects whose gravity is so large that light could not escape. But even the revolutionary Einstein was not prepared to accept such an outlandish notion. He felt that perhaps his equations would break down when gravity becomes infinite. His view was that Nature had to be ingenious enough to spare us from such bizarre creatures. He wrote papers in which he came out decidedly against the existence of black holes. As it turned out, he was wrong. Black holes do in fact exist. Astronomers have found them in the heart of galaxies. Even our own Milky Way harbors one. They are ubiquitous in modern astrophysics and often make it to the cover of magazines. Einstein did not live long enough to see black holes take center stage in astronomy. He passed away in Princeton, New Jersey, in 1955. But long before that, he had lost any interest for these bizarre compact objects that thumbed their nose at common sense in matters of space and time. There were several reasons for this lack of interest. First off, the 1920s had witnessed the birth of a new theory of matter—quantum mechanics. This theory interprets reality in terms of chance and probability, which Einstein, inveterate determinist as he was, found not at all to his taste. "God does not play dice," he was fond of saying. He spent the last thirty years of his life trying to develop a great theory that would unify the forces of gravity and electromagnetism, and would obviate this profoundly offensive indeterminism. Moreover, the demands that unfortunately come with fame consumed a lot of his time. During World War II, at the prodding of the Hungarian physicist Leo Szilard (1898–1964), Einstein wrote to the U.S. president, Franklin D. Roosevelt, urging him to launch a program aimed at building an atom bomb. Einstein was a dedicated pacifist, but his own experience of Nazism (he had to flee Germany to escape it) convinced him that there really was no choice. It was of paramount importance to beat Adolf Hitler to the finish line. After Hiroshima and Nagasaki, though, he lobbied tirelessly for a ban on all nuclear weapons and for a rapprochement with the Soviet Union. In 1952, David Ben-Gurion offered him the presidency of the state of Israel, but he declined.

A lone and revered figure, Einstein became increasingly removed from issues of contemporary physics. He displayed little interest in the great discoveries that revolutionized particle physics in the 1950s, and never relented in his opposition to quantum mechanics, despite its repeated successes in accounting for the behavior of atoms. As for exploring these bizarre black holes, he left it to his junior physicist colleagues, and took no part in it from the 1940s on. As they stepped into the world of black holes, his younger colleagues were to discover one fantastic phenomenon after another.

WHAT IF YOU BECAME A BLACK HOLE?

How does one find a black hole? To answer this question, it is necessary to first learn the "manufacturing recipe" for black holes—that is to say, how to create a gravitational field so intense that not even light can escape. In principle, any object can become a black hole if it is compressed sufficiently. If you weigh 160 pounds, and if a giant hand were to shrink your size to 10^{-23} cm, or 10 billion times smaller than an atomic nucleus, you would become a black hole. In fact, the size down to which an object has to be compressed is proportional to its mass. If you want to turn a 1,600-pound elephant—that is, 10 times more than your own mass—into a black hole, you will have to squash it down to 10^{-22} cm, or 10 times larger than in the case of your own distinguished person. But compressing things by hand is quite a challenge. The electromagnetic force, which binds atoms and molecules together, gives our bones their strength, and is responsible for the cohesiveness and shapes of the things of life, strenuously opposes any attempt at compression. This is good—otherwise, people would be going around trying to get even with their enemies by turning them into black holes.

As a matter of fact, to overcome the resistance of the electromagnetic force, it is necessary to call on the force of gravity for assistance. As Newton taught us, that force depends on the mass of the object involved. The more massive it is, the greater its gravity. Thus, to find black holes, the smart thing to do is to look for large masses. And what better place to look than in stars? But a note of caution is in order: Not any star will do. Take our Sun, for instance. Its has the enormous mass of 2×10^{33} grams (2 followed by 33 zeros). As large as this mass is, the Sun does not collapse under the effect of its own gravity, because it furiously generates energy in its core heated to more than 10 million degrees. Inside this cauldron, hydrogen nuclei (i.e., protons) collide, millions of nuclear reactions take place each second, and the Sun converts a very small amount (0.7 percent) of its hydrogen mass into radiation. This radiation is what makes the Sun shine and prevents its inward collapse. By making its way to the surface through the upper layers of matter, the radiation opposes the effect of gravity, which would otherwise cause it to implode. An equilibrium between the forces of gravity and radiation set in 4.6 billion years ago, when the Sun was born. This equilibrium will persist for another 4.5 billion years, the Sun being now only halfway through its life. At that time it will have exhausted all its fuel. The nuclear reactions will come to a stop, and there will no longer be any radiation to oppose gravity. The Sun will then collapse inward. Will it become a black hole?

The answer is no. Given its mass, it would have to shrink from its current radius of 700,000 km to less than 3 km. But physics mandates that the collapse should stop dead in its tracks at roughly 6,000 km, about Earth's radius, a far cry from the required 3 km. Why will it stop there? Because gravity will encounter another force that opposes it. This time around, it is not radiation that will stand in the way of gravity, but the collective action of electrons within the Sun. Solar matter heated to more than about 10 million degrees will collide furiously, freeing electrons from atoms. The collapse of the Sun will confine these electrons within an ever-shrinking space. Because they hate being piled up against one another, they will exert an opposing force strong enough to stop the collapse.[4] A new equilibrium will be reached between the force of gravity and that due to the electrons. The Sun will turn into what is called a "white dwarf." "White" because of the color of the emitted light, the energy generated by the collapse being converted to radiation, and "dwarf" because of its small size compared to a normal star. Matter is so compressed in a white dwarf that one cubic centimeter of this stellar corpse weighs as much as a ton. It is as if you crammed an elephant into a soup spoon. The white dwarf will keep radiating and cooling for another several billion years before it extinguishes completely and becomes a "black dwarf."

CELESTIAL BEACONS

Our search for black holes in Sun-like objects failed to produce any result. In fact, all stars up to 1.4 times the mass of our Sun will share the same fate. When they die, they will become not black holes but white dwarfs. This means we have to start looking for stars with a mass greater than 1.4 times that of the Sun. Once again, we must be cautious, as not all such stars will fit the bill. Stars with a mass between 1.4 and roughly 3 times that of the Sun never make it to the black hole stage either. Their gravitational collapse is arrested not by electrons but by neutrons within the star. Neutrons are elementary particles of matter with a mass comparable to that of protons, 1,836 times that of the electron, but they possess no electrical charge, as their name indicates. They, too, have a distinct aversion to being crowded in together.[5] They are, however, somewhat more tolerant to it than electrons, so that the collapse proceeds to a more advanced stage and does not stop until the radius has shrunk to 10 km. The resulting stellar remnant is a star composed entirely of neutrons and of a size comparable to the New York metropolitan area. Like an ice skater spinning faster by pulling his arms in, a neutron star rotates at breakneck speed about itself, sometimes several hundred times a second, like a genuine celestial top. It emits radiation, fed by the collapse energy, in the form of two light beams sweeping space much like a lighthouse by the sea, and as regularly as a metronome. Every time one of the beams sweeps by Earth, radiotelescopes detect a very short signal repeating itself at every revolution. It looks as though the star were pulsating. Hence the name "pulsar" (Figure 47). Matter is even more compressed in neutron stars. One cubic centimeter weighs one million billion grams. This time, it is like cramming a billion elephants into a spoon.

Since stars less than about three solar masses cannot satisfy our quest for black holes, we must start looking for even more massive stars. This time, we are on the right track. These stars are so massive that nothing can stand in the way of their collapse, neither the aversion of electrons to being crowded in, nor that of neutrons. The remains of such stars are neither white dwarfs nor neutron stars, but genuine black holes.

AN ATLAS OF BLACK HOLES

In order to discover the magical and phantasmagorical world of black holes, let us rejoin our friend Jules the explorer on another of his voyages. He has just been commissioned by the daily *The Universe* and by the International Society of Explorers to file a news report on "occluded celestial objects." As usual, his sedentary friend Jim will be his point of contact on Earth. So as to remain in constant communication, Jules has installed aboard his spacecraft a television camera intended to constantly send video and sound signals

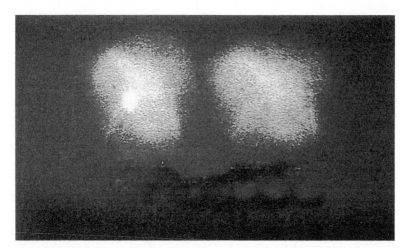

Figure 47. *A blinking neutron star.* At the center of the Crab nebula, remnant of a massive star that exploded in a violent death, the news of which reached us in the year 1054, there is a star that appears and disappears every 33 milliseconds. It is the Crab pulsar, a neutron star, 10 km in radius, which spins 30 times a second like a giant top. The star appears to turn on and off (hence its name, "pulsar") because it radiates not over all its surface, but only in two narrow beams of light that sweep the sky during the star's frenzied rotation. Every time one of the beams sweeps by the Earth, the star seems to light up *(left)*; and when the beam points away from Earth, the star seems to turn off *(right)*. (Photo courtesy of NASA)

back to Jim during the entire mission. Likewise, Jim will transmit nonstop news from Earth to Jules by means of a camera mounted in his living room.

The first thing Jules has to do is decide on a destination. He consults the Black Hole Atlas, published by the International Society of Astronomers. One of the most interesting black holes is the one in Centaurus. It is also one of the nearest, some eight light-years from Earth. It constitutes the stellar remnant left behind after the explosive death, about 5,000 years ago, of a star ten times more massive than the Sun. Jules calculates that if he travels at 80 percent of the speed of light, the round-trip will take 20 Earth years. If he leaves in the year 2000, he will be back in 2020, according to Earth calendars. Of course, because of the accelerations and decelerations of the rocket, he himself would age by only 12 years; his onboard calendar will indicate 2012 when he returns. That is entirely reasonable, on the scale of a human lifetime.

But there are many other candidates in the Black Hole Atlas to pique Jules's curiosity. For instance, there is the black hole in Sagittarius, which astronomers discovered at the heart of the Milky Way, some 30,000 light-years away from Earth. This black hole is far more massive, being equivalent to roughly 3 million Suns. It is believed to result from the coalescence of tens of thousands of smaller black holes, themselves the final product of the death of a multitude of massive stars. Jules calculates on his computer the time required for a trip to the center of the galaxy and back. He specifies the conditions required for the trip: a very high cruising velocity, as close as possible to the speed of light, and a rocket acceleration or deceleration just right to re-create inside the spacecraft the gravity on Earth (such an acceleration or deceleration is called "1 g"). The flight plan is as follows: Jules would travel with a constant acceleration of 1 g until the midpoint, then turn the rocket around and continue on with a constant deceleration of 1 g all the way to the black hole. With an Earth-like gravity in the spacecraft, Jules would feel as if he never left terra firma, which would make the trip extremely comfortable. The computer goes through the calculations and spits out the answer: The trip will take 30,002 years by Earth standards to get to the destination, and the same duration to come back. But since Jules would be traveling at an average speed of 99.99997 percent that of light, the whole trip would last only 20 years from his own vantage point.

Another black hole that attracted Jules's attention is located in the heart of the quasar 3C273. Its mass is more than a billion Suns, and it is some 2 billion light-years away from Earth, a thousand times farther than the Andromeda galaxy, and much beyond the supercluster of galaxies that the Milky Way belongs to. Quasars are part of the strange and fascinating zoo of modern astrophysics. They emit such a phenomenal amount of energy that they shine as brightly as foreground stars, even though they are located at enormous distances—billions of light-years, or almost as far as the edge of the observable universe. Hence the name quasar, which is a contraction of *quasi-star*. This fantastic energy, equal to that of an entire galaxy containing 100 billion Suns, is emitted in a volume scarcely larger than our own solar system. Astrophysicists believe that this enormous energy has its origin in the gluttony of an extremely massive black hole at the center of a galaxy. Tearing apart and devouring the unfortunate stars of the host galaxy that happen to venture too close by, the black hole makes the quasar blaze with dazzling brightness. The computer informs Jules that, with an acceleration and deceleration mimicking Earth's gravity on board the spacecraft, it would take him 2 billion Earth years to reach his destination, but only 42 years in his own time frame of reference. As fascinated as he may be with the black holes in Sagittarius and 3C273, Jules must be realistic: If he is to see his friend Jim alive at the conclusion of his trip, he cannot afford the luxury of missions requiring 60,000 Earth years, much less 4 billion. True, Jim could try to slow his rate of aging back on Earth by freezing his body and going into hibernation, but that technology is not quite fully developed yet. Besides, in that case, who would communicate accounts of Jules's adventures to the newspaper *The Universe?* Jim could leave instructions to his descendants to fulfill his task, but he could never be sure that his wishes would be respected to the letter. That settled it: Jules's destination would be the black hole in Centaurus.

A RAINBOW-COLORED DISK OF GAS

After a six-year interstellar journey according to the onboard calendar—10 years by Earth's standards—the black hole in Centaurus is in sight. Through the window of his spacecraft, Jules discovers a magnificent sight. The black hole manifests its presence by the attraction it exerts on the surrounding interstellar gas. Composed of three quarters hydrogen and one quarter helium by mass, this gas constitutes the remnants of the envelope of stars blown out into interstellar space during their explosive death (such explosions are called supernovae). The gas is not very dense. Containing on average one hydrogen atom per cubic centimeter, it is tens of billion billion times less dense than the air we breathe. However, being sucked in from all directions, it becomes increasingly dense and moves faster and faster as it gets closer to the black hole. The gaseous matter does not fall straight into the gaping mouth of the black hole. Rather, dragged by the rapid rotation of the black hole, it spirals in gradually until it eventually plunges in. The centrifugal forces due to the rotation cause the gas to settle in a flattened disk similar to a fried egg, with the black hole occupying what would be the yellow center. Because matter from the disk is accreted up by the black hole, it is referred to as an "accretion" disk.

From amorphous and lethargic far from the black hole, where gravity and attraction are weak, the gas becomes furiously agitated in the vicinity of the black hole due to the much stronger gravity and attraction prevailing there. The gas atoms collide and heat up. Starting from a chilly -260°C far from the black hole, the temperature reaches millions of degrees near it. Because of this huge temperature range, the disk emits a full palette of radiation. At the outer edges of the disk, far away from the black hole, the light emitted by the cold gas is in the form of radio waves. As the distance to the black hole decreases, the light becomes increasingly energetic and takes on successively the form of microwaves (the same type as in microwave ovens), infrared, visible, ultraviolet radiation, and X rays, all the way to gamma rays. Of course, Jules's eyes are sensitive to visible light only. Through the space-craft window, he sees a kind of luminous ring around the black hole, display-ing all the colors of the rainbow, from red at the outer edge of the ring to violet at its inner edge, passing through orange, yellow, green, and blue. To detect the other types of light that his eyes cannot see, Jules relies on the sophisticated instruments his spacecraft is equipped with.

At the center of the disk is a region that emits no radiation at all. That is the famous black hole Jules has come to explore. He is well aware that the black hole has a horizon surface beyond which he had better not venture if he wants to stay alive. Past this surface boundary, gravity is so strong that nothing, be it radiation or matter, can come back out. Jules knows that this limit has a radius of about 30 km, since the radius of no return is 3 km for the Sun and the mass of the Centaurus black hole is 10 times that of the Sun, the no-return radius being proportional to the mass of the black hole. In order to explore the properties of the black hole without risk, Jules decides to place his spacecraft in a circular orbit safely away from the no-return boundary. He chooses an orbit 160,000 km in radius, way beyond the fateful distance of 30 kilometers. Once in orbit, Jules shuts down his rocket engines to save fuel. He knows that for the time being he no longer needs them, since the centrifugal force due to the rocket's circular motion opposes the force of gravity and keeps the spacecraft in a circular orbit, out of reach of the black hole's tentacles. Under these conditions, the rocket completes a 1-million-kilometer-long revolution in roughly six minutes.

THE ULTIMATE SINGULARITY

Jules begins to ponder the properties of the black hole that he must relate to his Earth-based readers. Thanks to his knowledge of physics, he knows that all its external characteristics—the intensity of its gravitational attraction, the curvature of the space it causes, the deviation of the light's path, the shape and size of its horizon surface—depend on three parameters only: the mass of the black hole, its spin motion, and its electrical charge. Any other infor-mation about the black hole is lost. Once formed, it loses any memory of its progenitor. The chemical composition, odor, shape, texture, color, and size of the object that gave birth to it are forever gone and vanished. As such, black holes constitute a sink of information in the universe (Figure 48).

Suppose, for instance, that a giant hand compresses you sufficiently to turn you into a black hole, and that the same fate befalls your friend who happens to have exactly the same mass as you. The two resulting black holes will be indistinguishable when viewed from the outside. They will have the same mass, the same spin motion (namely, zero, if neither you nor your friend was spinning at the time of your mishap), and zero charge (since the

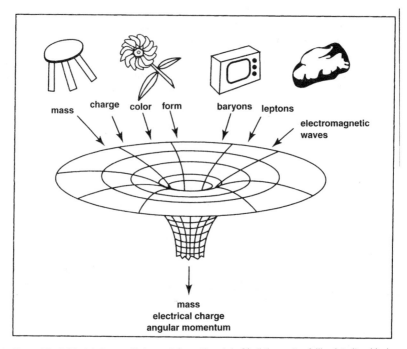

Figure 48. *A black hole constitutes an information sink.* Of all the matter falling into it, a black hole retains a memory of three quantities, and three only: mass, angular momentum (or spin), and electrical charge. Any other information, such as chemical composition, odor, color, shape, and size, is forever lost the moment this matter crosses the black hole's radius of no return (the horizon radius). Even if an observer could enter a black hole, survive, and carry out measurements, he could never communicate the results back to the outside world.

positive and negative charges exactly balance each other out in your bodies). An observer comparing these two black holes from the outside would have no way of telling which one of you produced which object. Information about your gender, your race, the color of your hair and eyes, your size, the clothes you were wearing, the shape of your body—all of this will have been erased the moment you turned into a black hole.

Jules knows the mass of the black hole. It is ten times that of the Sun. Its electrical charge is zero, because if the black hole had a positive or negative charge, it would have long ago attracted particles from the interstellar gas with the opposite charge to offset its own. As for the black hole's spin motion, Jules can already gauge how strong it is based on the furiously swirling mass of interstellar gas spiraling into the gaping black mouth, drawn by its irresistible gravity. In fact, Jules fully expected that this black hole would rotate rapidly since the massive star that gave birth to it was itself spinning fast. As it collapses inward, the resulting stellar corpse can only increase its rotation speed further, just as an ice skater spins faster as he pulls his arms in alongside his body. The rotation speed increases in proportion to the inverse of the square of the radius of the collapsed star. Because of this spinning motion, the horizon surface is not perfectly spherical. Instead, it is slightly flattened at the poles and bulging at the equator, much as Earth is subjected to the centrifugal forces due to its rotation.

Jules goes through a quick calculation. If ten solar masses are uniformly distributed in a more or less spherical volume of radius 30 km, the density of matter will be about 2×10^{14} g/cm3, which is 200,000 billion times greater than the density of water. However, Jules knows that matter inside a black hole is anything but uniformly distributed. Actually, relativity theory says that all the matter must be concentrated in an infinitesimally small region of space only 10^{-33} centimeter in size, or 10 million billion billion times smaller than an atom. Physicists call such a region a "singularity." Aside from a thin flux of interstellar gas pouring into the black hole, everything inside is essentially a vacuum—a vacuum extending from the horizon surface to the singularity.[6]

TIME FROZEN BY A BLACK HOLE

Leaving the spacecraft orbiting at a safe distance of hundreds of thousands of kilometers from the black hole, far from its grip, Jules transfers to a small capsule specially designed for exploring the vicinity of the horizon surface. He knows that he has to proceed with extreme caution. One false maneuver, and he could end up in free fall toward the gaping mouth of the black hole, never to be seen or heard from again. His plan is to approach the horizon in a spiral pattern, guiding his capsule along increasingly tighter circular orbits. To that end, he initiates a series of small braking thrusts. The first thrust slows him down very slightly from his initial orbital motion—the one the capsule shared at first with the spacecraft—and enables him to drop into a lower circular orbit, closer to the black hole. With a second slight braking thrust, the orbit becomes tighter still, and so on.

As he descends toward the black hole's horizon surface, Jules remains in contact with his friend Jim on Earth. At first, far from the black hole, all appears normal. Time flows at the same rate for both friends. Jim can verify that it is so by watching the TV images of Jules's clock inside the space capsule. But as Jules gets closer and closer to the black hole, the signals he sends to Jim have an increasingly hard time breaking free of the space distorted by the strong gravity. The radio signals lose more and more energy to escape the black hole's grip. The pictures sent by Jules become less frequent and are updated less often on Jim's television monitor. Jim has the distinct impression that Jules's time is slowing down more and more. Just to shave, Jules first takes a minute, then an hour, then a day, then a year by Jim's calendar. Jules keeps on displaying the same youthful cockiness, the same vibrant head of black hair, and the same smooth face, while old age is wreaking havoc on Jim, his hair turning white, his face being invaded by innumerable wrinkles. As far as Jim can tell, everything concerning Jules proceeds at a snail's pace. Even his mental faculties seem to have slowed down. It takes Jules an eternity to make the slightest decision.

Jules's perception of the situation is not different from his friend's. He also observes that Jim's time flows much faster than his own. The television images sent by Jim from Earth, captured by the black hole's strong gravity, reach the space capsule faster and faster and are refreshed at a dizzying rate on the screen aboard the spacecraft. While Jules spends no more than a few days descending toward the black hole's horizon, he watches Jim on the screen celebrate his fortieth, fiftieth, sixtieth, seventieth, etc., birthday. With a tinge of sadness and a heavy heart, he sees his old friend become less and less alert and increasingly prone to health problems. Time flies faster for Jim than for Jules because the situation is not symmetrical between the two friends. Jules experiences the unpleasant gravitational effects of the black hole, while back on Earth, Jim is shielded from them.

There are other consequences of the fact that Jules's time slows down. As he gets closer to the black hole, his speech transmitted to Jim stretches out and his voice becomes deeper and slurred, not unlike the cavernous sound we hear at the movies when the film breaks and the projector grinds to a stop. Jules, on the other hand, does not notice anything abnormal. From his own vantage point, everything proceeds normally. He ages at the same rate as before, his mental faculties are just as sharp, and his voice retains the same pitch. On the contrary, it is Jim's voice that seems grossly distorted. It becomes increasingly shrill to the point of almost hurting his ears; the words pile up at such a rapid pace that they become incomprehensible, much like the sound from an audiotape being fast-forwarded.

TIDAL FORCES STRETCH YOUR BODY

Jules may age less rapidly than Jim, but his body pays dearly for this prolonged youth. In fact, the gravitational effects of the black hole on his body are extremely unpleasant. Indeed, they can be downright painful. As he descends toward the black hole's horizon, Jules experiences increasingly strong forces on both his head and his feet. Since his feet are closer to the center of the black hole than his head by six feet—Jules's height—they are drawn more forcefully toward the singularity. These differences in gravitational pull acting on the various parts of Jules's body create what is called a "tidal force." This is by analogy with the difference in the gravitational attraction exerted by the Moon on the oceans and the center of Earth, which gives rise to the tides that topple the sand castles left by children on the beach. Every one of us living on Earth is subject to similar tidal forces, since our feet are closer to the center of the planet and are therefore more strongly attracted to it than our head. If we do not feel discomfort as a result, it is because the gravitational pull of Earth on our feet and head varies by less than one part in a million.

Things are quite different for our friend Jules when he gets close to the horizon of the black hole. The difference between the gravitational pull on his head and feet keeps increasing relentlessly. The tidal forces exerted by the black hole tend to stretch him like a strand of spaghetti, and his body is aching more and more. At a distance of about 13,000 kilometers from the horizon surface, Jules experiences a tidal force equivalent to one-fourth of Earth's gravity; at 8,000 kilometers, the force reaches 1 g, and at 3,000 kilometers, 15 g. Jules does the best he can to put himself into a fetal position, trying to pull his knees up to his head and make his body as small as possible in an attempt to counteract the tidal forces (ideally, if Jules's head and feet could merge at a single point in space and the size of his body could shrink to zero, the tidal forces would vanish). But it is a losing battle. The tidal forces are so powerful that they force Jules to unwind his legs and his body instantly falls pray to the assault of tugging gravitational pulls. Jules is in such pain that he can barely stand it. He decides to turn around. There is no hope for him to ever reach the horizon surface of the black hole without his body breaking apart. Jules fires the thrust engines of his capsule, which sends him spiraling out on ever-larger circular orbits to eventually rendezvous with the mother spaceship.

GRAVITATIONAL LENSES

Jules had to turn back because his body could no longer withstand the tidal forces exerted by the black hole. We are left with the frustration of not knowing what happens on the other side of the horizon surface. We would dearly love to find out the end of the story. Just to satisfy our curiosity, we shall pretend to be an omnipotent god and give Jules the superhuman power to resist the overwhelming tidal forces, which will allow him to make his way right up to the ultimate singularity.

In the revised script, Jules keeps on spiraling down all the way to the horizon surface. He cannot stop admiring the magnificent spectacle of the starry firmament all around him. Only a very small part of the sky is obscured by the opaque disk of the black hole immediately below him. Against a pitch-black background, innumerable points of light shimmer, revealing the stars of the Milky Way in all their splendor. With the help of his onboard telescope, Jules can also admire more diffuse and less dazzling objects; those are galaxies similar to the Milky Way, collections of hundreds of billions of stars bound by gravity, among hundreds of billions of other collections populating the observable universe. At first sight, it would seem that the black hole should block out all the stars and galaxies located behind it. That would indeed be the case if light propagated in a straight line. But, as we now know, it does not. The gravitational field of the black hole acts effectively like a lens that bends light rays. Light emitted by a point source on the far side of the black hole skims over the latter's horizon and reconverges on the near side to produce a faint ring framing the obscuring disk. The black hole acts very much like the lens of your eyeglasses, which bend light to correct for your nearsightedness. In this particular case, however, the lens is made not of the usual glass but of space distorted and curved by the gravity of the black hole. Hence the name "gravitational lens" (Figure 49). Within the ring surrounding the black hole, Jules can make out multiple images of each star blocked out by the black hole. A first image results from light deviated to the right of the black hole; a second one results from deviation to the left; a third results from light circling once completely around the black hole, and so on. Light emitted by a star takes several different paths through the black hole's gravity

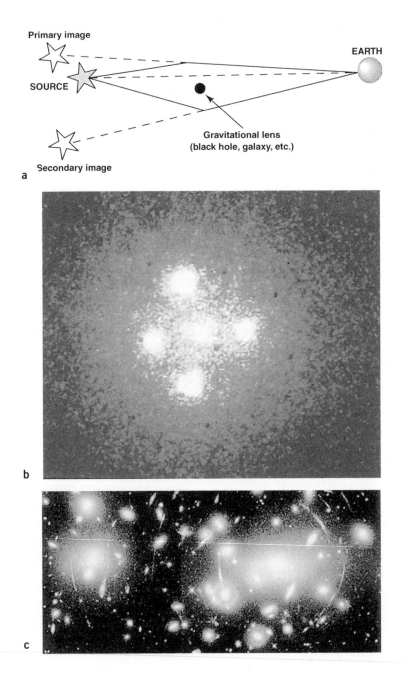

field, thus creating multiple images. Jules sees not the actual star, which is obscured by the black hole, but multiple mirages of it.

Einstein had predicted the phenomenon of gravitational lensing as early as 1936. That was one of the many unexpected ramifications of his theory of general relativity. Like all great scientific theories, this one, too, is a fabulous and inexhaustible treasure trove that never ceases to deliver new riches to physicists. Einstein had calculated that when two stars are exactly aligned with Earth, the more distant one should produce two images: a normal point-like image and, surrounding it, a second ring-shaped image, which is a grav-itational mirage caused by the gravity of the closer star (Figure 49a). Einstein believed that such an alignment was highly unlikely and that no one would ever see a cosmic mirage in the sky. Once again, he was too far ahead of the times. Not until 1979 was the first gravitational lens identified, although the actors involved in this saga were quite different from what Einstein had envi-sioned. The distant star was actually a quasar, one of those objects that emit such a tremendous amount of energy that they appear as bright as a closer star, in spite of being located near the observable limit of the universe. The closer object playing the role of gravitational lens was an entire galaxy, rather than a single star. As the light coming from the distant quasar approached the galaxy's gravitational field, it took two different paths to get around it, one to the right and the other to the left, thus producing two absolutely identical images on either side of the galaxy. As a matter of fact, it was pre-cisely the proximity and great similarity of these two images that alerted astronomers. Such a striking likeness could not be accidental, but could result only from multiple images of one and the same object.

Since then, the hunt for gravitational lenses has continued unabated. The effort has been amply rewarded. About two dozen of them have been discovered. Galaxies are not the only objects capable of acting as gravita-

Figure 49. *(a) A gravitational lens.* A black hole, a star, a galaxy, or a galaxy cluster produces, by virtue of its mass, a gravitational field that warps space around it. Light from distant celestial objects is bent when it goes through that space. Celestial objects creating gravity fields that bend light have been dubbed "gravitational lenses," because they deflect light in much the same way as the lenses in a pair of glasses do to correct for nearsightedness. The bending of the light can create gravitational mirages, just as warm air in the desert, by deflecting light, creates mirages and fools thirsty travelers. Depending on the relative alignment of the distant object, the gravitational lens, and Earth, the distant object can split into two, three, or even four images, or produce circular arcs.

(b) Einstein's cross. This picture, obtained by the Hubble space telescope, shows split images of a distant quasar located in the direction of the constellation Pegasus. The quasar is at a distance of 8 billion light-years from the Earth; it has split into four images, forming a kind of cross, which was named in honor of Einstein, who was the first to predict such gravitational mirages. The galaxy responsible for bending the light is aligned with the quasar and Earth, but it is much closer to us, being at a distance of only 400 million light-years. It is visible as the diffuse object in the center of the cross. (Photo courtesy of NASA.)

(c) Circular arcs around a galaxy cluster. In this picture, also taken by the Hubble space tele-scope, one can see the galaxy cluster Abell 2218 (each diffuse image in the photo represents a galaxy—that is, a collection of hundreds of billions of stars bound by gravity) acting as gravitational lens and distorting the light emitted by a distant galaxy in the background into fragments of circular arcs. These arcs can be seen around a large elliptical galaxy to the right of the photo. From the number, shape, and position of these arcs, astronomers can infer the total amount of matter, both visible and invisible, contained in the cluster. (Photo courtesy of NASA)

tional lenses. Galaxy clusters—collections of thousands of galaxies bound by gravity—can do quite nicely. Images of certain galaxies or distant quasars have been observed in sets of two, three, and even four (Figure 49b). If the alignment of the distant object, the lens, and Earth is perfect, images then take on the shape of a ring, just as Einstein predicted. With the slightest deviation from perfect alignment, the ring breaks up into multiple luminous arcs (Figure 49c).

Astronomers spend a great deal of effort studying gravitational lenses because they believe it may help them resolve one of the great mysteries of modern astrophysics—namely, the dark mass in the universe. We happen to live in a universe that could be likened to an iceberg. In excess of 90% of the matter in the universe does not emit any radiation at all. The matter that shines (stars and galaxies) is but the tip of the iceberg. There is, however, a fundamental difference between an iceberg and the universe. While we can be sure that the submerged part of an iceberg is made of ice, the amount and nature of the dark mass remain a complete mystery. Until some fresh light can be shed on the issue, astronomers are literally groping in the dark. They know that gravitational lenses create cosmic mirages through their gravity field, which in turn depends on the total mass, both visible and invisible, contained in the galaxy or galaxy cluster acting as a lens. Therefore, gravitational mirages are like signposts of the dark matter and can help determine its amount.

THE TUNNEL TO THE BLACK HOLE

Jules continues his spiral descent toward the horizon of the black hole. The splendor of the starry sky fills him with awe. Far enough from the black hole, the gas disk surrounding it forms a kind of dark floor directly beneath the rocket, while the sky above is completely clear and glitters with innumerable stars. But as Jules gets closer to the black hole, the blackness, which at first stretched over the disk only, spreads over the entire space around the capsule. It is as though darkness was closing in on him. Now he can see the starry sky only through a narrow circular opening above his capsule. He feels as if he is plunging into a long tunnel whose opening gets increasingly narrower. The reason darkness spreads its veil over Jules is that the black hole's intense gravity field captures light rays from the outside world and concentrates them into a narrow beam converging above the space capsule. Instead of following horizontal or oblique trajectories, light rays are forced by gravity to proceed along vertical paths. Visible light is turned into X rays. Jules can no longer admire his favorite stars with his own eyes, but he can still easily see them with his onboard X-ray telescope.

THE BLACK HOLE AND THE END OF TIME

While Jules continues his descent toward the horizon of the black hole, Jim sees his friend's time slow down more and more. Jules's capsule looks increasingly faint, as the light waves carrying the video images expend more and more energy to escape the gravitational grip of the black hole. Jules tries to remedy the situation by turning all the capsule's signal lights, but to no avail. By losing energy, visible light shifts more and more toward the red. On Jim's television screen, not only does Jules's capsule appear fainter but it also becomes redder.

Jules now reaches the point of no return. He knows that once he crosses the horizon, he will irrevocably fall toward the singularity. No matter how powerful his engines, he will never be able to come back. This is a one-way trip. But Jules has the soul of a genuine explorer and decides to press on. He is determined to discover the secrets of the black hole, even though he will never be able to share his findings with anyone else, since radio waves are irremediably imprisoned by the black hole's gravity. Jules may gain the satisfaction of knowing, but the black hole will forever lock in its secret. The singularity at its heart will never become accessible to our curiosity. Its truth will never be revealed. As the British mathematician Roger Penrose (1931–) put it, the black hole imposes a kind of "cosmic censorship" on our knowledge.

Back on Earth, Jim sees the image of Jules's capsule disappear from his television screen the moment his friend reaches the horizon of the black hole. Everything turns to darkness. The spaceship Jules was traveling in is engulfed in an endless night. As far as Jim can tell, Jules's time has become frozen. There is no updated image to come refresh the screen. Jules will forever flash that tense look and forced smile revealing his anxiety at disappearing into the gaping mouth of the black hole for all eternity. Jules, on the other hand, sees events unfold very differently. As far as he is concerned, time as measured by his onboard clock continues to flow quite normally. Instead of being stuck forever at the horizon, as his friend Jim believes, his space capsule does move on. In fact, already at 1.5 times the radius of the horizon, the rocket could no longer follow a stable circular orbit in spite of Jules's valiant efforts to retain control of his capsule. The gravitational attraction is so strong that no centrifugal force can compete with it. Jules tries to rev his engines to the limit in an attempt to boost his speed and increase the centrifugal force, but he is wasting his time. Even if the capsule could orbit around the black hole at the speed of light (which is impossible, because it would take an infinite amount of fuel), gravity is in total control. Jules goes into a free fall straight toward the black hole. He makes one last attempt to slow this fall ever so slightly, but to no avail. Because rocket, interstellar matter, and everything else fall so rapidly toward the singularity, vacuum is almost perfect inside the black hole.

By stepping inside the black hole, Jules went beyond time. Had he been able to stop his space capsule right at the horizon, he would have seen all eternity flash before his eyes in less than a microsecond on his watch: Jim's death; the passing of his children, grandchildren, great-grandchildren, and of all subsequent generations; the Sun burning its entire supply of hydrogen and inflating to one hundred times its current size; Mercury and Venus vaporized in its fiery envelope; the evaporation of Earth's oceans and the

scorching of its surface by forest fires; the transformation of the Sun first
into a white dwarf and eventually into a black dwarf after further cooling; all
the stars of the universe getting extinguished one after the other after running out of fuel; galaxies turning dark; galactic regions littered with stellar
remnants (black holes, neutron stars, black dwarfs); an eternal night spreading over an increasingly cold universe. . . .[7]

   In reality, Jules would not get to see eternity pass before his eyes, because
gravitational forces are so overpowering that, even if he engaged the full
power of his engines, his spacecraft could not remain fixed at the horizon of
the black hole. Since news from Earth and the rest of the universe takes a
finite amount of time to reach him, he will have plunged into the claws of
the black hole long before the news would make it to the horizon surface. In
any event, Jules will never be able to escape the black hole, because he will
have gone beyond time in the outer world. Reentering that world would be
tantamount to returning to the past, which is forbidden by physics as we
know it.

APPROACHING THE SINGULARITY

Near the singularity, the space-time pair goes haywire. Each oscillates in a
wildly chaotic manner, generating tidal forces that are even more chaotic,
much like the turbulent and frenzied motion of the foam on the surface of
the ocean during a raging storm. These tidal forces play havoc with poor
Jules's body. They execute a kind of diabolical dance, stretching his head
toward the north while jerking his right foot toward the east and twisting his
left one toward the southwest. It is as though two giant hands were tearing at
his body from everywhere and every which way, with different frequencies in
different directions. Since his body is not made of putty, Jules's pain is excru-
ciating. The tugging becomes so wild and powerful that even the atoms in
the body of our fearless explorer get deformed to the point of being unrec-
ognizable.

   The tidal forces are so overwhelming because the black hole in which
Jules plunged is relatively young. In time, they should become less intense
until they die out completely.[8] But the interstellar gas falling into the mouth
of the black hole invigorates them and keep them going, like a wild animal
replenishing his strength by feeding on fresh meat. With their renewed
strength, they pursue their tugging and twisting action with even more fieri-
ness. Less than a microsecond after Jules reaches the singularity, the tidal
forces become infinite, first at his feet, then his chest, and finally his head.
Despite the superhuman powers we bestowed on him earlier, the electromagnetic forces maintaining the atoms in his body can no longer withstand
runaway tidal forces. Electrons, protons, and neutrons—the building blocks
of atoms—as well as quarks—the building blocks of protons and neutrons—
stretch like spaghetti. At this point, Jules's body breaks apart, and it is
the end.

   As the singularity is reached, space-time as we know it ceases to exist.

Space shrinks to less than 10^{-33} cm and events take place in less than 10^{-43} second. Einstein's equations simply break down. The singularity constitutes a sort of wall to our knowledge, often called "Planck's wall," after the German physicist who first thought about the problem. To get past this wall, it is necessary to first unify the theory of relativity, which describes gravity, with quantum mechanics, which describes the subatomic world, in a grand synthesis called quantum gravity. The resulting theory is still in its very primitive stages. The few courageous physicists who have ventured into it tell us that in the singularity the space-time pair, so strongly intertwined up until then, split up. Time ceases to exist. We can no longer say that a particular event occurs before or after another because the concepts of before, now, and after become meaningless. Separated from its partner—time—space is reduced to a kind of froth with no well-defined shape. Its curvature and topology become chaotic and can only be described in terms of probabilities (Figure 50). Everything becomes random. It no longer makes sense to say that space spends a certain amount of time in a particular form, since time itself no longer exists. Physicists even theorize that these random and fleeting shapes of space in the singularity might give birth to new universes—that is to say, new regions of space-time, as perhaps happened to our own universe some 15 billion years ago. All these concepts are highly speculative. Quantum gravity remains shrouded in a thick fog that is not about to lift. The path toward a theory of quantum gravity remains long and arduous, and obstructed with many obstacles.

PARALLEL UNIVERSES AND WORMHOLES

Some physicists have even speculated that a singularity never actually comes into existence when a black hole forms. The question is relevant precisely because relativity theory breaks down when space shrinks to less than 10^{-33} cm and is powerless to describe what goes on beyond that limit. It is conceivable that space-time might continue to exist and that Jules could survive after all. He still would never be able to return to the universe he has just left, since he would have traveled beyond time in that universe. What he might do is emerge in another universe connected to ours through the black hole. In this new universe, he might even look for yet another black hole and through it reach a third universe, and so on. He could thus visit an infinite series of parallel universes without the possibility of ever returning to one he had already explored. For the time being, such a scenario remains in the realm of science fiction. There is no compelling reason to believe that such parallel universes do exist and that we may gain access to them via black holes without being irremediably torn to pieces by unforgiving tidal forces.

Some adventurous physicists have even dug deep into the equations of relativity to explore the possibility of entering a singularity in one corner of the universe and reemerging through another singularity somewhere else in the same universe. Both singularities would be connected by a kind of tun-

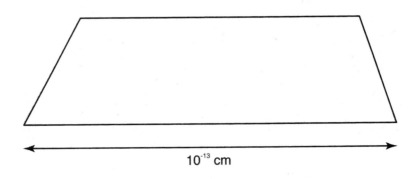

10^{-13} cm

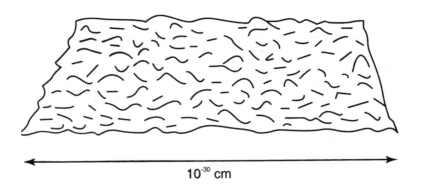

10^{-30} cm

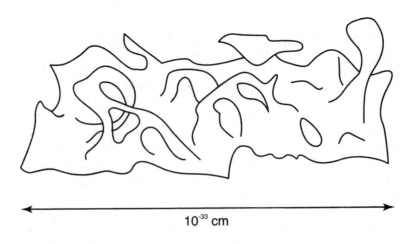

10^{-33} cm

A

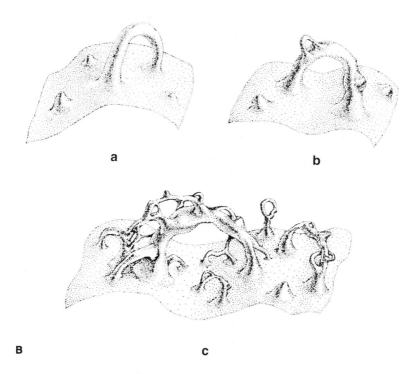

a   b

B   c

Figure 50. *The foam of space. (A)* On the scale of everyday life, the geometry of space appears perfectly smooth. But if we had access to a microscope capable of looking at things on scales as infinitesimally small as Planck's length (10^{-33} cm), we would see space as a kind of quantum foam in perpetual motion and constantly changing.

(B) The geometry and topology of space could no longer be described deterministically, but only in terms of probabilities. As shown in the figure, space would have, for instance, a probability of 0.2 percent of being in configuration *(a)*, 0.05 percent of being in configuration *(b)*, or 0.09 percent of being in configuration *(c)*.

nel existing not in ordinary space, but in a "hyperspace" resembling an underground passage connecting two holes dug in the ground by earthworms. The American physicist John Wheeler (1911–), who has always had a knack for inventing names apt to capture the public's imagination—he is the one who introduced the term "black hole" into the language of physics— dubbed such a pair of linked singularities a "wormhole" (Figure 51). A wormhole is much like a black hole, except that it does not have a horizon surface that, when crossed, precludes turning back. Whereas traveling into a black hole is strictly a one-way proposition, transit through a wormhole is allowed in both directions. You are free to enter, and nothing prevents you from getting back out and communicating with the rest of the universe if you so desire. Wormholes are naked and exposed for all to see. They exert no cosmic censorship. Most of all, they possess a remarkable property that makes physicists and science-fiction writers drool: They make it possible to travel through time. Enter a wormhole one way, and you will go into the

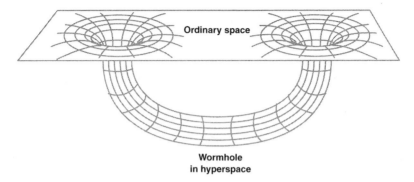

Figure 51. *A wormhole in space-time.* The mathematics of general relativity is so rich and complex that it allows for the existence of the most bizarre entities. As an example, they predict the existence of wormholes in space-time. Each such wormhole would have an entrance and an exit within our normal space, although the "tunnel" connecting the entrance and exit would be in a space outside our universe, called "hyperspace." These wormholes have an extraordinary property. They can serve as shortcuts to travel from one corner of the universe to another, or to enter a parallel universe. They even make it possible to go back in time. Thus, a traveler could slip through the entrance to a wormhole at noon and reemerge at the exit at 10 A.M., two hours earlier. Such trips into the past could play havoc with our familiar notion of causality, since a traveler able to go back into the past could in principle prevent his parents from ever meeting and render his own birth impossible. Fortunately, Nature seems to protect herself from such logical absurdities by making sure that a wormhole never stays open long enough for anyone to slip through. The "tunnel" snaps shut in the infinitesimally short time of 10^{-43} second. Some physicists have tried to figure out physical mechanisms to keep the wormhole open, but thus far no scenario appears promising.

future. Proceed in the opposite direction, and you will return to the past. Instead of letting you travel in space like an astronaut, wormholes allow you to become a time traveler—in other words, a "chrononaut."

But don't rush out to buy tickets just yet: Many problems remain to be solved, including some very daunting and perhaps insurmountable ones, before we know how to use wormholes as time machines. To begin with, it is not quite clear how to make them. We know that black holes result from the gravitational collapse of massive stars running out of fuel. But what causes the singularities at the entrance and exit of a wormhole? To answer the question, we need help from two protagonists. The first is quantum mechanics, the physics of the infinitely small, which is responsible for a number of most startling phenomena. The second is gravity, with its ability to warp space. The alliance of these two players results, as we have seen, in a theory called "quantum gravity." The laws of quantum gravity predict the existence of microscopic wormholes 10^{-33} centimeter in size, living for the extraordinarily short time of 10^{-43} second. A photographic flash lasts 10 million billion billion billion billion times longer! Wormholes appear and disappear in a frenzied cycle of life and death, like a kind of "quantum foam" floating in the space around us. If you had a microscope with the unlikely magnifying power of 10^{33}, space would no longer appear smooth, continuous, calm, and tranquil, but, on the contrary, filled with irregularities and discontinuities (Figure 50b). You would see it seething and boiling with activity, twisting and bending every which way to give birth to ephemeral singularities popping into existence only to vanish almost instantly. From time to time, two singularities would approach one another and link up to form a wormhole in space. As such, the existence of wormholes is determined not by deterministic laws but by the laws of probability.

Once a wormhole is created, how could one make use of it to travel through time? Two formidable obstacles must be overcome, which appear insurmountable for now. First off, you must act very, very quickly, since you have only the infinitesimally small time of 10^{-43} second to slip into the entrance of the wormhole and reemerge at the other end. Wormhole travel is not for those who enjoy taking their time to admire the scenery. Second, since the wormhole entrance is infinitesimally small (10^{-33} cm), one has to find a way to enlarge it so that a human being can fit through it. The few physicists who have worked on the problem believe that a wormhole would self-destruct in a violent explosion the moment you try to enlarge its entrance. That is apparently how Nature protects herself against noncausality and nonsense. Indeed, if a human being were able to travel back in time, he could in principle prevent his parents from ever meeting, making his own conception and birth impossible, which is patently absurd. This bizarre situation is referred to as "the mother paradox" (or grandmother, or any other distant ancestor, since a wormhole would allow you to go as far back into the past as you want and cut your genealogy tree arbitrarily high). It is also called the "matricide" or "patricide" paradox, referring to the more shocking and morally distasteful possibility that the time traveler could kill his own parents outright. Thus far, this self-protection against nonsense seems to be holding up well. As the British physicist Stephen Hawking (1942–) reminded us pointedly, we have not noticed hordes of tourists visiting us from the future. In any event, time travel remains for the time being the stuff of science fiction, despite some serious scientific work on the subject. It will remain so until Planck's wall is breached and the theory of quantum gravity finds a more solid footing.

BLACK HOLES GIVE THEMSELVES AWAY BY THEIR GLUTTONY

Relativity theory predicts that nothing can stop the collapse of massive stars running out of fuel, and that these end up as black holes. You may have thought that these objects with outlandish properties are nothing but theoretical entities existing only in the wild imagination of physicists. That is not at all the case. Astrophysicists have actually detected black holes in the depths of space. Of course, we cannot see the singularity they guard jealously in their midst, but they signal their presence by their gluttony and the havoc they inflict on their surroundings. Let us see how.

Just like human beings, stars like to live in pairs. The bond that unites a pair of stars is not love and tenderness, but plain gravity. Nearly half the stars in the Milky Way form a binary system, a technical term referring to a pair of stars orbiting each other. In general, the two members of a pair do not have the same mass, one being typically much larger than the other. The more

massive star requires large amounts of energy to power its prodigious luminosity. It consumes fuel (hydrogen, helium, carbon, and more) at a phenomenal rate, burning the candle at both ends. In a few tens of millions of
years—a mere eyeblink compared to the 15 billion years the universe has
been in existence—it goes out and turns into a black hole. Its less massive
companion, on the other hand, is far more frugal. Because it shines less
brightly, it uses energy more parsimoniously and conserves fuel, enabling it
to live much longer than its partner. After the massive star dies out, the
smaller partner continues to revolve around the stellar corpse as if nothing
had changed. Its orbit remains approximately the same since it is dictated by
gravity, depending only on the total mass of the system, which has hardly
changed at all. It is true that in its death throes, the massive star exploded
and ejected its outer envelope, but that involved relatively little material.
Almost all of the mass of the dying star ends up in the black hole.

The black hole's gravity causes gases to pour in from the upper layers of
the live star toward the singularity. Because of the spin of the black hole, all
of the surrounding material is dragged in a huge swirl. As a result, the gas
falls not straight in but along an increasingly tighter spiral. The centrifugal
force makes the gas settle into a flattened disk around the black hole, far
beyond the horizon. The mass of the disk increases as it keeps drawing in
gas from the companion star. We have already mentioned that it is called an
accretion disk. Heated to millions of degrees as it falls toward the black hole,
the gas emits high-energy radiation. The accretion disk emits copious
amounts of X rays. X-ray observatories orbiting high above Earth's atmosphere (since the atmosphere absorbs X rays, telescopes designed to detect X
rays have to be sent into outer space) have detected in the constellation
Cygnus a bright star with an invisible companion orbiting around it and
emitting X rays. Astronomers believe that this invisible companion object is
a black hole with about ten times the mass of the Sun. It gave itself away by
its gluttony, as its gravitation helps it to swallow up the gaseous envelope of
its luminous partner (Figure 52). This illustrates how black holes resulting
from the death of massive stars in the Milky Way are more likely to reveal
their presence when they form a pair with a live star. Solitary black holes,
such as the one explored by our friend Jules, are also adorned with an accretion disk made of the surrounding interstellar gas. But this disk is not dense
enough (its density is billions of billions of billions of times less than the air
we breathe) to produce X rays with sufficient intensity to be detected by our
space observatories. Such isolated black holes are lost for eternity in the
dark immensity of space, waking up only from time to time to devour an
unfortunate interstellar gas cloud passing too close by, or perhaps an occa-
sional reckless astronaut.

A BLACK HOLE AT THE HEART OF THE MILKY WAY

Binary systems are not the only objects in the Milky Way where astronomers
think they have seen evidence of black holes. They claim to have discovered

Figure 52. *A gluttonous black hole.* In the beginning, there were two stars orbiting each other as a binary system. The two stars did not have the same mass. The more massive of the two (with about 10 solar masses) quickly used up its nuclear fuel and died, collapsing in on itself and becoming a black hole. It then continued to orbit around the less massive star, which still has plenty of nuclear fuel left and continues to shine. A stellar wind of gases pours out from the star to the black hole, attracted by the latter's gravity. The gas swirls in an accretion disk around the black hole before falling into its gaping mouth. As they speed into the black hole, the gas atoms collide with one another and heat up to temperatures of millions of degrees, so that the inner edge of the accretion disk emits tremendous amounts of X rays. In the constellation Cygnus is an intense source of X rays that seem to come from an invisible source orbiting around a star. Astronomers think that this invisible object is a black hole and named it Cygnus X-1.

one at the very heart of our galaxy, some 30,000 light-years from Earth. The telltale sign, this time, is not copious amounts of X-ray radiation emitted by an accretion disk, but the frenzied movements of interstellar gas clouds in the vicinity of the galactic center. Their pace should be quite placid in the absence of a massive object, but much more brisk if one is there, since high velocities are required to counteract the gravitational attraction of a large mass. As it happens, gas clouds at the center of the Milky Way orbit at very high velocities, which can only be explained by assuming the presence of a central object with a mass of 3 million Suns. Moreover, astronomers have also detected in the same central region an extremely compact source of radio waves, no larger than our solar system. This radio emission could come from electrons stripped loose from the accretion disk of a black hole and accelerated to high speeds in a very intense magnetic field. This 3-million-solar-mass black hole might have been formed by the fusion of innumerable smaller black holes resulting from the explosive death of tens of thousands of stars.

Does the black hole at the center of the Milky Way present a danger to us? Could it someday devour Earth and us along with it? If so, that will not happen for what amounts to an eternity. The horizon of the black hole has a radius of 9 million kilometers, or only 30 light-seconds. Since Earth is located at 30,000 light-years from the galactic center, we are quite far away from the black hole's claws. It could, of course, extend its range of action by devouring all the stars that happen within its reach, thereby increasing its mass. The radius would increase in proportion to the mass. But even assuming that the black hole has a boundless hunger and devours all the stars in the Milky Way, its radius would increase to no more than 3,000 billion kilometers, which is still 90,000 times less than the distance between the galactic center and our solar system. At any rate, it would take billions of billions of years to consume the entire galaxy, or 100 million times longer than the current age of the universe. In short, even if the Sun's orbit were to change during this exceedingly long period, the prospect of being devoured by the black hole at the center of the Milky Way should be the least of our concerns. For those who cannot help worrying about the end of the world, the transformation of the Sun into a red giant in 4.5 billion years from now constitutes a far more real danger.

Astronomers have extended their hunting grounds for black holes to other galaxies as well and have not come back empty-handed. They have uncovered even more massive black holes (reaching one billion solar masses) at the heart of certain galaxies that eject matter (Figure 53), or at the center of quieter nearby galaxies such as Andromeda. Once again, the telltale signs are the extremely high velocities of stars near the galactic center, which provides evidence of a large mass, as well as a spectacular luminosity, which implies a large concentration of shining stars captured by the black hole's gravity. And that's not all. Astronomers believe that black holes with several billion solar masses—thousands of times more massive than the black hole at the center of the Milky Way—fuel the phenomenal energy of quasars by frantically devouring all the stars in the galaxies containing these quasars. It is in fact this very gluttony that will cause the quasar's demise. In a few billion years, the supply of stars in the host galaxy is exhausted and there is nothing left to satisfy the voracity of the monster whose appetite can never be satiated, at which point the quasar will extinguish itself. The quasar population, which flourished more than 10 billion years ago (or 2 to 3 billion years after the big bang), has already dwindled to almost nothing at present.

THE FUSION OF BLACK HOLES
AND THE VIBRATIONS OF SPACE-TIME

To explore the mysteries of black holes, astronomers have used their armada of telescopes, both ground-based and in space, to capture the light emitted by the accretion disks surrounding them. They have analyzed light in every region of the electromagnetic spectrum, from radio waves, the least energetic, to gamma rays, the most energetic, including, in increasing order of energy, infrared, microwave, visible, ultraviolet, and X-ray radiation. But the curiosity of astronomers is not easy to satisfy. Because electromagnetic radiation comes from the accretion disk, it probes regions located well outside the horizon of black holes. What astrophysicists really would like to learn

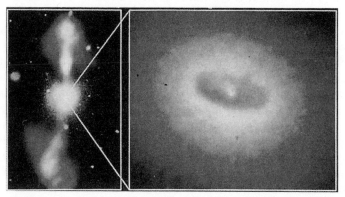

Figure 53. *A black hole at the heart of an elliptical galaxy.* (a) The photo at left shows the elliptical galaxy NGC 4261, located in the Virgo galaxy cluster, some 45 million light-years from Earth.
The galaxy ejects matter from its center into two diametrically opposite jets. These jets are seen not
as visible light but only as radio waves. The ejection of matter suggests very violent and high-energy
phenomena taking place at the heart of the galaxy.

(b) The photo at right shows the central part of the galaxy as seen by the Hubble space telescope. Able to resolve remarkably fine details, Hubble has detected a dust and gas ring, approximately 400 light-years in diameter, surrounding an extremely compact object. This object is most
likely a supermassive black hole, with a billion solar masses, in the process of devouring the stars of
the elliptical galaxy. The ring is composed of the debris of stars ripped apart by the tidal forces
exerted by the black hole, and the radio jets ejected perpendicularly to the plane of the ring derive
their energy from the black hole's gluttony. (Photo courtesy of NASA)

more about is what goes on very near the horizon. Are there other means
besides light to force black holes to reveal their secrets? Like priests consult-
ing with the oracle at Delphi, astronomers went to query Einstein's equations, and once again they were not disappointed. These equations do turn
out to provide a means of studying black holes without having recourse to
light.

To understand how that comes about, imagine yourself traveling to the
heart of a distant galaxy, and focus your attention on a pair of black holes
orbiting each other. Einstein's equations tell us that, during their frenetic
dance, the black holes create deep depressions in the fabric of space-time
and generate waves that propagate outward at the speed of light, just as the
waves created by a stone tossed in a pond propagate toward the shore. But
instead of peaks and valleys at the surface of the water, the waves we are concerned with here involve crests and troughs in the curvature of space.
Because the texture of space is sculpted by gravity, such disturbances in
space curvature are called "gravitational waves." While propagating outward,
they steal some kinetic energy from the two black holes. As a result, the
black holes spiral in on each other in an orbital motion that keeps on accelerating until it reaches nearly the speed of light. The horizon surfaces of

each member of the pair come closer and closer until they finally merge. At that point, the two black holes fuse and form a single one whose mass equals the sum of the masses of the pair members. This fusion is accompanied by a fresh burst of gravitational waves.

These gravitational waves make it back to Earth like the sounds of a faint melody in the distance. Just as sound waves carry musical notes to our ears and captivate us with the melody of a sonata by Chopin, so do gravitational waves carry the signature of events in the past life of black holes. Musical notes cause our eardrums to vibrate; these vibrations are transmitted by the auditory nerve to be decoded in our brain, and, as if by magic, we find ourselves enchanted by a fugue by Bach. Likewise, if we learn to detect and decipher gravitational waves, they are sure to thrill us with the history of black holes. They could tell us volumes about their mass, their spin, the shape of their orbit, or even the epic of their fusion and the resulting vibrations of space-time. By providing information about the warping of space-time, they complement electromagnetic waves that tell us something about the properties of matter surrounding black holes (such as temperature, density, and magnetic field intensity). Since gravitational waves do not interact with interstellar matter, they are neither distorted nor absorbed on their way to us. As such, they deliver untainted evidence directly from regions very close to the horizon of black holes.

THE MELODY OF SPACE

But what is the secret for detecting and deciphering gravitational waves? How does one build a giant eardrum to hear them? How does one listen to the melody of space?

The problem is formidable, because massive black holes that generate gravitational waves are not exactly next door to us, but very far away. The known black hole pairs with a few tens of solar masses scattered in the Milky Way are at least a few tens, if not thousands, of light-years away. We have seen that the black hole at the center of the Milky Way, weighing in at 3 million solar masses, is some 30,000 light-years away. The supermassive black holes of several billion solar masses that lurk in the midst of quasars are found only at distances of several billion light-years. Such enormous distances do not make it easy to search for gravitational waves. Just as the sound of an orchestra fades away with distance, the melody of space created by a pair of black holes spiraling in on each other and merging together becomes increasingly faint as they are farther away from Earth. The intensity of the music of black holes varies in inverse proportion to their distance to us. Move 100 times farther away, and the music you will hear will be 100 times more feeble.

The effect is quite spectacular, however, if you have a front-row seat. Our friend Jules the explorer can testify to that. During one of his many expeditions, he went on a mission to study the final fusion of the pair of black holes in the constellation Pegasus. Jules arrived just one week before the fateful event. Each one of the black holes had a mass about twenty times that of the sun. Separated by some 30,000 kilometers, they were engaged in a furious waltz, orbiting each other once every ten seconds. During the week following Jules's arrival, this vertiginous pace kept increasing as the black holes spiraled in on each other. The distance between them shrank to 20,000 kilometers, then 3,000, then 2,000. The 10 seconds required to complete a single orbit became 5, 2, and then 0.2 second. In the 10 seconds immediately before fusion, Jules's rocket began to shake violently, signaling an onslaught of gravitational waves. The vibrations became more and more intense, reaching a paroxysm at the precise instant of the fusion that unleashed a new burst of gravitational waves. Jules found himself alternately pulled and compressed from head to toe, as if some thoughtless giant were playing yo-yo with his body. The disturbance then moved on and the shaking subsided until everything returned to normal. Jules heaved a sigh of relief, as his poor body could hardly stand it anymore. The train of gravitational waves went away in the immensity of space and will someday reach the solar system and pay our planet a visit.

Back on Earth, physicists are pulling out all the stops in preparation for the arrival of gravitational waves and to give them the hearty welcome they deserve. Their challenge is to build instruments sensitive enough to detect them and wrest from them the story of black holes. As we can appreciate from what happened to our intrepid friend Jules, gravitational waves bear a family resemblance to the tidal forces exerted by the Moon on Earth. At the spot on Earth closest to the Moon (as well as diametrically opposite), the lunar gravity attracts the oceans and causes high tides, just as gravitational waves were tugging on Jules's body. But there are some profound differences between the two situations (Figure 54).

To begin with, tidal forces always pull on water, whereas Jules's body was successively stretched and compressed. It was pulled when the crest of a gravitational wave was passing by, compressed when a trough arrived, and stretched again at the next crest. By contrast, water withdraws from a particular area of Earth at low tide not because it is being compressed but because Earth's rotation takes that area farther away from the Moon.

Next, there is a huge intensity difference between the two effects. Because the Moon is very close by (it is only slightly more than one light-second away), it can raise the level of the oceans on Earth by a meter or more, which enables waves to come lap at the legs of the deserted lifeguard chair at high tide. By contrast, after traveling interstellar distances of hundreds, if not thousands of light-years, followed by intergalactic distances of billions of light-years, gravitational waves have lost much of their punch by the time they reach Earth. The ferocious forces that shook Jules's rocket so violently and tortured his body so ruthlessly have become mere shadows of themselves. The gravitational waves generated by the merging of a pair of black holes, each with about ten solar masses and located a billion light-years away, would raise the oceans by the infinitesimally small height of 10^{-12} cm: That's one-thousandth of a billionth of a billionth (10^{-21}) of Earth's diameter, or

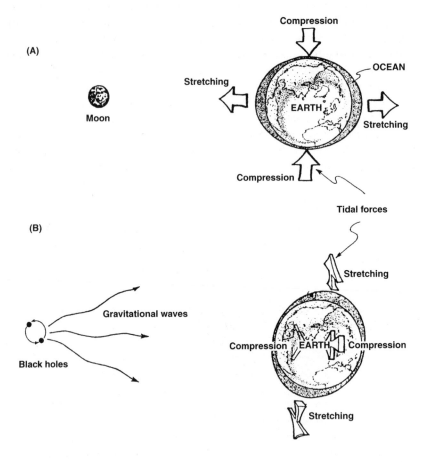

Figure 54. *Tidal forces produced on Earth by a passing gravitational wave.* Just as the Moon is responsible for the tides on Earth *(a)*, so does a passing gravitational wave created by two black holes merging into one also affect the water level in the oceans *(b)*. There are, however, three important differences between the two situations. First, gravitational waves propagate in space at the speed of light, as do visible light or radio waves. Hence, their effects come and go quickly. Tidal forces exerted by the Moon, on the other hand, are constant and unchanging. Second, whereas tidal forces due to the Moon can occur in all spatial directions, both along the Earth–Moon axis (longitudinal direction) and perpendicularly to it (transverse direction), gravitational waves can act only in a plane perpendicular to their propagation direction. The ocean levels go through periodic expansions and compressions as a gravitational wave goes by. When the crest of a wave arrives, expansion occurs in an up-down direction, while compression takes place in a front-rear direction. When a trough arrives, the reverse happens: front-rear expansion and up-down compression. The third difference has to do with the intensity of the effects. The Moon can raise or lower the level of oceans by several meters; by contrast, gravitational waves resulting from the merging of two 10-solar-masses black holes located at a distance of 1 billion light-year change the level of oceans by only 10^{-12} cm, which is 10,000 times less than the size of an atom.

10,000 times smaller than the diameter of an atom! Obviously, we are not about to notice the arrival of a gravitational wave on Earth by intently scrutinizing the level of the ocean. Physicists have to resort to all their ingenuity to devise gravitational wave sensors capable of detecting changes much smaller than the size of an atom.

A VIBRATING CYLINDER AND SINGING CRYSTALS

The American Joseph Weber (1919–) was the first to try to build such a detector, starting in 1959. He used a large block of aluminum, shaped like an enormous cylinder 2 meters long and 0.5 meter in diameter, and weighing a

Figure 55. *A gravitational wave detector.* This photograph shows the American physicist Joseph Weber working in 1973 on his gravitational wave detector—a large cylindrical bar of aluminum. The idea was that gravitational waves arriving from space would cause minute expansions or compressions of the cylinder's ends. To detect such tiny effects, Weber installed small piezoelectric crystals near the central part of his cylinder. When subjected to successive compressions and expansions as a gravitational wave passes by, these crystals generate an electrical voltage that can be measured. Unfortunately, this type of instrument has never been able to reach the sensitivity required to detect the gravitational waves produced by astrophysical phenomena, such as the fusion of two black holes or a supernova signaling the explosive death of a massive star.

ton (Figure 55). The idea was the following. If a gravitational wave were to pass through, the aluminum block would start vibrating just as Jules's spacecraft did. The aluminum block has a particular resonant frequency, just as a champagne glass has a natural tone, which is the lovely crystalline sound you hear when you give it a slight tap. All you would have to do is measure the vibrations of the cylinder to decipher the message conveyed by gravitational waves. It sounds straightforward enough, but it is easier said than done. Because the cylinder is only 2 meters long, which is 6 million times less than the diameter of Earth, the amplitude of each vibration of the cylinder will be 6 million times smaller than the lift of the ocean level on Earth by the gravitational waves. The vibration's amplitude will be about 10^{-19} cm, or one millionth the size of an atomic nucleus. Are such minute changes even measurable? The technological challenge was enormous. Weber had the idea of attaching to his cylinder small crystals that have the property of generating an electrical voltage when subjected to vibration-induced stresses (physicists call this phenomenon piezoelectricity). It was as though these crystals started singing to announce the arrival of gravitational waves. With many crystals hooked up in series, the induced voltage could be amplified and measured. Thanks to these magic crystals, Weber succeeded in measuring vibrations with an amplitude as small as 10^{-14} cm, or only one-tenth the size of an atomic nucleus. Marvelous as this technical prowess was, it was still 100,000 times short of what was required to detect gravitational waves created by the fusion of two black holes with ten solar masses located 1 billion light-years away.

There were other difficulties. The aluminum cylinder had to be isolated from ground vibrations and from the changing gravity of Earth's atmosphere. The best solution would have been to place the cylinder in space orbit, but the cost would have been prohibitive. Moreover, even if totally isolated from its environment, a cylinder could detect gravitational waves only in a very narrow range of frequencies centered about the block's natural resonance. To reproduce the full richness and beauty of a symphony by Beethoven, one needs an entire orchestra with dozens of different instruments, each of which is producing its own tone. Likewise, reproducing the symphony of gravitational waves would require thousands of cylinders with different sizes and weights, each tuned to a different frequency, which would be both expensive and impractical.

Another difficulty stood in the way—one of a theoretical nature. It concerns a fundamental principle of quantum mechanics, known as the uncertainty principle, first stated by the German physicist Werner Heisenberg. Heisenberg established that one cannot determine accurately the position of an atom without perturbing it with our measuring instrumentation, making its velocity random and unpredictable. The more accurately the position of an atom is known, the more uncertain its velocity. This uncertainty is built into Nature. It depends neither on the way you perform the measurement nor on the sophistication of your hardware. When it comes to detecting vibrations whose amplitude is a million times smaller than the size of an atomic nucleus, the uncertainty principle comes into play and reminds us that such a precision cannot be achieved without creating parasitic vibrations in the aluminum cylinder, which would completely obscure those due to gravitational waves.

A LASER INTERFEROMETER

Faced with such overwhelming difficulties, physicists turned toward other types of detectors. The currently most promising instrument is a laser interferometer. The idea is to suspend three weights at each end and at the right-angle corner of a huge L-shaped structure stretching over several kilometers (Figure 56). The arrival of a gravitational wave would slightly change the length of each of the two branches. As a crest passes by, one of the branches would contract and the other would expand. The opposite would take place when a trough arrives. These infinitesimally small changes in length would be monitored with a system of laser beams traveling back and forth along each of the two branches, reflected by mirrors positioned at each extremity. If one of the branches of the L-shaped structure expands and the other contracts, the first laser beam will take longer to complete a round-trip than the second. In other words, one beam will have a phase lag with respect to the other. The two beams are combined in a device called an "interferometer," the purpose of which is to measure such phase lags. In the absence of any gravitational wave, the distances separating the three weights are fixed. The two laser beams then take precisely the same time to complete their round-trip and the phase lag is zero. Thus, measuring the phase lag induced by gravitational waves provides us with a window on the history of the fusion of black holes. The uncertainty principle no longer presents an insurmount-

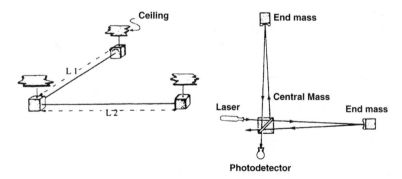

Figure 56. *A laser interferometric gravitational wave detector*. The instrument consists of two arms arranged in the shape of a large L. The distances L_1 and L_2 separating three masses placed at the corner and ends of the L change very slightly as a gravitational wave goes by. For example, the distance L_1 increases when L_2 shrinks. These minute changes are monitored by a system of laser beams traveling back and forth along the arms of the L. A beam covering the arm L_1 would take slightly longer than one covering L_2, introducing a slight phase difference between the two laser beams. The properties of gravitational waves can be studied by measuring this phase difference with the interferometer.

able obstacle in this approach because, interferometers being thousands of times longer than Weber's cylinder, the amplitude of the changes being measured are themselves thousands of times larger. Since the demands on accuracy are considerably relaxed, the parasitic perturbations introduced by the measuring apparatus become negligible.

The plan is for eight such interferometer systems to be built around the globe by the year 2007. Two of those are to be located in Pisa, Italy, the city in which Galileo studied gravity by reportedly dropping objects from the top of the Leaning Tower nearly four centuries ago. Galileo would surely be pleased that the same city is to be the site of an instrument capable of detecting vibrations in space-time caused by his beloved gravity, one of his favorite topics of research. The other six interferometers are to be scattered elsewhere around the world, two in Louisiana, two in Washington State, in the United States, and the last two in Japan. A global network is required to make sure that gravitational waves are indeed of cosmic origin. If so, all eight instruments should detect the same event at the same time. Otherwise, if the source of vibration is a local perturbation, it should register on one instrument only. With the help of this global network, astronomers in the twenty-first century may well be able to witness almost live the fusion of black holes and neutron stars, as well as the birth of the supermassive black holes that are responsible for the fire of quasars.

In the meantime, do not think that gravitational waves are pure abstract inventions living only in the overly fertile imagination of physicists. The existence of a system of two neutron stars, also called "binary pulsar"—as we have seen, neutron stars are also known under the name "pulsar" —have convinced astrophysicists that these waves are for real. The two pulsars in question, separated by a distance of some 700,000 kilometers, or roughly the radius of our Sun, orbit each other once every eight hours in a dizzying waltz. By monitoring this frenzied dance year after year, astronomers have noticed that the movements have been speeding up very slightly, by a relative amount of 2.7 billionths per year. This acceleration can be understood only if the binary pulsar behaves the same way a pair of black holes does, with each member spiraling in on the other in a process that results in the emission of gravitational waves. These waves steal energy from the pair of pulsars, causing them to get closer to each other and accelerate their orbital motion.

THE PATH TOWARD UNITY

Thus ends our exploration of the phantasmagorical world of black holes. By letting themselves be guided by aesthetical principles of beauty and elegance, by allowing themselves to be seduced by notions of harmony and symmetry, physicists were able to perceive the unity of Nature. They discovered intimate connections between phenomena that had no a priori similarity. As physics progresses, the unity of Nature is more and more evident. The dream of a unified description of Nature seems almost within reach.

In the seventeenth century, Newton blazed the trail toward unity by discarding the Aristotelian notions that Heaven and Earth are governed by separate laws, and by insisting that the movement of the Moon around Earth and the fall of an apple in an orchard are determined by the same universal gravitation. He unified Heaven and Earth. Some time later, when it became recognized that sound is but an oscillating motion of air, the science of acoustics was unified with Newtonian mechanics. The nineteenth century saw the great synthesis of electricity and magnetism by Maxwell. He proved that both types of phenomena are actually two facets of one and the same reality. By discovering that electromagnetic waves are in fact none other than light waves, he unified optics under the banner of electromagnetism. At the beginning of the twentieth century, by insisting that light always travels at the same speed and that the laws of physics remain invariant regardless of the motion of an observer, Einstein unified time and space. As he was pondering the dilation of time and the contraction of space, his physicist colleagues were making their first forays into the world of atoms. They discovered (with a little bit of help from Einstein himself) a world in which laws fly in the face of common sense and even defy the theory of relativity—a world from which determinism is banished and chance reigns supreme.

Our next order of business is to explore this strange world of atoms.

« 5 »

The Unbearable Strangeness
of Atoms

A WORLD OF COLORS

We live in a world of colors: the cerulean blue of the sky, the delicate pink of a rose, the luxuriant green of a leaf, the fiery red of a sunset. These colors soothe our souls and move our hearts. They bring a touch of gaiety to our existence and make life more worth living. To what do we owe this festival of hues, this exuberance of tones? Until the early part of the twentieth century, no one had a clue. Only with the discovery of the structure of atoms did we begin to have an answer.

The idea that matter is made of atoms now seems quite natural and self-evident. Yet it took a long time to be accepted. Like so many other concepts that now form the basis of Western civilization, the conviction that natural phenomena can be explained in terms of elementary constituents was born in Greece. In the town of Miletus of Ionia, around the sixth century B.C., Thales (ca. 625-547 B.C.) believed that fundamental substance to be water, while Anaximenes (ca. 550-480 B.C.) professed it was air. They were both wrong. Another century would pass before Democritus (ca. 460-370 B.C.) and Leucippus (ca. 480-ca. 420 B.C.) introduced the revolutionary idea that all matter is composed of indivisible and eternal particles, which they named "atoms" (from the Greek *atomos*, meaning "that which cannot be divided"). In the absence of any experimental evidence, things stood there for the next twenty-one centuries. The concept of atom faded away in favor of the four fundamental substances—water, air, earth, and fire—promoted by Aristotle (384-322 B.C.). Then came a long series of events, which included the annexation of Greece by the Roman Empire; the decline of Greek thought; repeated invasions by barbarian hordes, notably the Goths and the Huns; and the collapse of the Roman Empire and the rise of the Islamic Arab Empire, which picked up the torch of civilization and science, to hand it in turn to Europe at the Renaissance.

It was not until the 1600s that the notion of atoms resurfaced. Inspired by *On the Nature of Things* written by the Roman poet Lucretius, which expounded the atomistic views of Leucippus and Democritus, the French philosopher Pierre Gassendi (1592–1655) stressed the need to conduct experiments to test the atomic hypothesis. His call did not fall on deaf ears. The Irish physicist and chemist Robert Boyle (1627–1691) concluded in 1662 that the only way to understand the law of compressibility of gases was to accept that they were made of atoms. Another step toward the proof of the existence of atoms was taken by the French chemist Antoine de Lavoisier (1743–1794). He elucidated the secret of the composition of water and air, demonstrating that they involved chemical elements that always combined in a constant ratio.[1] In 1808, the English chemist John Dalton (1766–1844) came to the conclusion that the properties of chemical elements could be understood only if they were made of different types of atoms, each characterized by its own "atomic weight."

ORDER AMONG THE CHEMICAL ELEMENTS—THE PERIODIC TABLE

In 1869, the Russian chemist Dmitri Mendeleev (1834–1907) managed to restore order among the bevy of chemical elements that seemed to multiply at will. He arranged them according to their atomic weight. As if by magic, elements with similar chemical properties lined up neatly in groups of seven in an array of columns. Such a regular pattern made sense only if each chemical element was made of one distinct type of atom. This array of columns forms what is known today as the "periodic table" of the elements, which hangs on the wall of every high-school chemistry classroom from one end of the globe to the other. At the time Mendeleev came up with his list, only sixty-three chemical elements were known out of the ninety-two catalogued now. Some slots had thus to be left empty. So sure was Mendeleev of his ordering scheme that he did not hesitate to assert that the missing entries did not mean that Nature had an aversion for certain elements, but that scientists had yet to discover them. He turned out to be right. In time, the empty slots were gradually filled as new elements were discovered. That made Mendeleev famous the world over, except in Russia, where the tsar took a rather dim view of his too liberal political views.

The evidence in favor of the existence of atoms continued to mount. First, in the second half of the nineteenth century, the English physicist James Clerk Maxwell (1831–1879)—of electromagnetism fame—and then, a little later, the Austrian Ludwig Boltzmann (1844–1906) and the American Willard Gibbs (1839–1903) began to work on thermodynamics, the science of heat. They demonstrated that the properties of gases, particularly their temperature, could be understood in terms of atoms. Heat, they claimed, simply reflected the degree of movement of the constituent atoms. They are violently agitated in warm air, and become very sluggish at low temperatures. The hot winds of the Sahara desert feel uncomfortable because air molecules, heated by the incandescent Sun, strike your skin at high velocity, while they move much more gently in a cool and refreshing breeze. From this perspective, matter cannot be chilled below absolute zero (-273°C), at which temperature the movement of atoms stops completely.

Next, the fact that chemical elements could be arranged in Mendeleev's periodic table according to their atomic weight suggested that atoms possessed different degrees of complexity, heavier atoms being more complex. Therefore, the atom could not be a basic building block, contrary to what Democritus and Leucippus had thought. They had to have some internal structure and to be themselves made of even more fundamental particles. But up to that time, no one had ever seen an atom, let alone peeked at its internal structure. The first experimental proof of an internal atomic structure came in 1897, when the English physicist Joseph Thomson (1856–1940) discovered the electron while studying electrical discharges in gases. The newly found particle was named "electron" after the Greek word for "amber," because the Greeks knew that amber had the mysterious ability to attract objects when rubbed against wool. Each electron carries an electrical charge, and different atoms contain different numbers of electrons. In fact, that number turned out to be equal to the atomic weight and to correspond to the sequential position of the chemical element in the periodic table. For instance, lithium, which has a total of three electrons, occupies the third slot; boron, with five electrons, falls in the fifth slot, and so forth. Such a correspondence could not be accidental. It had to mean that the electron was a fundamental constituent of matter.

A NUCLEUS AS HARD AS IRON

Efforts to ferret out the secrets of the structure of matter redoubled in the early part of the twentieth century. To that end, what better strategy is there than to shatter matter to pieces by hitting it with high-speed projectiles? In 1910, the English physicist Ernest Rutherford (1871–1937) got inspired to bombard thin gold foils with high-energy particles. The results stunned him. While the vast majority of the particles went straight through the foil, a very small fraction (1 in 8,000) were reflected and came back to their starting point. It was as though a cannonball bounced off a piece of tissue paper. At the time, physicists thought that atoms were arranged in a solid object much like soft oranges stacked in a crate, occupying most of the space and leaving only small interstitial gaps between them. If so, none of the particles hurling at high speed toward the gold foil should turn back. They would all be expected to go through the target as easily as a bullet piercing an orange. Rutherford was forced to accept the evidence: There had to be inside the atom a nucleus dense and hard enough to reflect particles. Furthermore, the volume of that nucleus had to be very small compared to that of the entire atom, since most particles missed it and went right through the gold foil unencumbered. We know today that the diameter of an atomic nucleus is a tiny one-tenth of a thousand-billionth (10^{-13}) of a centimeter; that is 100,000 times smaller than the diameter of an atom (10^{-8} cm). An atomic nucleus occupies only one million-billionth of the volume of the entire atom (Figure 57). It can be compared to a grain of rice in a football stadium. All the objects around us—the book in your hands or the chair you are sitting on—are thus almost entirely vacuum.

On the strength of his experimental results, Rutherford went on to propose a model for the atom: It resembled a miniature solar system, with the nucleus playing the role of the Sun and the electrons that of the planets. Unfortunately, such a model cannot conform to reality, because the atom would not survive very long. The laws of classical physics tell us that, as the electrons orbit around the nucleus, they must radiate light, which means

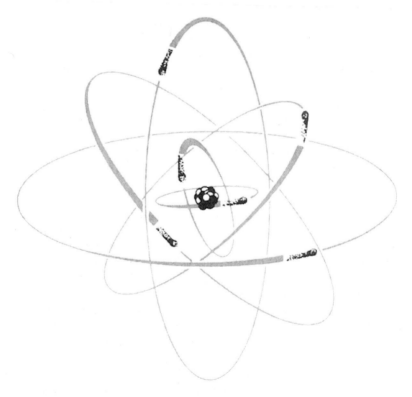

Figure 57. *Rutherford's model of the atom*. Electrons orbit around a nucleus whose size is 10^{-13} cm and which contains most of the mass of the atom. The space through which the electrons move is 100,000 times larger than the nucleus, stretching to 10^{-8} cm. Since the volume of the entire atom is one million billion times larger than that of the nucleus, matter is almost entirely made of void.

that they should gradually lose their kinetic energy and quickly spiral into the nucleus. If so, the atom would collapse on itself in one-millionth of a second, much faster than an eyeblink. Of course, we do not see material objects implode one after the other. They all seem to enjoy placid lives year in and year out. What, then, keeps electrons in place? What prevents them from losing energy and falling into the nucleus? What gives solids their cohesiveness? Completely new physical principles were called for.

GRAINS OF LIGHT

A twenty-seven-year-old Danish physicist named Niels Bohr (1885–1962) picked up the challenge in 1913. To solve the mystery of the stability of atoms, he seized upon a momentous discovery made by the German physicist Max Planck in 1900, in a fitting inauguration of the twentieth century. Planck was working on the seemingly simple problem of the radiation emitted by a heated material body. He immediately hit a formidable obstacle. The equations of classical physics kept telling him that such a body should emit an infinite amount of radiation in the ultraviolet part of the spectrum. That obviously did not make sense. Our kitchen ovens do not blow up like an atomic bomb the moment we turn them on. Yet they should do just that if the amount of ultraviolet radiation were indeed infinite. To resolve the paradox of the "ultraviolet catastrophe," as the problem was called, and bring theory in alignment with reality, Planck came up with a brilliant solution: He postulated that radiative energy was not a continuous quantity but, rather, had a discrete structure. It would come in the form of "grains" or "quanta." Each such grain of light would carry an energy equal to the product of a fixed number (known today as "Planck's constant") and the radiation's frequency (which is the number of crests of the light wave passing at a given point in space per second). After all, Planck argued, if matter has a corpuscular structure and a discontinuous nature, why should light be different?

Einstein, the quintessential revolutionary in science, did not hesitate a minute. He immediately latched onto Planck's idea to explain the photoelectric effect, which is that property of metals to eject electrons from their surface when exposed to a light beam. Curiously, the energy of the ejected electrons did not depend on the intensity of the illumination, but only on its frequency. High-frequency light, such as ultraviolet radiation, ejected electrons with a much higher energy than did low-frequency radiation, such as radio waves. This baffling result made sense only if the light absorbed by the metal was composed of discrete bundles, which we now call "photons." The photons absorbed by the metal gave up their energy to the ejected electrons. Since high-frequency light contained more energetic photons (the energy being proportional to the frequency), it stood to reason that it should eject electrons also with a higher energy.

THE STAINED GLASS WINDOW OF ATOMS

In a supreme act of creativity, Bohr combined Rutherford's planetary model of the atom with Planck's quanta. He accepted Rutherford's idea that electrons orbit the atomic nucleus in an orderly fashion, but he added to it Planck's concept of discreteness. Not only was light discontinuous, but so were the orbits themselves, which would take on "quantized" attributes. No longer would electrons be free to revolve around the nucleus wherever they pleased; they were now forced to remain in well-defined orbits, at specified distances from the nucleus. How does one prove the validity of such a model? How can one verify that electrons really do behave that way? If the art of science is to see connections that were not obvious at first, Bohr was an artist in the purest sense of the word. He realized that there existed a window through which it was possible to study the behavior of atoms. This window had panes with motifs as colorful and rich as those magnificent stained glass windows gracing the walls of Gothic cathedrals. The stained glass window of an atom was its emission spectrum. It is revealed by decomposing the light emitted by an atom into its different energy components, or colors.

We have all admired the beauty of rainbows, those multicolor arches that sometimes spring up in the sky on the side opposite the Sun after a rain shower. A rainbow is simply the spectrum of sunlight. Raindrops decompose white light into all its constituent colors, producing this marvelous festival of red, orange, yellow, green, blue, indigo, and violet, in order of increasing energy. The transition from one color to the next is continuous. It takes place gradually, with no gap interrupting this gorgeous luminous pattern. Things are quite different for light emitted by an atom. Its spectrum is not continuous at all, but appears broken up by numerous sharp vertical lines. For instance, the visible light from a hydrogen atom—the simplest, lightest, and most abundant chemical element in the universe—is characterized by three bright lines: a red one, a blue-green one, and a blue one. Bohr argued that the discontinuous character of this spectrum is related to the discontinuity of electronic orbits in the hydrogen atom itself. Each one of the lines is the result of a release of energy. A grain of light (or photon) is emitted by a hydrogen atom every time its lone electron executes a quantum jump from a distant orbit to one closer to the nucleus. The energy of the emitted photon is precisely equal to the difference in energy between the initial and final orbits. For instance, the red line is observed when the electron jumps from the third to the second orbit, the blue-green line when it jumps from the fourth to the second orbit, and the blue line when it jumps from the fifth to the second orbit. The arrangement of the lines in the spectrum thus becomes an accurate reflection of how electron orbits are arranged inside the atom. This arrangement is unique to each chemical element. It constitutes a fingerprint of sorts, or an identity card, as it were, for each element. Much as Sherlock Holmes can zero in on the identity of the criminal with the help of the fingerprints the latter inadvertently left behind, astrophysicists have learned to identify the chemical elements present in the atmosphere of stars by analyzing the structure of their emission spectra. It is by relying on these spectra that they have been able to determine the chemical composition of stars, galaxies, indeed of the entire universe.

Just as stained glass windows allow daylight to filter through inside a Gothic cathedral, spectra enable us to peek inside atoms and learn the secrets of their structure. The new world of atoms turned out to be governed just as much by mathematical laws as Newton's world of classical mechanics, unveiled two centuries earlier. Mathematics continued to conform to Nature's way, even on the smallest scales. But in the world of atoms, the era of the continuum had come to an end. From here on, discontinuity and quantum principles would be the rule. That was a radical revolution. It was as though a dictatorial government had suddenly decreed that its citizens would be allowed to take steps in lengths of only 20, 30, or 35 centimeters, and that any other length would be prohibited under penalty of death. By forbidding electrons to occupy any arbitrary orbit, Bohr prevented them from spiraling in on the nucleus and saved atoms from imploding. But while his atomic model did account for atomic spectra, it also raised many new questions difficult to answer. His model was a peculiar mixture of classical and quantum physics. As long as they uneventfully orbited around an atomic nucleus, electrons observed the laws of classical mechanics as they were known to Newton and Kepler. They behaved much like planets in the solar system. But as soon as they decided to jump from one orbit to another, they would suddenly obey the brand-new rules of quantum mechanics. It all seemed to lack consistency.

WHY IS A ROSE PINK?

Despite these conceptual difficulties, Bohr's atomic model can account for a number of properties of matter, notably the festival of colors all around us. Why is a rose pink, a poppy field red, and chalk white? The answer lies in the atomic structure of the materials that form them. We interact with the universe through the interplay of light and matter. Because we are made of electrons and protons, we are electromagnetic creatures that communicate with the external world by means of electromagnetic waves. A large part of the human experience derives from sunlight reflected off the objects populating our surroundings. Thus, the soft light of the Moon reflected off the tranquil surface of a lake starts as particles of light produced deep inside the Sun in the frenzy of nuclear reactions. About a million years later, these light quanta emerge from the Sun's surface and travel to the Moon to interact with the atoms of silicon and oxygen making up the rocks on the arid lunar surface. In turn, they pay a visit to a lake here on Earth, grazing its calm surface and interacting with the electrons in the atoms of hydrogen and oxygen bound together in water molecules. They finally make their way to our eyes, where they stimulate the clouds of electrons bound to the protein chains in our retinas.

Just as the surface of a lake reflects the image of the Moon, so does a rose charm us with its delicate pattern by redirecting sunlight into our eyes. But, you might object, sunlight is white. Why, then, is a rose not white too? The answer is to be found in the atoms that make up the rose. As we have seen in connection with rainbows, white light is a mixture of colors spanning the range from red to violet. A rose absorbs the blue and violet, but reflects the red, which then mixes with white to produce the color pink. Why this preference for blue and violet? The answer has to do with the arrangement of electronic orbits in the atoms and molecules the rose is made of. In order for an atom or a molecule to absorb light, an electron must make a quantum jump from a low-energy orbit, close to the nucleus, to a higher-energy orbit, more distant from the nucleus. The energy difference between these two levels determines the energy (or color) of the light that will be absorbed. It just so happens that some of the electron orbits in the atoms of a rose are configured in such a way that their energy difference corresponds to the colors blue and violet; therefore, it is those particular colors that are being absorbed. On the other hand, there are no orbits whose energy difference corresponds to the red, so that color is not absorbed; instead, it bounces off unaffected and goes on to ultimately tease our retinas. And that is why roses look pink.

Likewise, chalk is white because the molecules that it is made of have electron orbits whose energy separations correspond to none of the colors of the rainbow. All the colors contained in white light are reflected equally, and we perceive chalk as white.

Thus it is that the great variety in the structure of atoms and molecules composing ordinary matter is responsible for all the colors that surround us and brighten our lives. Neither the orange and mauve tints of Cézanne's apples nor Monet's azure-blue sky would exist if not for the amazing diversity of chemical elements in the paint these marvelous artists used to create their magic. The world would be a depressingly dull place if all atoms had the same structure. Imagine what would happen if all matter conformed to the blueprint of chalk atoms. We would live in a world of white. Not only would chalk be white, but so would poppy fields; butterfly wings would lose their dazzling colors and fade into a bland and monotonous white.

ELECTRON WAVES

The next actor to enter the stage in the quantum epic was the French physicist Louis de Broglie (1892–1987). Einstein had endowed light with a dual nature: Besides its wave characteristics, it could also take on the attributes of a particle. If light can have such a split personality, why not matter also? de Broglie asked himself. In 1923, he proposed that electrons, accepted by everyone as particles, could also take on the appearance of waves. He postulated that the size of an electron orbit in an atom had to be such that an integral number of waves could fit in. For instance, the smallest orbit, the one closest to the nucleus and with the smallest energy, had to have a circumference just large enough to accommodate one wavelength (a wavelength being the distance between adjacent crests). The second orbit had to be able to accommodate two wavelengths, the third three, and so on.

De Broglie's brilliant intuition was given a solid mathematical footing in 1925, when the Austrian physicist Erwin Schrödinger (1887–1961) derived his famous electron wave equation, which came to be known universally as Schrödinger's equation. Although it had been worked out for electrons, it would prove equally applicable to any other particle within an atom or molecule. Armed with this equation describing how the wave associated with a particle evolves as a function of time, physicists began to frantically calculate all sorts of things—energy levels, molecular structures, and anything else they could think of. An avalanche of results ensued, which, to everyone's amazement, turned out to agree remarkably well with observations. That proved conclusively that matter indeed had wavelike properties. It was the golden age of quantum mechanics. The times were heady.

If Schrödinger's wave equation was undeniably capable of describing the behavior of atoms and molecules, no one—not even the fathers of the theory, such as Schrödinger and de Broglie themselves—had the slightest idea what this wave meant. A wave is described at any instant by a set of numbers. For a sound wave, for instance, a relevant number would be the air pressure everywhere the wave propagates. For a light wave, the set would include the intensities and directions of the electric and magnetic fields at every point in space. But what in the devil is the meaning of a number characterizing an electron wave? What is it in this case that oscillates like a wave in the ocean? The answer came from a theoretical study of the behavior of an electron beam, described by a wave packet, when it is launched at high speed against an atom. When applied to this situation, Schrödinger's equation tells us that, at the moment of impact, the wave packet breaks up into several wavelets that are scattered in all directions, much as water from a garden hose splashes everywhere when aimed at a wall. What is the proper interpretation of this result? If electron waves truly describe matter, a wave packet splitting into many wavelets would imply that an electron breaks up into pieces spreading in all directions. Such a conclusion is evidently absurd, since electrons cannot split. They remain whole and preserve their identity. If so, electron waves cannot represent waves of matter. But then, what are they? The solution to the conundrum came from the German physicist Max Born (1882–1970). He proposed in 1926 that the wave described by Schrödinger's equation was not a concrete wave, made of actual matter propagating through space, but a rather more abstract entity—a probability wave.

Picture yourself playing pool. With the cue, you hit one ball and send it rolling toward another. Upon colliding, the two balls rebound in different directions. These directions can be predicted with Newton's laws of classical mechanics if you know the force and direction with which you set the initial ball in motion. Indeed, the determinism of the ball trajectories is what makes the game of pool possible in the first place. Things are quite different in the world of atoms. When you smash an electron against an atom, the electron's trajectory after the collision is no longer determined unambiguously. Rather, it can rebound in *any* direction. Yet indeterminacy does not mean complete unknowability. This is where Schrödinger's wave equation comes in. It tells us the probability of finding an electron moving in a particular direction and at a particular location after the collision. Born showed that this probability is equal to the square of the amplitude of the wave function. The chance of finding an electron is highest at the crest of the wave function, and lowest at the trough. But even at the crest, there is never complete certainty that an electron will be found there. It will show up perhaps three times out of four (a probability of 75 percent), or nine times out of ten (a probability of 90 percent), but the probability will never be 100 percent. In the world of atoms, the boring certainty and stifling determinism of classical mechanics are thrown out. Taking their place are the exhilarating uncertainty and liberating indeterminacy of quantum mechanics.

AN INDETERMINATE WORLD

Neither Schrödinger nor de Broglie, both of whom were determinists at heart, were too thrilled with this probabilistic interpretation of their beloved invention. Yet the very next year, in 1927, a young German physicist named Werner Heisenberg (1901–1976) brought fresh ammunition in support of indeterminism. While thinking about the relation between observer and object observed, he came to the stunning conclusion that uncertainty was inherent to the subatomic world, and that nothing could alleviate that fact.

Suppose we want to determine both the position and speed of an object. That is quite routine in everyday life. When you drive around in your car, all you have to do is look at the roadway signs to figure out where you are, and glance at your speedometer to know your speed. In principle, you could determine both with any precision you want. This ability goes by the wayside the moment you step into the subatomic world. Heisenberg proved that there exists a fundamental limitation to our knowledge. To determine the position of a particle, you have to illuminate it in order to see it. As it happens, light can delineate the contours of an object only to the extent that its wavelength (the distance between two successive crests or troughs) is comparable to the size of the object in question. When the object is smaller than the wavelength of the light, it becomes fuzzy. If you use a radio wave with a wavelength of a few meters to illuminate someone's face, all you will see is an indistinct blur. Only when you drop the wavelength down to a few centimeters will the outline of the ears, for example, begin to take shape. Decrease the wavelength further to less than a millimeter, and you are now working with infrared radiation (Figure 46). A pair of slanted eyes, a slightly upturned nose, and the outline of a faint smile all start to become recognizable. Continue to drop the wavelength down to a few tenths of a thousandth of one millimeter. We are now in the range of our beloved visible light, which our eyes can see. Beauty spots, wrinkles, and even fine hair now come into focus. Keep going and shrink the wavelength to less than one-tenth of a thousandth of one millimeter, and we now penetrate the ultraviolet region. Under an ultraviolet microscope, the cells of our skin can be magnified 3,500 times, and you can make out individual chromosomes. One more step toward shorter wavelengths, to only a few hundredths of millionths of a millimeter, and you step into the domain of X rays. This type of energetic radiation can pass through our bodies with ease. It constitutes a precious gift of Nature to physicians, because it enables them to image our skeleton. It also allows your dentist to examine the cavities and fillings in your teeth, and customs agents to scrutinize the contents of your suitcases without having to open them.

To pinpoint the position of a subatomic particle, we proceed in the same way as we did for the human face. We illuminate it with light whose wavelength is comparable to the size of the atom, or one-hundredth of a millionth (10^{-8}) centimeter. Light with such a minuscule wavelength belongs in the X-ray range. But by using light with such a high energy, we inevitably disturb the particle we are trying to pinpoint by imparting a momentum to it. That completely changes the velocity the particle had just before the observation. We are thus faced with a dilemma: Either we determine the position of the particle with the greatest accuracy possible by illuminating it with light of extremely short wavelength, which has a very high energy and therefore causes a large perturbation, and we have to resign ourselves to the fact that there is very little we can know about its velocity (for instance, if we desire to locate an electron to within one-hundredth of a millionth of a centimeter—roughly the size of an atom—the uncertainty on its velocity would be such that one second later, the electron could be anywhere within a radius of a thousand kilometers, which is more than the size of the state of Texas); or we decide from the outset that we are interested only in the particle's velocity, in which case we would illuminate it with light causing the least possible amount of disturbance—in other words, light with little energy and a very long wavelength. But, under those circumstances, the position would become a total blur.

Atoms impose a basic limitation to our knowledge. There is no hope of ever measuring both the velocity and the position at the same time with any arbitrary accuracy. Heisenberg's uncertainty principle forces you to take the plunge and make a choice. Uncertainty is inherent to the world of atoms. Regardless of what we might do to increase the sophistication of our instrumentation, we will always run into this fundamental hurdle.

Quantum fuzziness has spread over the subatomic world, pushing aside the determinism Laplace had extolled so passionately. Nature asks us to be tolerant and to renounce mankind's age-old dream of absolute knowledge. The degree of tolerance is quantified by a number called "Planck's constant." Heisenberg taught us that the product of the uncertainty on the position and that on velocity can never be smaller than Planck's constant divided by 2π. If Planck's constant were equal to zero, positions and velocities could, of course, be determined simultaneously with any desired accuracy. But Nature decided otherwise. Planck's constant is in fact not zero: In a system of units known as *cgs* —in which lengths are measured in centimeters, weights in grams, and time in seconds—the constant is equal to 6.626×10^{-27} erg-sec (where the number 6626 comes after 27 zeros). As small as this number may be, it imposes a fundamental and absolute limit to what we may and may not know.

All of this raises a question. If quantum fuzziness is so prevalent in the subatomic world, how come we seem to be shielded from it in our everyday life? We are, after all, made of atoms. Why does this uncertainty, which affects the behavior of atoms, not manifest itself on the scale of ordinary objects? The answer lies in the mass of these objects. Because they are typically big and have a large inertia, they are not easily perturbed when illuminated. High-energy X rays may pass through our bodies unimpeded, but they do not slam us against a wall. The impulse that light imparts on ordinary objects is all but negligible, and that is why their speed can be measured as precisely as you want at the same time as their position. Indeterminism disappears. This is just as well. We have our hands full enough as it is, dealing with the trials and tribulations of life, without having to also worry about uncertainties concerning the behavior of things all around us.

PARTICLES WITH THE GIFT OF UBIQUITY

Human beings do not have the ability to be in two places at the same time. They are not endowed with the gift of ubiquity. Michael cannot be both on Fifth Avenue and in Greenwich Village, and Carol cannot be having dinner in Chinatown and at her cousin's on the Upper West Side at the same time. Michael and Carol are either here *or* there. They cannot be both here *and* there. Yet that is exactly what quantum mechanics says can happen in the case of subatomic particles. This property of ubiquity manifests itself in a particular experiment involving electrons, duplicating one conducted in the eighteenth century by the English physicist Thomas Young (1773–1829), who was using light. Young had demonstrated the wave nature of light by illuminating two parallel slits with a single light source. He projected the image of the two slits on a screen located immediately behind them. Young noticed that the image was not simply two bright parallel bands, as would be expected if light propagated in a straight line. Instead, what he saw was a series of bright bands, spaced regularly and separated by dark bands. This pattern of dark and bright striations could be explained only if light behaved like a wave. In such a picture, light was no longer constrained to propagate in a straight line when forced through slits narrower than the separation between two adjacent wave peaks. Light waves passing through both right and left slits could superimpose. Wherever the two waves arrived at the screen with the same phase, the crests of each wave would reinforce each other and produce a bright band. Where they arrived out of phase, the crest of one wave would superimpose on the trough of the other, and the two waves would cancel each other, resulting in a dark band. Such alternating bright and dark bands created by light interfering with itself are called "interference fringes."

All this seems to make perfect sense. But things really get confusing when we repeat Young's two-slit experiment with electrons rather than light. Imagine that we replace the light source by an electron gun similar to the type found in an ordinary television set, and the screen by an array of electron detectors. Let us now examine the behavior of electrons as they pass through the slits. Our intuition tells us that there should be no interference pattern—i.e., no bands of maximum electron counts interspersed with bands of minimum counts—since the electron gun ostensibly fires electrons in the form of particles, not waves. We are in for a surprise. As it turns out, the detectors record precisely a series of maxima and minima in the number of electron hits, exactly like what was observed with light. The conclusion is inescapable: Electrons must have undergone a radical metamorphosis. They may have left the electron gun as particles, but they have turned to waves before reaching the slits. Moreover, the electron wave must have passed through both slits at the same time, since an interference pattern can be generated only by the interaction of two separate waves. In other words, the wave nature of an electron gives it the ability to be simultaneously in two places; indeed, it can be anywhere at once. Its trajectory is no longer defined. The electron has become subject to "quantum fuzziness." Just as yin and yang go together, electrons exhibit two aspects that complement each other: They are both particles and waves. This "complementarity principle," as Bohr called it, joins Heisenberg's uncertainty principle in governing the subatomic world.

ALL ROADS LEAD TO ROME

Because of quantum fuzziness, Bohr's model, in which the atomic world is populated by electrons following well-defined orbits around the atomic nucleus, like planets around the Sun, becomes meaningless. The notion of trajectory from point A to point B goes by the wayside. We may say that a particle is at A or B, but we would be at a loss to describe how it goes from A to B. An automobile traveler may drive from New York to Los Angeles by way of St. Louis. His itinerary is well-defined. He chose it so as to minimize the time required to reach his destination. But should he ever travel in the subatomic world, he would be hard-pressed to tell which route he picked to reduce travel time. He would be forced to describe reality in terms of probabilities. He might say that he went from New York to Los Angeles with a 65 percent probability of passing through St. Louis, 20 percent through Chicago, or 15 percent through Dallas. But he would never be 100 percent certain of having gone through any one of those cities. All roads lead to Rome, and electrons take them all. In Young's two-slit experiment, we will never be able to tell which particular slit an electron went through.

To say that randomness lies at the heart of matter does not necessarily mean that all knowledge is out of reach or that the laws of physics no longer apply. On the contrary, quantum mechanics predicts many properties of matter, always in perfect agreement with observations. The only catch is that such predictions never apply to individual events, but only to a collection of many events. When you toss a coin up in the air, the laws of probability do not tell you whether, on the very next throw, it will land on heads or tails. All they tell you is that when you throw the coin repeatedly, on average it will land on heads half the time, and on tails the other half. The same goes for the subatomic world. An individual event is not causally determined, but the behavior of a whole series of similar events is.

PARTICLES THAT CAN PASS THROUGH WALLS

The fuzziness of trajectories in the subatomic world can lead to situations that defy common sense. If you were to throw a rock at a window in real, everyday life, one of two things might happen. Either you threw it softly enough and the rock would just ricochet, or you did throw it hard enough and the window would shatter to pieces, causing the homeowner to come out and yell at you. Imagine now that we step into the subatomic world, replacing the rock by an electron and the window by an array of atoms acting as an electron barrier. In most cases, the electron behaves quite similarly to the rock. When it does not go fast enough, it simply bounces off the barrier and rebounds in the direction it came from. When it has plenty of energy, on the other hand, it goes right through the barrier of atoms. So far, nothing unusual. But from time to time some bizarre happenings, typical of the subatomic world, rear their head, and the electron does something most unusual. It may turn around even though it has more than enough energy to get past the barrier. But what is even more bizarre is that it can miraculously show up on the other side of the barrier even though its energy is not nearly sufficient to do so. It is as though a rock thrown way too softly, instead of rebounding against the window, made its way across without even breaking the glass. The French fiction writer Marcel Aymé, who wrote a novel about a character who could walk through walls, would have been thrilled to learn that quantum mechanics enables perhaps not living beings, but at least electrons, to perform the trick. It is as though the electron dug itself a kind of tunnel to get through to the other side of the wall. As it happens, physicists have dubbed this magic trick precisely the "tunnel effect."

Heisenberg's uncertainty principle provides an explanation for why electrons can pass through walls and cross barriers that seem insurmountable. Just as we cannot simultaneously know the position and velocity of a subatomic particle, so does quantum mechanics make the energy of an elementary particle fuzzy at any given time. This fuzziness allows particles to borrow energy from Nature, which then plays the role of a bank, as it were. It is this additional energy that enables particles to cross barriers. But there is a catch. This energy loan is not meant to be forever. The greater the loan, the faster it has to be repaid. Thus, the particle must act very quickly if it is to derive any benefit from borrowing energy. Most particles receive loans that are insufficient to cross the barrier. Those simply turn back. From time to time, though, a particularly lucky particle is the beneficiary of an unusually large loan and makes it across. Do not think that the "tunnel effect" exists only in the unchecked imagination of physicists. As you enjoy your favorite concerto being played on your hi-fi system, you should be thankful to the tunnel effect for making your listening pleasure possible. It is indeed that effect which makes some of the components in consumer electronics function.

THE OBSERVER CREATES REALITY

In our experiment involving electron beams passing through parallel slits, we have no way of telling which slit an electron went through. We are forever condemned to speak in terms of probabilities, or "to play dice," as Einstein put it. You might retort that all you have to do is place some detectors just behind each slit to monitor the passage of electrons. But the very act of spying perturbs the system so that we cannot have access to reality as it was before an observation. As long as we do not observe it, a subatomic particle can be here, there, and everywhere. It then dons its wavelike habit; waves add up or cancel out, and interference fringes appear on the screen behind the slits. It is only when we try to observe it that the particle decides to be either here *or* there, that it materializes as a particle, and that interference fringes vanish. In other words, it is the observing itself that creates reality. Prior to the observation, the electron behaves as a wave and remains pure potentiality. That potentiality becomes actualized only after the observation. Whereas for Laplace the world was a well-oiled machine that kept on running on its own without any divine or human intervention, quantum mechanics reinstated the observer to his preeminent status. In a manner of speaking, the external world is defined by the questions we ask ourselves about it.

If the role of the observer is so important for crystallizing reality at the subatomic level, is it also true in everyday life? Do the elements of reality we perceive around us while we go about our ordinary business—a bed of multicolored flowers, a tree-lined avenue, a desk cluttered with papers—also depend on the observer? Do they not have an existence of their own? Is the Moon there only when we look at it? Does a tree falling in the forest not make any noise if no one is around to hear it? Must we abandon the objectivity of the world, the idea that it exists independently of ourselves, whether we are observing it or not?

Those are all legitimate questions, given the fact that ordinary things in life are also made of atoms. If reality in the subatomic world is so subjective, why should it be any different for people and objects around us? We are all aware that our behavior is apt to be affected by someone looking over our shoulder. We act differently when we are alone and when someone is observing us. The natives in an Amazonian Indian tribe adjust their behavior in subtle ways the moment an anthropologist steps off a canoe to study them.

There is no doubt that observation can influence human psychology. The question is whether it can also alter reality in everyday life.

Erwin Schrödinger, who invented the wave function describing all the potentialities of an elementary particle—that is, all its possible movements and positions—had a difficult time accepting that reality should not be independent of any observation. He was so unsettled by the bizarre properties of quantum mechanics that he once cried out to Bohr: "I am sorry I ever got involved with quantum theory." To demonstrate the kind of paradoxical situation a probabilistic interpretation of reality can lead to, he imagined the following scenario. Lock a cat inside a box containing a flask of cyanide. Hanging above the flask is a hammer controlled by a radioactive substance, a material that decays spontaneously after a certain time. As soon as the first decay reaction occurs, the hammer comes crashing down and breaks the flask; the poison is then released and the cat dies. It all sounds straightforward so far. Things get wacky, though, the moment we try to predict the fate of the cat. Quantum mechanics tells us that we cannot know precisely when the first decay will happen. We can describe the situation only in terms of probability. For example, an hour into the experiment, there is a 50 percent chance that the nucleus will have decayed and that the cat will be dead. But there is also a 50 percent chance that nothing will have happened and that the cat remains alive. As long as we do not open the door of the box to check inside, the best we can say is that the state of the feline is a combination of 50 percent dead cat and 50 percent alive cat. Does it mean we can manufacture cats that are both dead and alive? Of course not. The moment you open the door and peek inside, you will see a cat that is either dead or alive, but certainly not suspended between the two. It is as though Nature waited for an observer before deciding between the two alternatives. But what goes on inside when no one is looking?

A MULTIPLE REALITY?

Schrödinger concocted the cat story to illustrate how a probabilistic interpretation of quantum mechanics can lead to strange, if not downright absurd, conclusions. But he was really waging a losing battle, as quantum mechanics continued to score one success after another in elucidating the behavior of atoms. So as to avoid situations involving cats suspended between life and death, the American physicist Hugh Everett proposed in 1957 an even more exotic solution—the so-called "parallel universes" theory. In this picture, the universe would split into two almost identical copies every time there is a choice to be made between two alternatives. The cat would be alive in one of these two universes, and dead in the other. Reality would thus multiply ad infinitum. Each parallel universe would come into existence with its own space-time completely severed from ours, in such a way that no communication between different universes, including our own, would be possible. As observers, we too would subdivide endlessly, cloning ourselves in each of the parallel universes. Each clone, however, would be able to perceive its own universe only and would be conscious of only a single reality. As you read these lines, there would be a multitude of clones of yourself sitting on the same chair, living in the same house, in a multitude of parallel universes differing from one another very slightly. All clones would have experienced the same past and accumulated the same memories. One could even envision an infinitude of Marcel Prousts sifting through the same past, recollecting the flavor of "madeleines" in an infinite number of slightly different versions of *Remembrance of Things Past*. But each of your clones would get to experience his own particular future, sometimes with only very slight differences, sometimes with more pronounced ones. Some clones might just go on quietly reading this book, others might get up and fix themselves a cup of tea, while still others might go out for a walk and stretch their legs. It is even conceivable that some in the latter group might get run over by a car while crossing the street, abruptly ending their existence. Copies of this book would also proliferate to infinity. Unfortunately, the writer's royalties would not pile up in kind, because there would also be an infinite number of clone-writers, each demanding his own share.

In short, quantum mechanics allows for the existence of a multiple reality. But since there is no way to actually observe parallel universes, these must, until further notice, remain purely hypothetical entities lacking any experimental proof, the product of the fertile imagination of a few inventive physicists.

Be that as it may, scientists continue to actively explore the twilight zone between the subatomic world and that of everyday experience. They try hard to understand how the transition from quantum fuzziness to classical determinism is made. They hope to explain why after assembling billions of billions of billions (10^{27}) of atoms to make a cat, we can no longer talk of a cat that is 50 percent dead and 50 percent alive, but only of a cat that is either dead *or* alive. The task remains daunting. The largest atomic clusters studied thus far do not exceed some 5,000 particles. That is still small enough for quantum fuzziness to prevail.

ERASING THE PAST

Quantum mechanics gives rise to many other peculiar situations in our ordinary world. One of them is the ability to erase the past. To see how that comes about, we must return to Young's two-slit experiment. You might recall that, as long as we do not monitor which slit a photon (or an electron) made its way through, a wave behavior prevails and interference fringes appear in all their glory on a screen placed behind the slits. But the moment we use detectors just past the slits to spy on photons and determine which slit they went through, the particle behavior takes over and interference fringes vanish. Quantum mechanics goes even further. It allows for photons to wait until *after* they go through the slits before deciding which of the two complementary aspects of reality—wave or particle—they will adopt, even though logic would dictate that the decision must be taken before.

To see how this comes about, let us set up detectors past the slits to record photons passing by. In such a situation, photons put on their particle-like hat and no fringes form. Nothing out of the ordinary here. But suppose now that, after the photons have gone through the slits, we change our mind and decide we are no longer interested in their actual path. We can set up before the screen (but after the slits) instruments whose function is to erase the path information. As soon as that information is erased, the inter-ference fringes reappear as if by magic. This means that photons have exchanged their particlelike hat in favor of their wavelike one. But what is truly remarkable is that they did it *after* going through the slits, rather than before. It is as though the photons knew ahead of time, even before going through the slits, that the information was going to be wiped out and had adjusted their behavior accordingly. In other words, a decision taken by human beings to activate instruments erasing information can influence the nature of reality retroactively. Quantum reality thus seems to possess some mysterious link with the past. But there are limits to this. While the actions of an experimenter or observer can help determine the nature of quantum reality in the past, under no circumstance can they interfere with the causal-ity of past events. You cannot, for instance, use quantum mechanics to send information back in time, stop your parents from ever meeting, and thereby prevent your own birth.

A GLOBAL REALITY

The quantum world thus appears to be endowed with a kind of global qual-ity (also called "holism") that transcends time. It also seems to transcend space, as shown by the following experiment imagined by Einstein.

Einstein, the magician of twentieth-century physics, was no more inclined than Schrödinger to accept that reality is not objective and that it cannot exist independently of an observer, as Bohr and his followers main-tained. Einstein felt that quantum fuzziness could not be inherent to Nature, but that it merely resulted from the lack of sophistication of our measuring instruments. An inveterate determinist, he firmly believed that reality was ruled rigidly by causal laws, and not haphazardly by the laws of chance and probability. The world had to resemble a round of pool, not a game of roulette.

Einstein spared no effort looking for a chink in quantum theory. He tried relentlessly to come up with scenarios proving that a probabilistic interpretation of subjective reality could lead only to absurdities. Over the years, he concocted a number of situations that he felt exposed a weakness in quantum mechanics. But Bohr always managed to fend them off. Kept in check, Einstein eventually gave up his resolve to prove that quantum mechanics was flawed, concentrating instead on the more modest goal of showing that it was incomplete and failed to give a comprehensive descrip-tion of reality. In 1935, he and his coworkers Nathan Rosen and Boris Podolsky devised the following thought experiment, commonly referred to as the "EPR experiment," after the initials of the three authors.

Consider, they argued, a particle spontaneously decaying into two grains of light denoted *A* and *B*. By virtue of symmetry, these must fly off in diametrically opposite directions. We can verify that it is so by deploying measuring instruments. If *A* turns up going west, we do indeed detect *B* going east. So far, nothing seems to be out of the ordinary. But that does not take into account the peculiarities of quantum mechanics. Before being detected, *A* exhibited the behavior of a wave, not a particle. But a wave is not localized, and *A* has a nonzero probability of going in any direction. Only after it has been detected does *A* metamorphose into a particle and "learn" that it was propagating west. But if *A* did not "know" ahead of time which direction to take before being detected, how could *B* "anticipate" *A*'s behavior and adjust its own behavior accordingly so as to be detected at the same moment in the opposite direction? This did not make any sense, unless *A* could *instantly* communicate its direction to *B*; but that would violate the theory of relativity, which precludes signals from traveling faster than the speed of light. "God does not send telepathic signals," argued Einstein. He described the concept of instant communication disparagingly as "spooky action at a distance," and concluded that quantum mechanics did not provide a complete description of reality. According to Einstein, before separating off from *B*, *A* had to already know which direction it was going to take and somehow share that piece of information with *B*. *A* had an objective reality that should in no way depend on the fact that it would later be recorded by the detector located west. That certain and deterministic reality was merely "hidden" behind the appearance of quantum uncertainty and indeterminism (physicists refer to this point of view as "hidden variable theory," the variables in question being the position and velocity of a particle).

Unfortunately, Einstein was wrong. The French physicist Alain Aspect of the University of Paris conducted in the early 1980s a series of experiments on photon pairs in order to test the validity of the EPR hypothesis. His results consistently validated quantum mechanics. There were in fact no "hidden variables." The flaws so desperately sought by Einstein in the theory remained conspicuously missing.

If so, how does one explain the fact that *B* always seems to know instantly what *A* is doing? This constitutes a problem only if we insist that reality is fragmented and localized at each particle. The difficulty goes away if we are willing to accept that *A* and *B* are part of a global reality. There is no need for *A* to send to *B* signals traveling faster than light, because the two quanta of light are in constant contact through some mysterious interaction. Quantum reality thus dismisses any notion of localization. The concepts of "here" and "there" become meaningless, as "here" is identical with "there." The universe is a vast system of particles all interacting with one another. Quantum mechanics had already conferred a holistic character on time. It now does the same to space.

THE ATOMIC NUCLEUS HAS ANOTHER CONSTITUENT

While Schrödinger, Bohr, and Einstein were grappling with the conceptual difficulties of quantum reality, physicists pressed on with exploring the depths of matter. They started to unlock more and more of its secrets by bombarding it with beams of particles accelerated to ever-greater speeds in increasingly larger machines, stretching over tens of kilometers. The population of elementary particles began to proliferate.

In 1930, Bohr's atom included a nucleus and electrons that orbited around it. This atom was made almost entirely of vacuum, so extremely compact was the nucleus composed of positively charged protons. There was one negatively charged electron for each proton, so that the total electrical charge of the atom was zero. The configuration of electrons around the nucleus determined the chemical properties of the corresponding atom and its place in Mendeleev's periodic table. But dark clouds were gathering on the horizon and threatened this comfortable picture. There existed in Nature some atoms with the same number of electrons, even though they had different atomic weights (they are called "isotopes"). For instance, neon occurs in two forms: neon-20 and neon-22, which have 20 and 22 times, respectively, the mass of a hydrogen atom; yet they have the same number of electrons. This difference in mass implied the existence of an additional constituent in the atomic nucleus. It was first suggested that it could all be explained with extra electrons. Thus, the atomic nucleus of neon-20 would contain 20 protons, while the neon-22 nucleus would have 22 of them, with two additional electrons to neutralize the excess positive charge due to the two extra protons. But Ernest Rutherford—the same one who had previously demonstrated the existence of the atomic nucleus—had a much better idea: Instead of attributing the weight difference to extra protons, why not postulate a new type of particle with the same mass as the proton? This new particle would have no electrical charge—it could be called "neutron"—which would do away with the need for extra electrons within the nucleus.

Rutherford's brilliant intuition was soon to be spectacularly confirmed by his colleague and former student, the Englishman James Chadwick (1891–1974). Chadwick discovered new particles that neither magnetic nor electric fields could deflect, which implied that they had no electrical charge. Yet, when used to bombard atomic nuclei, they caused protons to be ejected, which meant that the two types of particles had to have similar masses. They turned out to be indeed the neutrons Rutherford had postulated. Neutrons thus joined protons and electrons as the third of the building blocks of matter responsible for the complexity and beauty of the world.

NEUTRONS ENSURE THE STABILITY OF MATTER

Why did Nature bother to invent neutrons? Was that a gratuitous act? Not in the least. As it turns out, without neutrons, atomic nuclei containing protons only would not be stable, and all the matter around us, the teacup you are holding in your hands, the walls that support the roof over your head, the roses exuding their fragrance in your living room, would simply disintegrate. The electromagnetic force causes particles with opposite charges to attract each other. By the same token, it also dictates that particles with like charges, whether positive or negative, repel each other. As a result, protons inside a nucleus want to fly apart, and nuclei would explode if not for a force preventing them from doing so. What offsets the electromagnetic force and keeps the constituents of the atomic nucleus locked in (these individual constituents are called "nucleons") is known as the "strong nuclear force." The term "strong" seems appropriate enough, since it turns out to be approximately 100 times more powerful than the electromagnetic force.

You might think that, given its strength, the strong nuclear force would make mincemeat of the electromagnetic force, and that the latter should become completely inconsequential. But Nature is subtle and did not put all its eggs in one basket. The strong nuclear force has an exceedingly short range. Its effect is felt by nucleons only when they practically touch each other. As soon as they move apart by the smallest of distances—much less than one-tenth of a thousand-billionth (10^{-13}) centimeter—the strong nuclear force loses all influence and potency. The electromagnetic force, on the other hand, may be much weaker, but it has a considerably longer range. To be sure, it too becomes weaker as the distance between nucleons goes up. But its strength decreases only as the square of the distance, rather than exponentially as in the case of the strong nuclear force. The situation is somewhat analogous to a boxing bout in which one of the combatants has a vicious right uppercut but short arms, while the other's blows are less devastating but can reach farther out.

Likewise, nucleons within an atom are bound together by the strong force, but only when they are virtually in contact. By contrast, electrical repulsion makes itself felt from one end of the nucleus to the other. That is why the electromagnetic force wins the battle hands down in large nuclei. A bulky uranium nucleus, for instance, which has 92 protons and 140 neutrons, is highly unstable and splits readily. The electrical repulsion between that many protons causes the nucleus to undergo fission, a process accompanied by the release of energy. Man has learned how to tame this energy, but he has not always used that knowledge wisely: The atom bomb that rained death and destruction on Hiroshima derived its incredibly devastating power from the fission of uranium. In the opposite case—small nuclei, such as hydrogen nuclei—the strong force predominates, causing nuclei to fuse together rather than split apart. This fusion process also releases energy and is ecologically far more benign, as it does not leave behind any radioactive by-products as in the case of nuclear fission. Scientists are trying to exploit fusion to produce energy in a controlled manner, but so far with little success.

What is the role of the neutron in all of this? It brings stability to matter and sees to it that objects in everyday life do not constantly decay or fuse all around us. Indeed, the neutron plays a crucial role in promoting a delicate balance between the strong nuclear and electromagnetic forces, making sure that neither wins the match by KO. Being electrically neutral, it does not contribute to electrical repulsion, but its presence helps the strong force in maintaining the cohesiveness of the nucleus. The nucleus of the simplest and most abundant chemical element in the universe—hydrogen—is composed of a single proton. Neither the electromagnetic force nor the strong nuclear force comes into play in this case, and the presence of a neutron is, therefore, not necessary. Things are quite different for helium, the next element in the periodic table, which also ranks second in terms of abundance in the universe. Its nucleus includes two protons and two neutrons. Absent the neutrons, the helium nucleus would disintegrate instantly. And that would be the end of helium, the gas that lifts the colorful balloons of little children high in the sky and gives us this funny high-pitched voice. More important, the Sun's heat, responsible for all life on Earth, would no longer come caress our faces. As we know, the fire of stars, including the Sun, comes from innumerable fusion reactions of four protons into a helium nucleus. Should helium ever lose its stability, the stars would stop shining and the universe would turn glacial and desolate. And since we are all made of stardust, we would not be around to commiserate about this sorry state of affairs. Only hydrogen clouds would float about, here and there, in a thoroughly dull and uniform universe. Chemistry and complexity would be absent. The fragrance of roses and the singing of nightingales would be no more. Most important of all, a conscience capable of apprehending the beauty and harmony of the world would not exist.

THE MUSTACHIOED TWIN OF THE PROTON

Nature has thus invented neutrons to prevent matter from flying apart. There still remains one puzzle to be solved. Measurements show that a proton and a neutron have nearly identical masses—the mass of the neutron exceeds that of the proton by a mere 0.14 percent, both being about 2,000 times heavier than an electron. Yet such a coincidence of masses is not essential for neutrons to perform their function of stabilizers of matter. Moreover, the strong force between two neutrons is virtually the same as that between two protons. In other words, a neutron is practically indistinguishable from a proton (to within 0.1 percent), except for the electrical charge. The proton and the neutron are like identical twins, with the difference that one wears a mustache and the other one does not.

In 1932, Werner Heisenberg, of uncertainty principle fame, advanced the notion that this nearly perfect similarity could not be accidental, but resulted from a principle of symmetry pervasive in Nature. As we have seen, symmetry in physics means that the reality of an object remains invariant when it is subjected to various operations, such as a rotation or a mirror reflection. Heisenberg extended the concept of symmetry further. Instead of restricting himself to concrete operations that can be visualized in real space, he asserted that the principle of symmetry applies just as well to totally abstract operations. In this picture, a proton could turn into a neutron, not by means of a concrete operation such as a rotation, but through a change in an abstract quantity called "isospin." Isospin is a property that characterizes particles subject to the strong force, much as electrical charge characterizes particles subject to the electromagnetic force. While one observer might see a proton, another, whose perspective is altered because of a change in isospin, will see a neutron. Since proton and neutron are but two aspects of a single physical reality, it should then come as no surprise that the strong force between two protons is the same as the one between two neutrons. But while a proton and a neutron are symmetrical with respect to the strong nuclear force, they are not with respect to the electromagnetic force. That explains the slight difference in masses between the two particles. What it all means is that Nature loves beauty and symmetry. By using symmetry principles to fashion reality in all its complexity, she ensures that aesthetical principles prevail at the very heart of matter.

A PLETHORA OF PARTICLES

The electron-proton-neutron trio was soon to be joined by hundreds of other new particles. Some came straight from the sky in the form of "cosmic rays," the flux of high-energy particles generated during the violent explosive death of massive stars (called supernovae) in the Milky Way, ultimately reaching the Earth after a long interstellar journey. They were given exotic names, such as "muon," "pion," or "tau." The vast majority of these newcomers, though, are created right here on Earth by sending particle beams accelerated to tremendous speeds smashing into targets of matter in the bowels of huge, monstrous machines. They typically live for only the very briefest of instants, one millionth of a second or less. A good fraction of the Greek alphabet was put to use to name them: rhos, sigmas, thetas, omegas, and others entered the lexicon of particle physicists. Still other particles were predicted by pure thought. We have seen how the English physicist Paul Dirac theorized the existence of the antielectron (or positron) in 1928, ushering in the physics of antimatter, when the equation he had derived to describe the behavior of electrons consistently gave two solutions: one for the electron itself, and the other for a similar particle with an opposite charge. The positron would eventually be discovered in 1932 in cosmic rays. We have also already encountered Wolfgang Pauli, who, in order to explain radioactivity, postulated the existence of the neutrino, a phantom particle interacting so weakly with ordinary matter that two decades were to elapse before it was actually detected experimentally in 1955.

The zoo of particles began to proliferate out of control, and just as zoologists strive to classify the innumerable living species populating the face of Earth, physicists resolved to restore some order in this plethora of particles. To describe this runaway population, they invented increasingly abstract mathematical entities. We are all familiar with the concepts of mass, charge, and spin of an elementary particle. Heisenberg took the first step toward abstraction by introducing the concept of isospin. Soon thereafter, the concepts of parity and strangeness appeared. They were even . . . well, stranger than the things they claimed to describe. Yet, despite this bevy of new concepts, a great confusion seemed to pervade the world of particles until the early 1960s. The situation was reminiscent of the one prevailing at the end of the eighteenth century with the proliferation of new chemical elements. A new Mendeleev was needed to restore some order in this chaos. The American physicist Murray Gell-Mann (1929–) took up the challenge. But whereas chemists had to wait until the discovery of the proton and the electron to understand the arrangement of chemical elements in the periodic table, Gell-Mann did not hesitate to invent whatever particles he needed to establish harmony in the subatomic world.

THE FLAVOR AND COLOR OF QUARKS

According to Gell-Mann, the existence of a certain category of particles could be understood only if they were composed of a fundamental particle, which he named "quark." A great admirer of the language inventiveness of James Joyce (1882–1941), Gell-Mann was drawn to the catchy sound of the phrase "Three quarks for Muster Mark," which appears in Joyce's *Finnegans Wake.* "Quark" appeared to Gell-Mann an eminently appropriate name for his newly invented elementary particle as, in his scheme, protons and neutrons were precisely made of three quarks.

Physicists quickly came to the realization that a single type of quark was not nearly enough to explain the stunning variety of particles populating the subatomic world. They found themselves forced to introduce several others, attributing to each of them two properties that they whimsically dubbed "flavor" and "color." While the term *flavor* was indeed introduced to evoke the various kinds of ice cream we enjoy on a hot summer day, the flavor of a quark actually has nothing in common with ice cream. Quarks are not to be sampled like fine wine. Here, flavor designates an abstract quality that has nothing to do with taste. Likewise, the term *color* does not have the same connotation as in "the red color of poppy fields." Both terms refer to abstract properties, as did the concept of isospin. Physicists might just as well have decided to call them "opulence," "obesity," "courage," or "harmony," although that may not have resonated as deeply with our emotions as the words *flavor* and *color*, which appeal to our senses of taste and vision, thereby providing a link, tenuous as it may be, between our ordinary world and the world of particles. The choice of the word *color* was not entirely arbitrary. It was picked because the rules specifying how quarks of different colors combine to form colorless protons and neutrons are reminiscent of the way the three primary colors mix to produce white light.

The menu at your friendly neighborhood physicist's café features three different families of quarks and a choice of two flavors per family, for a total of six different quarks. The first family includes quarks whose flavor is either "up" or "down." The "up" quark is the lightest, with only $\frac{1}{235}$ the mass of the proton, while the "down" quark has a mass equal to $1/135$ that of the proton (the designations "up" and "down" have no particularly deep significance; they simply reflect where in a table the corresponding flavors are usually listed, the "up" being at the top, and the "down" at the bottom). The second family includes quarks with flavors that are either "strange" or "charm." Both kinds have one-sixth the mass of a proton. The third and last family includes quarks with flavors designated "bottom," with 5.2 times the mass of a proton, and "top," with 170 times the mass of a proton (here again, there is no special meaning associated with the terms *bottom* and *top*, except for their traditional places in a table). Physicists often call the last two flavors "beauty" and "truth," respectively, which further emphasizes the abstract and mathematical nature of a quark's "flavor." Furthermore, for each "flavor," you are entitled—at no extra cost—to choose between three different "colors," which brings the total number of distinct quarks to 18. The available choices are the same as the three primary colors of light— yellow, red, and blue. The quarks of a given "flavor" all have the same mass, regardless of their "color." It is as if you ordered scoops of ice cream for dessert. As long as the size of the scoop remains the same, you can request any color you wish for the same price. The cost of the coloring ingredients is constant. The fact that quarks of different "colors" have the same mass is another manifestation of Nature's penchant for symmetry. Quarks of a given "flavor" but different "colors" correspond to the same physical reality viewed from different perspectives.

Finally, each quark possesses a fractional electrical charge determined by its "flavor." In particular, the "up," "charm," and "top" quarks all have a positive charge equal to $2/3$ that of a proton, while the "down," "strange," and "bottom" quarks have a negative charge equal to $1/3$ that of an electron. Quarks in a given family can change "flavor." For instance, an "up" quark can mutate into a "down" quark, and vice versa. They can also change family, although with somewhat greater difficulty. But if they do so, they can only mutate into quarks with a different electrical charge. For instance, an "up" quark with charge $2/3$ can transform into a "strange" quark with charge $-1/3$, but the mutation from "up" to "charm" is forbidden because the charge is $2/3$ in both cases.

Ordinary matter—your body, the book you hold in your hands, the sofa you are sitting on, the walls of your living room, etc. —is made of quarks of two flavors only: "up" and "down." That is so because all ordinary matter is made of protons and neutrons. Furthermore, a proton results from the combination of two "up" quarks and one "down" quark, which gives it a positive charge equal in magnitude and of opposite sign to that of the electron. A neutron, by contrast, is made of two "down" quarks and one "up" quark, resulting in zero electrical charge. Protons and neutrons have no "color." The primary "colors" of their three quarks combine so as to cancel out, somewhat the way the primary colors of light blend into white. The other "flavors" of quarks are involved in more exotic matter found only in cosmic rays and in particle accelerators. Protons and neutrons, as well as some other particles also made of quarks, belong in the family of "hadrons" (meaning "strong" in Greek), which includes all particles subject to the strong nuclear force.

THE LEPTON FAMILY

The other fundamental constituent of ordinary matter is the electron. It belongs in the family of "leptons" (meaning "weak" in Greek), which includes all the particles on which the strong nuclear force has no effect. As was the case for quarks, Nature offers a menu of six lepton "flavors," also arranged in three families. You have a choice between the ordinary electron and its sidekick the *electron neutrino* (we have already encountered it—it is often called simply "neutrino"), which form the first family. Next come the *muon* and its counterpart the *muon neutrino*, which form the second family. Finally, the *tau* particle and its partner the *tau neutrino* constitute the third family. Once again, leptons of different "flavors" have different masses. The electron is the lightest of the bunch, its mass being only $1/1836$ that of the proton. Next comes the muon, whose mass is intermediate, at $1/9$ that of the proton. The heavyweight record is held by the tau particle, weighing in at 1.9 times the mass of the proton. As for the three types of neutrinos, their mass is either extremely small or zero. The mass of the electron neutrino, for example, is only one-millionth that of the electron. Both the muon and the tau particle have the same electrical charge as the electron. All three neutrinos, on the other hand, are electrically neutral.

While electrons last forever—fortunately for the stability of matter around us and luckily for us—the muon and tau particles have extremely short lifetimes, on the order of a millionth of a second or less. As a result, they do not play much of a role in everyday life, and they show up only in cosmic rays and in particle accelerators. In distinct contrast to quarks, which are not overly picky about whom they associate with, there is a rigid demarcation line between the three families of leptons. The two members of a given family stay to themselves and never fraternize with others. Thus, an electron can transform into an electron neutrino, and vice versa, or a tau particle into a tau neutrino, but do not expect to see an electron changing into a muon neutrino or a tau neutrino.

QUARKS ARE NEVER FREE

Matter is made of quarks and leptons. But while all six lepton "flavors" have been discovered in Nature or detected in cosmic rays or particle accelerators, the same cannot be said of quarks, which have never deigned to reveal their existence directly to us. Physicists cannot be blamed for not trying hard enough to bring them out into the open. As early as 1968, they used the 3.2-kilometer-long Stanford University linear accelerator (or SLAC) to send extremely high-energy particle beams smashing into protons, hoping to shatter them and liberate quarks. It did not work, despite the fact that the way the particles bounced off the protons confirmed that they indeed had a substructure composed of three pointlike entities. Why, then, did we fail to free the quarks from their prison? The answer lies in their bizarre properties. When two quarks are very close to each other, they completely ignore each other and act as if the other does not exist. They experience no force whatsoever. It is as though they are completely free and the strong nuclear force had vanished altogether (indeed, physicists refer to this situation as "asymptotic freedom"). But as soon as quarks move apart, the opposite occurs: The strong force manifests itself again and they become strongly attracted to each other. This behavior is reminiscent of that of some lovers or spouses: As soon as they are apart, they profess their undying love for each other and cannot wait until they reunite; but the moment they are back together, their amorous fervor turns to indifference and they barely speak to each other.

The way the strong nuclear force acts on quarks is most peculiar and counterintuitive. Its behavior is markedly different from other interactions. The intensity of the electromagnetic force, for instance, decreases when the distance between two electrons increases. When two magnets are moved apart, they interact less and less. Likewise, the force of gravity decreases as the square of the distance between two masses. How, then, can we understand this bizarre behavior of the strong force? It is helpful to picture quarks as being tied together by a string. When they crowd in together, the string is loose and the quarks experience no force. When they drift apart, however, the string becomes taut and prevents them from moving freely. Suppose we pull on the string harder and harder. Our intuition tells us that the string will eventually break and the quarks will be freed. But that ignores the fact that by pulling on the string, we give it energy. At some point, that energy will exceed the mass energy of a quark-antiquark pair (the energy in question is equal to the mass of the pair times the square of the speed of light, as Einstein taught us). When the string breaks, what happens is not the liberation of individual quarks, but, rather, the creation of quark-antiquark pairs bound together in the form of particles called "mesons" (Figure 58). So quarks can never be seen in isolation. We will never be able to see one in all its glorious individuality. Trying to extract a quark from a proton is like trying to isolate the north pole of a magnet: You can always cut a magnet in two, and you will end up not with isolated north and south poles, but with two separate smaller magnets, each complete with its own north and south poles.

PHANTOM PHOTON MESSENGERS

The world is no longer simply a huge collection of inert and isolated particles subject to mechanistic and deterministic laws, the way Newton and Laplace had imagined. Quantum mechanics tells us that localized reality has no meaning and that these particles are part of a whole. The universe is unified into a vast network of connections and interactions. The idea of a non-localized reality is not exactly new. It had already surfaced in the nineteenth

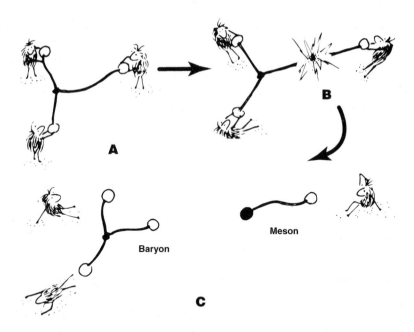

Figure 58. *The captivity of quarks.* Physicists believe that quarks are the most fundamental build-ing blocks of matter. Yet no one has ever seen a quark in a free state. To understand why that is, consider a baryon (a proton or a neutron) made of three quarks. These are bound together by the strong nuclear force depicted by strings in drawing *A*. Suppose that, like the cartoon characters in drawing *B*, we try to free the quarks by pulling on the strings. One of the strings breaks, releasing a considerable amount of energy. As shown in drawing *C*, the quark attached to the ruptured piece of string does not gain its freedom. Because of the equivalence of energy and matter, the released energy gives birth to a quark-antiquark pair (the antiquark is depicted in black in the drawing). The newly produced quark replaces the one that had just separated to reconstitute the original baryon, while the antiquark combines with the lone quark to form a new particle called a "meson." In other words, by attempting to release a quark, the only thing the cartoon characters have managed to do is produce a meson. Quarks will never be seen free.

century with Michael Faraday. As we have seen, in order to explain the action at a distance of electric and magnetic forces, Faraday had envisioned immense fields of line forces emanating from an electrical charge or a mag-netic pole and stretching across space. An electron is a material point resid-ing at the center of a vast halo of invisible electromagnetic energy. As a sec-ond electron approaches, it senses the influence of the electromagnetic field, and will experience a force of repulsion. It is as though the first elec-tron had sent to its companion a message by way of the field to tell it: "Go away!" Maxwell believed that the message was transmitted by electromag-netic waves propagating at the speed of light through the field, much as waves propagate on the surface of the ocean.

In the twentieth century, physicists picked up on Faraday's notion of field and incorporated it into quantum mechanics to construct a field the-ory called "quantum electrodynamics." This theory has been extremely suc-cessful in explaining the behavior of subatomic particles and appears to conform remarkably well to the way Nature acts. It says that the information is transmitted not by waves, as Maxwell believed, but by discrete particles of light, called "photons." Picture one electron approaching another. Its trajectory is going to be deflected as the electron is being repelled. As far as Maxwell was concerned, the repulsion of two electrical charges of like sign was due to the electromagnetic force. A modern physicist, on the other hand, would describe the same process as the exchange of a messenger photon: A photon is emitted by the first electron and absorbed by the second. It is as though the first electron had fired a cannonball and recoiled under the effect of the shot, while the second one was knocked off course by the impact of the projectile. Electrons can also exchange two or more photons, rather than just one, although this type of interaction is less probable and not nearly as important.

Photons are very handy messengers, but where do they come from? They require energy in order to exist. The question is: What is the source of this energy? That is where the quantum fuzziness of energy comes in. Thanks to Heisenberg's uncertainty principle, photons can borrow energy from the "Bank of Nature" in order to materialize. However, the loan is subject to very strict conditions imposed by the uncertainty principle. The more energy borrowed, the shorter the loan. Nature demands to be promptly repaid in full. As a result, the photon returns its energy and disappears after an infinitesimally short time by being reabsorbed by another electron. Thus, ephemeral photons (they are called "virtual") come and go in a frenzied cycle of births and deaths. Quantum uncertainty sees to it that every electron is surrounded by a throng of virtual photons, like a swarm of bees buzzing around its hive. Photons that are closest to an electron have the most energy, because it takes them very little time to pay off the energy loan and be reabsorbed. Those that are farther out do not benefit from as large a loan, since it takes them longer to refund the bank. In this picture, electrons are immersed in an ocean seething with virtual photons, that ocean being none other than the electromagnetic field postulated by Faraday. Whenever a second electron wanders into this bubbling effervescence, it can absorb one of the many photons teeming around the first. The absorbed photon acts as a messenger and tells the second electron what to do; thus it is that we see it deviate from its original path. Messenger photons are condemned to remain forever in a virtual state; we will never be able to see them because their lifetime is far too short. But their existence is indispensable for the theory of quantum electrodynamics to account for how Nature behaves.

THE STRONG FORCE AND GLUONS

Now that we know that the electromagnetic force is transmitted by virtual photons, what about the other forces? Besides the electromagnetic force, the universe is governed by three other forces that we have already encountered: the strong nuclear force, which binds protons and neutrons, forming atomic nuclei; the weak nuclear force, which is responsible for the decay of atomic nuclei (the phenomenon of radioactivity); and lastly, the gravitational force that keeps us firmly planted on Earth, maintains the planets in orbit around the Sun, gathers the one hundred billion or so stars in the Milky Way, and prevents galaxy clusters from dispersing away. Is it possible that these three other forces are also transmitted by messenger particles? The answer is yes for the strong and weak nuclear forces. The jury is still out as far as the gravitational force is concerned.

The strong nuclear force is believed to be transmitted by eight messenger particles called "gluons." The number eight keeps recurring in particle physics. It results from a particular symmetry principle in the subatomic world (the American physicist Murray Gell-Mann called it the "Eightfold Way," in reference to the eight paths advocated by Buddhism to attain Nirvana: right comprehension, right thought, right speech, right action, right living, right effort, right intention, and right concentration). A total of eight gluons is required to account for the different "flavors" of quarks. To account for the quark colors, each gluon, like a quark, is assigned one of three primary "colors"—blue, red, and yellow. Gluons keep shuttling back and forth between quarks; they are emitted by one quark, only to be absorbed by another, like tireless diplomats constantly carrying messages from one head of state to another, or like a go-between of yesteryear endlessly relaying billets-doux between lovers. Thanks to this nonstop to-and-fro hopping, gluons hold quarks captive within protons and neutrons as in a prison, and act effectively as the glue confining protons and neutrons inside atomic nuclei. Like quarks, gluons are prisoners for life and will never be set free. And like photons—the messengers mediating the electromagnetic force—gluons have neither mass nor electrical charge and travel at the speed of light. Nature may be made of quarks and electrons, but she relies on gluons and photons as messenger particles. Without photons, we would not be able to communicate with distant stars and galaxies and receive their message of splendor, harmony, and universality. Without gluons, the fragrance of roses and the glorious sight of sunsets would be absent from our lives.

THE WEAK FORCE AND PHOTOGRAPHIC PLATES

Thus, the electromagnetic force and the strong nuclear force both have their own messenger particles. How about the weak nuclear force? It made its debut on the physics stage in 1896, when the French physicist Henri Becquerel (1852–1908) discovered serendipitously that a photographic plate fogged up in the presence of a sample of uranium. Marie Curie (1867–1934) and her husband Pierre (1859–1906) demonstrated that the intensity of the effect did not diminish with time; they further showed that another element—radium—had the ability to heat up and vaporize an entire block of ice without noticeably changing appearance. Evidently, uranium and radium were emitting a new type of radiation produced by some force that could be neither gravitational nor electromagnetic. Scientists soon recognized that these two chemical elements—uranium and radium—were radioactive substances containing unstable nuclei. By rearranging their internal structure, these nuclei emitted high-energy particles that had the ability to fog up photographic plates and vaporize ice blocks. Physicists went on to use those radioactive substances as sources of high-energy particles for bombarding matter and studying its structure, before the advent of large accelerators.

The weak force is mediated by messenger particles, just as the other forces are. There are three such particles, which have been given the rather unpoetic names of W^+, W^- (the first letter in the word *weak*), and Z. Unlike gluons, forever imprisoned out of sight, the three particles in question have been detected in 1984 in the huge accelerator at CERN (European Center for Nuclear Research). The weak force is weak because its range is extremely short, even more so than that of the strong force, which is itself quite limited. In fact, so short-range is the weak force that two particles are highly unlikely to ever come close enough to each other to experience it. That is why this particular force is unable to bind particles together and act as a glue, unlike the other forces. All it can do is render some particles short-lived. In particular, it is responsible for the instability of all quarks, with the exception of the lightest one—the "up" quark—as well as of all leptons carrying an electrical charge, here again sparing the lightest one, the electron. As a result, quarks and leptons end up decaying to adopt the most stable configurations possible—namely, those of "up" quarks, electrons, and neutrinos. Neutrinos owe their stability to the fact that, even though they are leptons, they carry no electrical charge. Virtually all particles are subject to the weak force. Photons and gluons are the only exceptions.

The infinitesimally small range of the weak force (one hundred thousand-billionth of a centimeter) compared to that of the electromagnetic force is related to the heavy mass of its associated messenger particles. W and Z particles are 85 and 95 times, respectively, heavier than a proton. Unlike photons, which have no mass and can therefore roam around all over space with great ease, W and Z particles are so massive that they never get a chance to venture very far away from where they were born. The weak nuclear force rarely manifests its presence in everyday life. We do not run into radioactive materials around every corner (fortunately for our health, because the high-energy particles they emit are hazardous to living cells and can induce genetic defects). Nevertheless, the weak nuclear force is responsible for our very existence. The reason is that the nuclear reactions that fuel the fire in the interior of stars, particularly the Sun, could not have gotten started without the participation of the weak force.

THE IMPROBABLE BUT REAL KINSHIP BETWEEN THE WEAK AND ELECTROMAGNETIC FORCES

At first sight, the weak force and the electromagnetic force have absolutely nothing in common. The former has an infinitesimally short range, much smaller than the size of an atomic nucleus, while the latter has a range that is unlimited. Photons, the messengers of the electromagnetic force, are as light as particles can get; with no mass at all, they zip around the universe at the greatest possible speed—the speed of light. By contrast, the W and Z particles, which carry the weak force, are bogged down by their heavy weight and can barely move. Yet there are strong similarities between photons and W and Z particles, particularly with regard to their spins. All particles have a spin. They rotate about themselves like tops. However, they cannot rotate any way they please. Quantum mechanics imposes some very strict rules on them. For instance, particles called "bosons" (in honor of the Indian physicist Satyendranath Bose [1894-1974]) have a spin restricted to an integral multiple (0, 1, 2, . . .) of a quantity equal to Planck's constant. Photons and W and Z particles all belong in the family of bosons with spin 1. While all messenger particles transmitting interactions are bosons, material particles (electrons, leptons, quarks) belong in another family known as "fermions" (in honor of the Italian physicist Enrico Fermi [1901-1954]). Particles in that family have a spin that can only be a half-integer multiple ($\frac{1}{2}$, $\frac{3}{2}$, $\frac{5}{2}$, . . .) of Planck's constant.

That photons and W and Z particles all have a spin of 1 confers on them a curious family resemblance, despite their obvious differences. It is as though we were dealing with two people with remarkably similar facial traits, except that one is tall and skinny as a pole, while the other breaks all records of weight and obesity. Deep down, could there be after all a kinship between the messenger particles of the weak and electromagnetic forces? The Pakistani Abdus Salam (1926-1996) and the American Steven Weinberg (1933-) were able to show that indeed there is. In 1967, they succeeded in unifying the two forces. To accomplish that feat, they invoked yet another symmetry principle. We have seen that the neutron is like the "mustachioed" twin of the proton: By modifying an abstract property called isospin, a proton can mutate into a neutron. Along the same line, Salam and Weinberg demonstrated that photons and W and Z particles are like identical triplets that were separated at birth and retained only a distant memory of their common origin. They no longer display any evident family resemblance, because we now live in a very low-energy universe that has become too cold.

It is now believed that some 15 billion years ago, a gigantic explosion—the big bang—gave birth to the universe, time, and space. Since then, a relentless march toward complexity has taken place. Out of a subatomic vacuum successively emerged quarks, electrons, protons and neutrons, atoms, stars, and galaxies. An immense cosmic tapestry was woven, composed of hundreds of billions of galaxies, each made of hundreds of billions of stars. In the suburbs of one such galaxy, on a planet close to a star, man appeared, capable of marveling at the beauty and harmony of the cosmos and endowed with a conscience and intelligence enabling him to comprehend the universe. As the universe aged, it expanded, creating more and more space between galaxies, all the while cooling down. The cosmic background radiation, which constitutes the remnants of the fiery heat of the creation, now stands at a temperature no higher than a frigid -270°C! Yet, early on, just 10^{-43} second after the big bang, the universe reached the unfathomable temperature of 10^{32} degrees. When the cosmic clock hit the one-millionth of a millionth (10^{-12}) of a second mark after the primeval explosion, the temperature of the universe was still millions of billions of degrees. During these first few fractions of a second, everything was seething in a primordial soup of elementary particles, including W and Z particles. Extreme temperatures mean enormous energies and particles moving at breakneck speed. The kinetic energy of the W and Z particles far exceeded their mass energy, which for all practical purposes was then negligible. Under these conditions, the W and Z particles behaved as though they had no more mass than photons. Thus, in a high-energy environment such as what prevailed in the primordial universe, the W and Z particles were truly the siblings of photons. "We are clearly part of the same family," they gush today to the photon. "We are just as efficient messengers as you are. It is just that at low energy, in a universe that has cooled off so considerably, we are now very much hindered by our mass and can no longer cover as much territory as you to deliver our messages."

By using abstract symmetry principles, Salam and Weinberg were able to show that at the very beginning of the universe—when its size had not yet reached one-thousandth of the diameter of a proton (10^{-16} cm) and the ambient energy was hundreds of times the proton's mass energy—the electromagnetic force and the weak nuclear force were one and the same; they were united into what has come to be known as the "electroweak" force. Only after the 10^{-12}-second mark did the kinetic energy of W's and Z's diminish sufficiently for their mass to start hampering their movements; at that point the union was broken, and the two types of forces went their separate ways.

THE GRAND UNIFICATION

Thanks to Salam and Weinberg, photons and W and Z particles rediscovered one another in a most unlikely reunion. After the emotion of the moment subsided, physicists turned their attention to the eight gluons, messengers of the strong nuclear force. Could there also be some unsuspected family links between them and the other messenger particles? If the unification of two forces had succeeded, why not three? After all, all these messenger particles belong in the same family of the twelve spin-1 bosons. "Impossible!" clamored the eight gluons in unison. "We have nothing in common with the other messengers. We are strong and powerful, whereas the others are puny and impotent!" The argument is certainly valid in a cooled-off universe, in which energies are low; but it does not hold in a hot and high-energy universe. The higher the energy, the more likely it is for particles moving at high velocity to get close to one another. We have already pointed out that when quarks (whose messengers are gluons) get packed like sardines, they become virtually free (remember the asymptotic freedom argument); the strong force becomes so small as to be practically negligible. The electromagnetic force, on the other hand, behaves in exactly the opposite way: The smaller the distance between particles, the greater the intensity of the force. Now, if one force increases and the other decreases with the energy of the universe, it is reasonable to expect that there is a particular energy where the two become equal. In other words, increasing the energy of the universe gives the electromagnetic force, which in our current universe is roughly 100 times weaker than the strong nuclear force, a chance to catch up to the latter and become its equal.

The key question is whether the energy where the electromagnetic and strong nuclear forces become equal is the same as that where the electromagnetic and weak nuclear forces merge into one. Only then would the "grand unification" be possible—"grand unification" is what physicists call the unification of the three forces, to distinguish it from the simpler "electroweak" synthesis. Nature has to arrange a triangular rendezvous. The situation is analogous to the one faced by three mountain climbers (Figure 59). The first, named "Strong," comes down the mountain, while the other two, named "Weak" and "Electromagnetic," go up. They must adjust their respective pace so as to arrive at an agreed-upon meeting spot somewhere between the base and the summit at exactly the same time. "Weak," who starts from the very bottom, must hurry if he is to catch up to "Electromagnetic," who

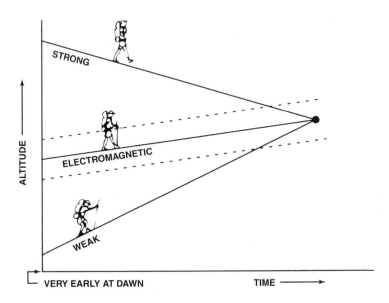

Figure 59. *A synchronized rendezvous.* The unification of the electromagnetic force and the two nuclear forces at some past temperature and epoch of the universe can be compared to the situation faced by three climbers, each starting from a different altitude on a mountain, and who have to meet at the same place and time. The climbers named "Electromagnetic" and "Weak" must go up (just as the two corresponding forces must become more intense as we go back into the universe's past), while the climber named "Strong" must come down (just as the strong force must diminish in intensity in the past).

begins his ascent from a higher location. "Electromagnetic" should not climb too fast, and "Strong" should not race down the mountain, else they would meet long before "Weak" can arrive. The synchronization required for a successful outcome is delicate and tricky, but it seems that Nature managed to do it. The relative intensities of the three forces (corresponding to the starting altitudes of our three mountain climbers) appear just right for the "grand unification" to occur. The three forces become one when we go back in time to the 10^{-35}-second mark, when the energy of the universe was a hundred thousand billion (10^{14}) times the mass energy of a proton, and its size was only one-thousandth of a billion-billion-billionth (10^{-30}) of a centimeter.

The lightweight photon, which once despaired of being an only child, suddenly finds himself with a brood of eleven brothers and sisters. When he first got to meet the two W particles and the Z, he had no idea that he had anything in common with such chubby characters. Then he found out that they, too, used to be massless once upon a time, when the energy of the universe was very high; he was their long-lost brother after all. Next came the eight gluons, so completely different by virtue of their strength and power. But it turned out that in an even higher-energy universe, gluons would lose their strength, while the photon, the W's, and the Z would gain some. The upshot is that at the very beginning, everybody started out as one and the same.

THE INFINITESIMALLY SMALL GIVES BIRTH
TO THE INFINITELY LARGE

Is it at all possible to ever verify the grand unification theory? Particle accelerators are called in for help. Such accelerators actually amount to giant microscopes. Physicists use them to bombard matter with beams of extremely high-energy particles, shattering the target into a thousand pieces. The higher the energy, the deeper physicists can probe into the heart of matter and scrutinize infinitesimally small dimensions. The unification of electromagnetic and weak forces occurs at a dimension of one-tenth of a million-billionth (10^{-16}) centimeter. The largest accelerators in existence today can reach energies high enough to probe that small a scale. That is how the accelerator at CERN was able to confirm the unification of those two forces by producing evidence of W and Z particles with masses in agreement with theoretical predictions. But the grand unification—the synthesis of the electromagnetic, weak, and strong nuclear forces—does not occur before the infinitesimal dimension of 10^{-30} centimeter is reached. This is 100,000 billion times smaller than the scale of the previous unification. We are not about to construct accelerators that can reach the phenomenal energies required to explore such tiny dimensions, particularly in these times of rampant unemployment and budget deficits. Not to mention that, barring a giant technological leap, such an accelerator would have to be larger than the diameter of Earth; it would actually stretch all the way to the nearest star, which is not especially practical. Faced with this situation, physicists have turned to the mother of all particle accelerators—the universe itself.

One of the most profound scientific developments in the last twenty years is without a doubt the marriage of particle physics and cosmology. Our immense universe started out in an unimaginably small, hot, and dense state. The temperatures and densities were so extreme that there is no chance to ever duplicate them here on Earth; they can exist only in the imagination of scientists. The ultimate experiment was conducted once and for all 15 billion years ago, and the best we can do now is contemplate the universe and try to reconstruct its history. The infinitesimally small has given birth to the infinitely large. It thus becomes quite natural for astrophysicists to join forces with particle physicists to decipher the secret melody of the universe. This alliance is all the more beneficial since astrophysicists cannot go back all the way to the origin of the universe with their instruments. They explore the universe's past with telescopes, which are effectively time machines. Since light takes a finite amount of time to reach us, we always see things with a certain time delay. We see the Moon the way it was a little more than a second ago, the Sun as it appeared eight minutes earlier, the nearest star as it was four years ago, the galaxy nearest to the Milky Way as it looked 2 million years ago, and so on. Seeing far is seeing early. If so, is it conceivable that by building ever-larger and more powerful telescopes we may someday "see" the moment of creation? The answer is no. The universe was so dense and hot during the first 300,000 years of its existence that light could not propagate then and the cosmos was completely opaque. This constitutes a natural barrier that no telescope will ever be able to overcome. The earliest image of the universe, taken by the COBE satellite in 1991, dates back to when it was already 300,000 years old.

Is there a way to get past this wall of opacity? While astronomers are helpless, particle physicists are able to come to the rescue. From the time of the primordial explosion to 300,000 years after the big bang, all matter in the hot and dense universe existed in the form of extremely high-energy particles. Indeed, the early universe was nothing but a giant particle accelerator. That is the reason why accelerators, such as the ring of several tens of kilometers in circumference at CERN, in Geneva, Switzerland, take over from telescopes and enable us to go back in time and study the history of the universe up to about 10^{-12} second after the big bang. Going back still further would require even larger and costlier machines, an unlikely prospect. To explore times between the primordial explosion and the first thousandth of a millionth of a second, we have no choice but to resort to the previously described theory, which physicists refer to as the "standard model."

Our ability to answer some of the most pressing and fundamental questions posed by modern cosmology will hinge on our ability to understand events as they unfolded during this first fraction of a second in the life of the universe. Why is the universe made primarily of matter, rather than of equal quantities of matter and antimatter? How did seeds of galaxies form and grow to produce the vast cosmic tapestry we see today? There is also the problem of the dark mass of the universe. It appears that we live in an iceberglike universe; 90 percent to 98 percent of its mass is not directly observable with our telescopes, because it does not emit any radiation. Yet, as we have already pointed out, there is a fundamental difference between an iceberg and the universe: We know perfectly well what the submerged part of an iceberg is made of, while the nature of the invisible mass remains a vexing enigma. While the simplest "standard model" describes ordinary matter (protons, neutrons, electrons), it also "predicts" the existence of massive particles, born in the first instants of the universe, and whose nature is completely different from the type of matter we are made of. Perhaps one of those will turn out to constitute this mysterious invisible mass? If so, Copernicus's ghost would strike yet again. Not only do we not occupy the center of the cosmos, not only do we represent a mere eyeblink in its history, but we may even turn out not to be made of the same matter as most of the rest of the universe.

Another problem has to do with the extremely precise fine-tuning of the universe. The evolution of the cosmos is determined by initial conditions (such as the initial rate of expansion and the initial mass of matter), as well as by fifteen or so numbers called physical constants (such as the speed of light and the mass of the electron). We have by now measured these physical constants with extremely high precision, but we have failed to come up with any theory explaining why they have their particular values. One of the most surprising discoveries of modern cosmology is the realization that the initial conditions and physical constants of the universe had to be adjusted with exquisite precision if they are to allow the emergence of conscious observers. This realization is referred to as the "anthropic principle" (from the Greek word *anthropos*, meaning "man"). Change the initial conditions and physical constants ever so slightly, and the universe would be empty and sterile; we would not be around to discuss it. The precision of this fine-tuning is nothing short of stunning: The initial rate of expansion of the universe, to take just one example, had to have been tweaked to a precision comparable to that of an archer trying to land an arrow in a 1-square-centimeter target located on the fringes of the universe, 15 billion light-years away! Will the standard model someday mature into a theory capable of explaining everything, including the values of physical constants and the initial conditions of the universe?

THE DEATH OF PROTONS

There is another way to test the grand unification theory without resorting to monster particle accelerators. It has to do with the death of protons. The "grand unification force," resulting from the unification of the electromagnetic and nuclear forces, has a messenger particle of its own, designated somewhat cryptically "X." Like any other messenger particle, the X comes into existence by borrowing energy from the ubiquitous "Bank of Nature." We are talking here about an unusually heavy particle, having a mass of one-tenth of a billionth of gram, which is 100,000 billion (10^{14}) times more massive than a proton. That means it can have only a very short existence (it is expected to return its enormous energy loan promptly to the bank), and its range can be only infinitesimally short. During its ephemeral appearance, it can travel no farther than 10^{-25} centimeter, a thousand billion times less than the diameter of a proton. Given the fact that there are only three quarks within the vast ballroom that is a proton, the likelihood that an X particle will be absorbed by a quark other than the one that generated it in the first place is exceedingly minute. In order for this to happen, two quarks would have to come to within less than 10^{-25} centimeter of each other. The probability of such an encounter is minuscule. It is about the same as the probability of two flies, out of a population of three, meeting each other by chance in a 10-million-kilometer-long hangar, a distance stretching over one quarter of the distance from Earth to Venus. Unlikely as the event may be, when it does happen, an X particle is exchanged between two quarks, with catastrophic consequences for the proton. The two quarks involved transform into an antiquark and a positron (or antielectron). The positron is ejected out of the proton, while the antiquark combines with the third quark to form a new particle, called a "pion." The pion is itself unstable and decays in a fraction of a second, producing photons. In short, when two quarks encounter each other inside a proton and exchange an X particle, the net result is the death of the proton, leaving behind one positron and several photons. Thus, protons do not last forever; they will die someday. No need to panic, though: The protons in your body are not on the verge of disintegrating before your eyes, and you are in no immediate danger of transforming into positrons and photons. The grand unification theory has yet to be verified experimentally; encounters of the type just discussed are so rare that the lifetime of a proton is estimated to be 10^{30} years, or hundreds of billion billion (10^{20}) times the age of the universe!

Does such a very long life expectancy mean that there is no chance whatsoever of observing a proton-decay event and verify the grand unification theory? Not at all. The life expectancy is only an average number. Quantum theory tells us that an individual proton can decay actually at any time. Since the proton lifetime is 10^{30} years, if we were to gather together 10^{30} protons in one place, we should expect to see one death per year on average. Better yet, round up even more protons, say 10^{33}, and several decay events should take place on average every day. To that end, American physicists have filled a large tank, 20 meters to the side, with 5,000 tons of water (water is an excellent source of protons, since each water molecule is composed of two atoms of hydrogen—a hydrogen atom includes one proton and one electron—and one atom of oxygen). They placed the tank deep underground in an abandoned salt mine in Ohio, in an effort to filter out cosmic rays that might trigger reactions mimicking real proton decays. The water must be extremely pure and clear so that photons signaling the death of protons can travel across several meters and be detected by an array of photomultipliers lining the walls of the tank. (If ocean water had the same purity and clarity, we would have no difficulty admiring the marvelous marine animal and plant life 20 meters deep.) Unfortunately, despite valiant efforts (there have been similar experiments in Japan and Europe), no one has ever recorded a proton decay. Could it be that protons are more stable than previously thought? Is it necessary to modify the grand unification theory? Physicists have not yet conceded defeat. The Japanese and the Italians are feverishly building detectors one hundred times more sensitive in the hope of catching at least one proton in the act of decaying.

GRAVITY KEEPS TO ITSELF

While the electromagnetic, weak, and strong forces have allowed themselves to be unified, the fourth character in the play—gravity—has steadfastly resisted: It is not joining in. Quantum mechanics is perfectly adequate for describing the subatomic world, in which gravity hardly matters; and relativity is well suited to account for the properties of gravity on the cosmic scale of planets, stars, galaxies, and even the entire universe, where nuclear and electromagnetic forces do not occupy center stage. But no one has been able thus far to unify the theories of the infinitely large and of the infinitesimally small into a single theory of "quantum gravity," which would describe the situation where all the forces are treated on an equal footing and, furthermore, which can be verified experimentally. This is in spite of sustained efforts on the part of physicists. The great Einstein himself spent the last twenty-five years of his life trying to unify gravity with electromagnetism, without success (because of his continuing opposition to quantum mechanics, Einstein never concerned himself with the two nuclear forces).

The problem is formidable. According to the standard model, as we have seen, the electromagnetic and nuclear forces are transmitted by messenger particles. It therefore seems logical that the same should be true of the fourth force. Quantum mechanics does predict that the force of gravitation has a messenger particle called a "graviton." Like the photon, the graviton would belong in the family of bosons, have no mass or electrical charge, and race across the universe at the speed of light. But there the resemblance would end. The graviton would revolve about itself twice as fast as the photon, having a spin of 2, rather than 1. While photons are extremely gregarious and just love to interact and toy with matter in order to convey to our eyes the color and beauty of the things of life, the gravitons are rather antisocial and limit their interactions with material particles to a bare minimum. They would go right through your body, indeed through the entire Earth, as if there were nothing to it, giving up less than 1 percent of their energy. That is why it is so difficult to detect gravitational waves with our sensors made of ordinary matter.

There is, however, an important fact we left out of this great story: No one has ever seen a graviton. For the time being, it is a purely mathematical entity existing only in the imagination of physicists.

GEOMETRIZING THE ELECTROMAGNETIC FORCE

As we said earlier, quantum mechanics requires that the force of gravity be transmitted by a messenger particle called graviton. The theory of general relativity, on the other hand, makes no such demand. It describes gravity not as a force, but as a consequence of the geometry of space. The Moon traces out an elliptical orbit around Earth. Newton explained that it is so because Earth exerts a gravitational force on the Moon. Einstein, however, asserted that the mass of Earth warps the space around it, and the Moon simply picks the shortest path in this curved space to go from one point to another; this path is called a "geodesic" and happens to be shaped like an ellipse. The fact that space is warped by a mass has been verified countless times. We have seen, for instance, that the light coming from a star is deflected in the curved space around the Sun.

Thus, force is synonymous with geometry in the theory of relativity. To unify quantum mechanics and relativity theory so as to meld the four forces into a single "superforce" in a scenario that physicists refer to, perhaps with a bit of excessive grandiloquence, as the "theory of everything," is no trivial task. It would require reconciling two points of view that a priori have nothing to do with each other: a force of gravity transmitted by messenger particles on the one hand, and a force generated by the geometry of space itself, on the other.

The first step toward such a reconciliation was taken as early as 1919, just four years after the publication of the theory of general relativity, by a little-known Polish physicist named Theodor Kaluza (1885–1954). Only the gravitational and electromagnetic forces were of concern back then; the two nuclear forces had yet to make their appearance onto the stage. Impressed by the ability of geometry to account for gravity in Einstein's theory, Kaluza resolved to "geometrize" the electromagnetic force as well. To do so, he proposed a most bizarre and unexpected solution. He managed to prove that electromagnetism too was due to the curved geometry of space, although it involved not our familiar three-dimensional space but, rather, a hypothetical space with an additional fourth dimension. It has been known since Einstein that we live in a space-time continuum with four dimensions—three dimensions for space, and a fourth for time. Kaluza did not pull any punches: He boldly assigned an additional dimension to space, thereby creating a space-time with a total of five dimensions. According to him, gravity in such an over-dimensioned space-time would be equivalent to ordinary gravity in a conventional four-dimensional space-time; but as a bonus—and that is the crucial point—it would encompass electromagnetism as well. In such a five-dimensional space-time, the radio waves carrying your favorite tunes to your receiver and the light waves that bring you daylight would all be propagating in the fourth spatial dimension.

THE HIDDEN DIMENSIONS OF SPACE

You may argue that this is all very well and that Kaluza's idea is quite clever, but it cannot possibly be right because there is no evidence whatsoever for this additional space dimension he is talking about. You can go forward or backward, up or down, and right or left. That's three dimensions, not four! Is there really another direction for us to move in? If indeed space has a fourth dimension, where on earth does it hide?

The answer was provided in 1926 by the Swedish physicist Oskar Klein (1894–1977). He proposed that the fourth dimension appears missing because it folded back on itself and shrank to such a small size that it became imperceptible. Picture a water hose (Figure 60). From far away, it looks like a skinny wiggly line; only when you approach it does it begin to reveal its cylindrical shape. What from a distance looks like a point with one dimension becomes a two-dimensional circle from up close, ringing the cylindrical surface of the hose. Similarly, said Klein, what we take to be a point in our normal three-dimensional space is actually a tiny circle in a fourth dimension. We do not notice it because the circle is infinitesimally

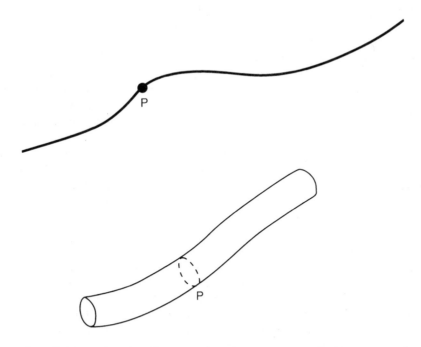

Figure 60. *Hidden dimensions.* What appears from afar to be a one-dimensional line can turn out to be a two-dimensional cylindrical tube when viewed up close. Each point on the line is in reality a tiny circle on the surface of the tube. One dimension is hidden because of the very minute size of the circle. Likewise, the physicist Oskar Klein proposed that what appears to us as a point in three-dimensional space is actually a very small circle in a space with four dimensions, the fourth spatial dimension being hidden from us.

small: Its circumference is a mere 10^{-33} centimeter, which is 100 billion billion times less than the size of an atomic nucleus. So much larger is an atom that there is no possibility for it to squeeze through this fourth spatial dimension and reappear in another corner of the universe. Science-fiction writers are fond of this type of stratagem because it enables their characters to move from one point to another in a three-dimensional universe without being limited by the speed of light, a tortoiselike speed on the scale of the universe. Needless to say, such travel in extra dimensions of the universe remains pure fantasy. At best, we may surmise that a fifth dimension might be hidden deep within atoms.

Despite its cleverness, the Kaluza-Klein theory remained dormant for the next 50 years. Physicists just could not take it seriously; they thought of it as an interesting mathematical curiosity devoid of any connection with reality. After the discovery of the strong and weak nuclear forces in the 1930s, a theory that unified only gravity and electromagnetism, while ignoring the other two forces, just did not command a whole lot of interest. But in the 1980s, against the backdrop of the grand unification theory and the dream of unifying gravity with the electromagnetic and nuclear interactions into a "theory of everything," the idea of "geometrizing" all forces regained popularity; the Kaluza-Klein theory was resurrected and overhauled.

A UNIVERSE WITH ELEVEN DIMENSIONS

Geometrizing the remaining two forces exacts a price. While just one extra dimension did the trick for the electromagnetic force, the strong and weak nuclear forces required several more each. As a matter of fact, to take care of all three, it took no fewer than seven additional dimensions, bringing the total to 10 to describe space alone. Adding one more dimension for time, we now face a space-time with 11 dimensions. Once again, we have to "hide" the seven additional dimensions to make them unobtrusive in everyday life. Following in Klein's footsteps, we have to assume that they fold back onto themselves. However, whereas there is only one way to fold a single dimension into an infinitesimal circle, there exist several choices when dealing with more than one dimension. For instance, a two-dimensional surface can be folded back into either a sphere or a torus, two objects with completely different topologies. With seven dimensions, the options are far more plentiful and varied. One of the simplest ways to accomplish the objective is to construct a sphere whose surface is not two-dimensional, as is usual, but seven-dimensional. Such an object is called a "7-sphere" and has been the object of mathematical studies for more than half a century. It possesses rather unique geometrical properties and symmetries not shared by the ordinary sphere.

A universe with 11 dimensions, seven of which are hidden away in a 7-sphere, makes it possible to unify the force of gravity with the other three forces, producing a single superforce. In such a universe, all the forces in Nature would be manifestations of the geometry and structure of space-time. The force of gravity would be due to the curvature of our familiar space with three dimensions, while the other forces would result from the curvature of space in the seven extra dimensions. From this vantage point, forces—the source of all change and evolution in the universe—would simply result from geometrical shapes carved and molded over immense stretches of space.

Given the fact that the seven hidden extra dimensions do not exactly jump out at us, is there any prospect of ever observing them with improved technology? As we have seen, the higher the energy of the particles used to bombard matter, the smaller the scale that we can probe. Does this suggest that we might ultimately be able to see these hidden dimensions? The answer is a categorical no. The theory predicts that the extra dimensions must fold back within a 7-sphere whose radius is one million-billion-billion-billionth (10^{-33}) of a centimeter. This infinitesimally small size is called "Planck's length," after the German physicist whom we have already met—he is the one who first recognized the quantized nature of light. Planck calculated the dimension threshold at which gravity, which is so weak as to be negligible on an atomic scale, would be equal in intensity to the other forces. (Gravity is by far the weakest of the four forces; if the strong force is assigned an intensity of 1, the electromagnetic force would be 10^{-2}, the weak force 10^{-5}, while gravity would be only 10^{-39}.) The results of his calculations yielded the unimaginably small dimension of 10^{-33} centimeter, which is 1,000 times less than the scale at which the "grand unification" of the two nuclear and the electromagnetic forces occurs. If you want to probe a scale this small, you will need an energy ten billion billion times higher than the mass energy of a proton; that is 100,000 times greater than the energy at which the "grand unification" occurs. You may recall that, short of a technological breakthrough, it would take an accelerator reaching all the way to the nearest star to test the grand unification. Probing extra dimensions folded in a 7-sphere would be even more challenging: A machine as large as the entire Milky Way, 100,000 light-years across, would be needed. That pretty much qualifies as an impasse.

Physicists, rarely short of imagination, may have worked out a theory unifying gravity with the other three forces, but for the foreseeable future there is no way to verify it experimentally.

FROM SYMMETRY TO "SUPERSYMMETRY"

Space must include extra dimensions if forces are to arise from geometry. Curiously, if you are fond of particles and take the view that forces derive not from the geometry of space but from the constant shuttling back and forth of messenger particles, you will still be forced to introduce extra dimensions in space. This conclusion comes from a theory that purports to unify not the forces themselves but the very substance of the universe—matter and light.

Ordinary matter is made of quarks and electrons, both of which belong in the family of fermions—that is, particles endowed with a spin of $\frac{1}{2}$. By contrast, the quantum of light, or photon, is a boson, which is a particle with a spin of 1. In the standard model, bosons and fermions are completely different, particularly when it comes to their behavior in groups. Fermions are fiercely individualistic. They cannot stand being crowded in with other fermions. This antisocial behavior is intrinsic; it has nothing to do with electrical repulsion. Indeed, not only are electrons loners, but so are neutrons and neutrinos, even though they carry no charge. You can cram fermions into a given volume only up to a certain point and no more. Beyond that, they will protest and resist. The fermions' aversion to crowding is expressed in the "exclusion principle," originally proposed by the Austrian physicist Wolfgang Pauli (the same one who postulated the neutrino). This principle states that no two electrons can be found in the same quantum state; it explains why electrons in an atom are not all packed on top of one another in the orbit closest to the nucleus, which has the lowest energy. Instead, electrons fill orbits around the nucleus in a systematic fashion, according to a very strict rule dictated by Pauli's exclusion principle. The rule is as follows: If n is the quantum number characterizing a particular orbit, the maximum number of electrons that can be found in it is $2n^2$. By way of example, the orbit closest to the nucleus has a quantum number of 1 and can accept at most two electrons. The next orbit has a quantum number of 2 and can accommodate eight electrons; the third can hold at most 18, and so on. If electrons did not adhere to such order and discipline, the periodic table of elements would not exist, chemistry would not be possible, and all the beauty and complexity of the world would disappear. Do away with Pauli's exclusion principle, and all stars would become black holes. We would then be deprived of the thrill of discovering white dwarfs (what our Sun will become when it runs out of nuclear fuel) or neutron stars. As we have seen, white dwarfs and neutron stars exist only because electrons in the former and neutrons in the latter cannot bear being too crowded in and resist the pressure of gravity.

Bosons, on the other hand, are extremely congenial. They have no problem sharing the same bed and take great pleasure in rubbing elbows with one another. They do not want to be bothered with any exclusion rule; they would rather work as a team and help each other than make life difficult for the other guy, the way fermions do. That is what enables them to cooperate and produce macroscopic effects perceptible to our senses, even though they themselves are of subatomic size. For instance, armies of photons can band together and form an electromagnetic radio wave that propagates at the speed of light to bring your favorite television show right into your living room. To be sure, electrons do have waves associated with them as well; but we will never get to see one of those waves propagate in space in routine, everyday life, precisely because of the electrons' refusal to work in unison.

To unify light and matter, fermions and bosons must somehow become coupled. It is necessary to find a symmetry principle linking them, one that is capable of transforming a fermion into a boson, and vice versa. In these times that do not shy away from hyperboles, the inventors of the symmetry in question dubbed it "supersymmetry," which, predictably, earned them the sobriquet of "superphysicists" doing "superphysics." Unfortunately, the initial effort, which attempted to link fermions and bosons of the types already known, ended in failure. Cornered, our "superphysicists" were forced to double the population of existing particles; they linked ordinary fermions to bosons invented from scratch, and ordinary bosons to yet-to-be-discovered fermions. According to "supersymmetry" (or SUSY, for short), each ordinary boson is associated with a fermionic "superpartner," and, likewise, each fermion to a bosonic "superpartner." A "superpartner" is identical in every respect to the original particle (same mass, same electrical charge, same color, etc.), except for its spin, which must always differ by $\frac{1}{2}$. Thus, photons with spin 1 have superpartners with spin $\frac{1}{2}$, and electrons with spin $\frac{1}{2}$ have superpartners with spin 0.

A PROLIFERATION OF SUPERPARTICLES

The sudden birth of a plethora of (as yet) hypothetical particles taxed the ability of physicists to come up with new names. The terminology they proposed and eventually adopted did not exactly turn out to be a shining example of elegance. So as to preserve the familiar, albeit peculiar, nomenclature of ordinary particles and make it easier to remember who is associated with whom, all fermion "superpartners" simply received the prefix *s* (the initial in the word "supersymmetry") tagged onto the traditional name. This procedure led to some rather dreadful names like *selectron, smuon, stau, slepton, squark, stop,* etc. As for the boson superpartners, they were given names generally more pleasing to the ear. The suffix *ino* (meaning "littler" in Italian) was simply added to the normal designation, even though the corresponding particles are not always as small and light as the suffix connotes. For instance, the photon's superpartner is the "photino," that of the graviton is the "gravitino," and so on. Occasionally, though, the recipe leads to rather bizarre and ugly names. For instance, the W, Z, and gluons have superpartners called Winos (a most unfortunate outcome), Zinos, and gluinos.

Physicists have begun in earnest to look for such superpartner particles in accelerators. But none has been sighted as of yet. If a superpartner to the electron really exists with the same mass, it ought to combine with nuclei and form atoms that should have been observed by now. The fact that no machine has been able to detect these superpartners means that either the idea of supersymmetry is flawed, or current accelerators do not have enough energy to create them. In the latter case, the superpartners would be more massive than their ordinary counterparts. That would imply a breakdown of the mass symmetry. Some astrophysicists have even floated the idea that these hypothetical massive superpartners might constitute the mysterious dark mass in the universe.

It did not take long for "superphysicists" to turn their attention to gravity. After all, supersymmetry exuded hints of geometry. Two successive supersymmetrical operations turned out to be equivalent to a common geometrical operation—a simple translation in space. Thus was born the theory of "supergravity." In this new framework, gravity is no longer transmitted just by gravitons of spin 2, but by a whole family of supersymmetrical particles as well—the gravitinos—each of spin $3/2$. In our ordinary world, we will never be able to separate the effect of gravitons from that of gravitinos. We can only experience their combined effects. It is their concerted action that causes us to fall when we trip. The most economical description of supergravity in mathematical terms also requires extra dimensions added to space. As you might have anticipated, we find ourselves, once again, face-to-face with a space-time with eleven dimensions.

In short, whether we start from general relativity, where forces are but manifestations of geometry, or quantum mechanics, which holds that forces are carried by messenger particles, we end up with the same result—namely, a universe with eleven dimensions, seven of which are folded back onto themselves so tightly as to become invisible. Should we read this as simple coincidence or profound truth of Nature? Only time will tell.

At any rate, none of the supersymmetrical particles predicted by the theory, including the gravitino, has ever been seen. They remain pure products of the creative imagination of physicists. There are, however, good reasons to hope that they might be detected in the not-too-distant future. The LHC (the acronym stands for Large Hadron Collider, a machine that smashes protons against protons), which is planned to come on line around the year 2005 at CERN, will be capable of energies seven times higher than in the current most powerful accelerator, located at Fermilab, in Batavia, Illinois, just outside of Chicago. That is 7,000 times the mass energy of a proton, enough to produce "supersymmetrical" particles—if the theory is right. Until then, supersymmetry will have to remain a theory in search of a reality to describe, much as the characters of Pirandello were in search of an author.

THE SYMPHONY OF SUPERSTRINGS

One of the major hurdles and thorniest problems standing in the way of unification has to be the problem of infinities, which are infinite quantities that keep popping up in the theory. That is a sure sign that the theory cannot be true to reality, since Nature abhors infinities. We have seen that the closer the force-carrying messenger particles are to the particles of matter that generated them, the higher their energy. There is no upper limit to how high this energy can go, since there is nothing to prevent the messenger particles of matter from getting arbitrarily close to the particle of matter until coinciding with it. In the standard model, particles of matter are mathematical points without any dimension. In an attempt to remedy the problem of infinities, it was suggested that a finite size be ascribed to these particles of matter. As long as this size is much smaller than what our instruments can measure, we would never be aware of it. That is how the so-called theory of "strings" came into existence.

From simple dimensionless mathematical points, particles of matter have evolved into infinitely thin, one-dimensional strings, somewhat in the image of extremely fine spaghetti. There had been previous attempts to describe material particles as tiny spheres, but they failed because they contradicted the theory of general relativity. Initial calculations based on the concept of strings were not very promising: They led to particles traveling faster than light, which is forbidden by relativity. The next step was to call for reinforcements in the form of "supersymmetry." Strings became "superstrings," which remained well-mannered and never exceeded the speed of light, and string theory was saved. As it turned out, supersymmetry brought a surprise guest nobody had expected—a massless particle traveling at the speed of light and with a spin of 2. You may have recognized it: It is none other than the graviton, the hypothetical messenger particle of the gravitational force. And so, quite unintentionally, since the original motivation was to get rid of infinities, physicists have let gravitation wiggle its way into string theory. Gravity, which up to then had proved remarkably uncooperative, all of a sudden decided to join the other forces without even being invited. It sneaked into the theory while nobody was looking. Physicists were ecstatic. Superstring theory seemed to hold great promise for the long-sought unification of the four forces of Nature.

The theory tells us that superstrings have the infinitesimally small length of 10^{-33} centimeter—our old friend Planck's length, which is the scale at which the intensity of gravity is on a par with that of the other forces. All particles of matter and light, as well as those that transmit the forces holding the world together and are responsible for all changes in it, are simply manifestations of the vibrations of these superstrings. As the strings of a guitar or violin vibrate, they delight our ears by producing varied sounds accompanied by all their harmonics (i.e., frequencies that are multiples of the fundamental). In much the same way, superstrings are the source of the diverse sounds and harmonics that fill Nature and register on our measuring instruments in the form of protons, neutrons, electrons, photons, and so on. Indeed, the mathematics developed in the nineteenth century to describe the vibrations of strings in musical instruments account quite well for the vibrations of superstrings. The energy in each vibration mode corresponds to a specific particle whose mass is equal to that energy divided by the square of the speed of light, as prescribed by Einstein's formula. In this picture, the proton is simply a set of three vibrating superstrings, each string corresponding to a particular quark. Just as a trio of cellos can delight us when playing a piece by Mozart, so do the combined vibrations of three superstrings produce the music of a proton, which translates into a mass, an electrical charge, and a spin of $\frac{1}{2}$ when the melody is detected by our scientific instruments. The atom, being a combination of protons, neutrons, and electrons, includes yet more musicians in its orchestra to create its music. In a molecule made of several atoms, the musicians are more numerous still, and the sound becomes even fuller and more majestic. All around us, superstrings vibrate and sing, and the world is but a vast symphony. Next time you listen to Beethoven's Emperor Concerto in your living room, keep in mind that, mixed in with the sounds of the violins, trumpets, and drums, which work together in producing such sublime and uplifting music, are perhaps (I say perhaps because superstring theory has yet to be verified experimentally) the vibrations of all the matter around you.

SUPERSTRING LOOPS

A given superstring can manifest itself under several guises, depending on its vibration frequency. At one frequency, it may appear as a graviton; by switching to another, it pops up as a photon. Superstrings do not lead quiet and sheltered lives. They constantly move, interact, link up, or subdivide. Their demeanor in groups is extremely complex. Through lengthy mathematical calculations, physicists are just beginning to understand their social behavior. The mathematics required is extremely complicated, and for the first time in the history of physics, mathematicians had not developed it ahead of time. To speed up progress, we can only hope that the experts will be able to devise better mathematical tools or find conceptually and technically simpler ways to approach the theory.

The tension of superstrings plays a determining role in their behavior. At low energy, such as in our current universe, the tension is high and pulls on both extremities, bringing them closer together; under such conditions, superstrings shrink down to points. That is why classical physics, which treats particles as points, was able to describe phenomena at low energy so accurately. By contrast, when the energy is very high, as during the big bang or in particle accelerators, the tension loosens up, in which case particles become true superstrings and behave quite differently. The two ends of a superstring can either remain free or hook up to form a loop. Two distinct superstrings can also combine to form a single loop. The loop configuration seems particularly well suited to account for Nature's properties, particularly the symmetry principles that physicists have discovered in the subatomic world.

Since "supersymmetry" and "supergravity" postulate a vast population of massive and invisible particles, some physicists have theorized the existence of a "shadow world," populated with particles similar to ours (quarks, electrons, neutrinos, and so forth) that would interact with our own world only through the bias of gravity. It would be possible, for instance, for someone made of this "shadow matter" to go right through your body without your being at all aware of it. The reason is that the gravity exerted by your body is extremely weak. But should the intensity of gravity increase, the consequences would become quite noticeable. For example, if a "shadow planet" happened to traverse our solar system, its gravity might knock our beloved Earth out of its orbit. For the time being, though, these ideas lack any hard experimental proof; shadow planets, shadow black holes, and shadow galaxies remain purely speculative.

Since gravity made its way into the theory of superstrings through the bias of supersymmetry, it should come as no surprise that additional dimensions of space manifest themselves as well. In one version of the theory, superstrings inhabit a universe with nine spatial dimensions, resulting in a 10-dimensional space-time. In another, the universe has 25 spatial dimensions, giving a space-time with 26 dimensions. Once again, we have to assume that the 6 or 22 (depending on which version you adopt) extra dimensions fold back onto themselves, collapsing down to a size so infinitesimally small (10^{-33} centimeter) that they become completely imperceptible. Physicists hypothesize that during the big bang all dimensions were on a par with one another, and that particles in the primordial soup were able to sense them all. But after 10^{-35} second, the universe entered a phase of

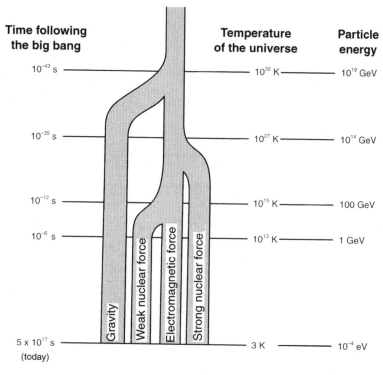

Figure 61. *The unification of the four forces of Nature.* The intensities of the four forces that govern Nature (the gravitational, strong nuclear, weak nuclear, and electromagnetic forces) are quite different from one another in our current universe. However, this intensity depends on the temperature of the universe. Physicists believe that these four forces had the same intensity, hence were unified into one and the same, when the universe was only 10^{-43} second old and its temperature was an unimaginable 10^{32} degrees Kelvin. The unification of the electromagnetic and the weak nuclear forces, resulting in the electroweak force, occurs at a temperature of 10^{15} degrees Kelvin, which prevailed 10^{-12} second after the big bang; it has been demonstrated experimentally. The "Grand Unification" theory postulates that the electroweak and strong nuclear forces were still merged into an electronuclear force at 10^{-35} second after the big bang, when the universe had a temperature of 10^{27} degrees Kelvin. As for gravity, it has thus far been a stubborn holdout and has resisted any attempt at unification. Physicists have yet to develop a quantum theory of gravity unifying the gravitational and electronuclear forces. A promising step in that direction is the theory of "superstrings," which holds that elementary particles are vibrations of infinitesimally small bits of string in a space-time with ten dimensions.

(a) End of nineteenth century

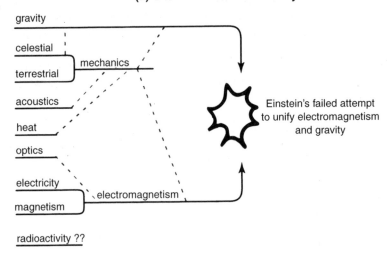

(b) End of twentieth century

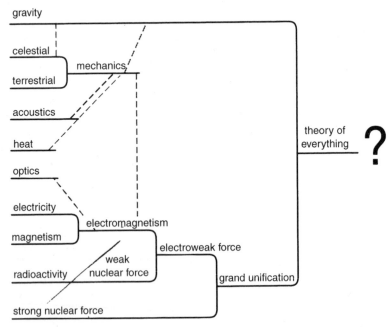

extremely rapid expansion called the "inflationary phase," which enlarged three of the spatial dimensions exponentially, while leaving all the others unaffected. As a result, the extra dimensions virtually dropped out of sight and no longer make their presence known to us except indirectly through the properties of particles and forces. Thus, according to this script, all forces originated in geometry, even though in our current universe gravity is the only force to remain associated in a perceptible way with the geometry of space-time.

In summary, the theory of superstrings presents several important advantages. By ascribing a nonzero size to particles, it gets around the problem of infinities. By encompassing gravity, it brings us closer to the much desired goal of devising a unified description of Nature (Figure 61).

THE MARCH TOWARD UNITY

We are at the end of our extensive journey into the heart of matter. It has been marked by a relentless march toward Unity (Figure 62). As physics progressed, intimate connections were discovered between phenomena once thought to be completely distinct. A unified description of Nature, sometimes referred to a bit too pompously as "the theory of everything," has become the holy grail of modern physics. Symmetry has constantly and reliably guided the physicists' first hesitant steps in this quest for Unity. It has taken on more and more importance, while at the same time becoming increasingly abstract. At the beginning, there were simply right-left and rotation symmetries. Now scientists routinely deal with "matter-light symmetry."

The march toward Unity began in the seventeenth century with Isaac Newton. He unified mechanics in the heavens and on Earth. Acoustics followed in joining the fold of mechanics. Next was the turn of the science of heat to connect with mechanics, when physicists discovered in the nineteenth century that heat was due to the frenetic agitation of the molecules making up ordinary objects. You sweat in the summertime because air molecules move rapidly and collide with your body with a lot of energy; you

Figure 62. *The unity of Nature.* As physics progresses, it keeps revealing unexpected connections. At the end of the nineteenth century *(diagram a)*, the whole of physics was united under the banner of two forces only—gravity and electromagnetism. Radioactivity, which had just been discovered, was not included. Einstein spent the last years of his life trying to unify these two forces, but without success. And for good reasons. The weak and strong nuclear forces had yet to be discovered. At the end of the twentieth century *(diagram b)*, four fundamental forces govern the world. Joining the gravitational and electromagnetic forces were the strong and weak nuclear forces. Steven Weinberg and Abdus Salam succeeded in unifying the electromagnetic and weak nuclear forces into the electroweak force. Grand Unification theories appear able to unify the electroweak force with the strong nuclear force, forming an electronuclear force. Physicists are hard at work trying to unite the electronuclear and gravitational forces into a single superforce. Such is the quest for the Theory of Everything, the holy grail of modern physics. One version of this next step is the theory of superstrings, which views elementary particles as vibrations of infinitesimally small bits of string in a space-time with 10 dimensions.

shiver in winter because these same molecules then move much more sluggishly and strike your body with far less energy. Another big step toward
Unity was taken the moment it was realized that mechanical interactions
between objects, such as friction (the phenomenon that causes the brakes of
your bicycle to heat up when they rub against the wheel), could be
explained as electromagnetic interactions between the atoms and molecules
in these objects.

Electromagnetism itself had made giant strides toward Unity. James
Maxwell had unified electricity, magnetism, and optics in the nineteenth century. By the end of that century, all movements of particles could be understood in terms of gravitational and electromagnetic forces. To the extent that
the type of science that studied such movements in those days was essentially
mechanics and nothing more, the claim could be made that all of physics
had been unified into two interactions—gravity and electromagnetism. The
old subdivisions of physics into separate fields, such as mechanics, heat,
optics, sound, electricity, magnetism, and gravity, which can still be found in
old books, were no longer relevant and fell into oblivion.

Progress toward Unity had been spectacular. Yet the ultimate goal
remained elusive and the path strewn with obstacles. During the first half of
the twentieth century, Einstein tried hard to unify his beloved gravity with
electromagnetism, but, sadly, without success. And for good reasons. New
actors had made their entrance and demanded to be reckoned with.
Radioactivity, discovered in the latter part of the nineteenth century, did not
fit in the then-prevailing picture. The strong nuclear force also needed to be
included in any unified theory. Salam and Weinberg met the challenge and
showed that the electromagnetic and weak nuclear forces were actually one
and the same—the electroweak force. Shortly thereafter, the grand unification theory made its appearance, attempting to unify the strong nuclear and
electroweak forces. The fourth character in the drama—gravity—continued
to resist stubbornly. The theory of superstrings then came along, with signs
that it might be able to overcome gravity's resistance. It described the world
as a vast symphony of vibrations of infinitesimal strings in a 10-dimensional
space-time. The world would then be governed by a single superforce, melding all four currently known forces, and whose rule would extend over the
entire universe.

Are we about to reach the goal? Will we get to see Unity in all its glory?
There are those who believe so. Personally, I am not convinced. It is, for the
time being, unlikely for the grand unification theory to be verified experimentally, since it would require phenomenal energies. The only chance to
verify it is indirectly, through the process of proton decay. Thus far, protons
have displayed a longevity surpassing initial predictions. The theory of
superstrings is even less amenable to direct experimental confirmation, as
the unification of everything is predicted at energies that defy human imagination (there remains an outside chance that supersymmetrical particles
predicted by the theory might be observed in upcoming machines). More-

over, the theory is shrouded in such a thick mathematical veil that it no longer has any connection with reality. And, as long as physics is not rooted in reality, it is no more than metaphysics.

Having marveled at Nature's prodigious inventiveness as she deals with inanimate matter, let us next focus on how she exercises her creativity to generate life.

« 6 »

The Creative Universe

ODE TO IMPERFECTION

By letting symmetry be their guide, physicists have been able to edge closer to Unity. They discovered that at the moment of creation, when the universe was extremely small, hot, and dense, forces could have been unified and Unity could have been perfect. Today, by contrast, after 15 billion years of evolution, we live in a universe of stunning complexity and variety, governed by four distinct forces of completely different intensities and characters. For instance, the strong nuclear force is 100 times more intense than the electromagnetic force, and 10^{39} times greater than the gravitational force. Matter itself displays a most amazing variety. Quarks assemble in groups of three to form protons and neutrons, but electrons do not combine into more complex structures. Electrons have both a mass and an electrical charge, while photons do not.

Nature delights in trying as many variations as possible. Why did she not choose to maintain the perfect unity that prevailed in the beginning? Why did she decide to vary the bricks of matter, rather than making them all the same? The answer is quite simple: A perfect universe is certainly possible, but it would be totally sterile and dull. Such a universe could not allow for the fragrance of roses, the singing of nightingales, fiery sunsets, or Monet's water lilies. Nor could it produce a conscious intelligence capable of wondering about where it came from and where it is going. Perfect unity, flawless symmetry, and absolute perfection would be synonymous with sterility and death. The principle of life wilts away in the face of the cold perfection of overly precise symmetry and absolute unity. We simply could not exist in a perfect and undifferentiated world. We are here because symmetry was broken.

The importance of being imperfect is evident in many different areas, from physics to biology, even in mathematics. Imperfections often prove beneficial, if not crucial. Minor errors, slight flaws, trace impurities, tiny exceptions to the rule—all are essential to the successful operation of a system. The following example, which deals with railroads in Japan, illustrates the point clearly.

Experts had long wondered why French catenaries sold to Japan seemed to wear out much faster in the Land of the Rising Sun than in France. It turned out that Japanese engineers, meticulous to excess, erected their pylons at precise intervals of 100 meters, while their French counterparts were perfectly satisfied installing them *roughly* every 100 meters. The Japanese practice turned out to generate large-amplitude vibrations, which caused electrical cables to rub excessively hard against the catenaries and induced premature wear. The problem was too much perfection built into the system; the solution was to introduce a small degree of imperfection in the spacing of pylons.

Perfection can be just as harmful in time as in space: Soldiers should never march in lockstep on a bridge, because repeated synchronized impulses can cause it to collapse.

Whereas perfection is sterile, imperfection is pregnant with possibilities. Without impurities added to crystals of silicon and germanium used in the manufacture of the transistors in your radio receiver, you would be unable to listen to your favorite station. These impurities introduce "surprises" in the otherwise infinite repetition of the crystal lattice making up these elements, and help electrical currents flow more readily in transistors. In pharmacology, highly pure products are sometimes less effective against microbes than impure natural mixtures. If living matter had been a perfect machine programmed to reproduce itself endlessly, permanently generating identical and self-similar replicas without the slightest variation, life could not have evolved. It would have forever remained at the same primitive stage. Only through imperfections in reproduction can new structures appear and different forms of life take hold. A gene transforming (perhaps by interacting with cosmic rays) or a chromosome reorganizing itself after being broken apart are sources of mutations and evolution. Imperfections in genetic reproduction are responsible for the extraordinary richness and variety of animal and vegetal species that populate our present-day ecosystem, from the delicate forget-me-not to the giant baobab, from parasitic bacteria to thinking human beings. Hindu temples are not quite symmetrical: The legend goes that too much perfection would make the gods jealous. But is the real reason not, in truth, that the "perfection" of a work of art stems from controlled imperfections in the details, just as happens in Nature? Is it not because true perfection is sterile, while measured imperfection begets novelty?

MAGNETS AND CHILLED SUPERCONDUCTORS
LOSE THEIR SYMMETRY

The history of the universe is a long trail of broken symmetries and controlled imperfections. Indeed, imperfection in the midst of perfection is what made it possible for us to come onto the scene and for dark night skies to be filled with planets, stars, and galaxies. The world basks in vibrant colors, but also abounds in sound and fury. The wind caresses our face and brings us the sweet smell of a rose. The waves of the oceans splash softly and lull us to sleep. The Sun floods the vast prairies with its blazing fire. If all this beauty that soothes our souls is at all possible, it is because the forces number four rather than one. The force of gravity attracted matter into galaxy seeds—tiny heterogeneities in the density of the universe that appeared during the first 300,000 years—and enabled them to grow into stars and galaxies. Later on, it also collected dust grains into planets and moons. By binding neutrons and protons to form atomic nuclei, the strong nuclear force gave birth to all matter within and around us. By binding atoms and molecules, the electromagnetic force led to the emergence of the twisted double helix of DNA and, ultimately, life. The weak nuclear force is responsible for the nuclear reactions fueling the fire within the stars. The four forces are responsible for the great diversity and variety of the things of life. Yet, as we have seen, they were in all likelihood melded into one in the beginning.

In order to create complexity and novelty, the universe had to break the perfect symmetry of the original single force. It did so spontaneously (physicists speak of "spontaneous symmetry breaking") by cooling down. Indeed, cold temperatures break symmetries. You can appreciate it by considering ice: At high temperatures, ice turns to water, which has no particular structure; liquid water is completely symmetrical. On the other hand, when the temperature drops below 0°C, water freezes into ice; symmetry is broken in the process, since ice crystals display a specific spatial orientation so that all directions in space are no longer equivalent. By causing liquid water to change to solid ice (physicists call this change of state a "phase transition"), cold is able to generate structure from a structureless state.

Another example has to do with magnets. At high temperatures, magnetism is not activated and a magnet does not attract nails. As it turns out, a magnet is made of microscopic components. When heated up, the tiny magnets experience a frenetic and completely chaotic agitation; they point in all possible directions and their effects cancel out. But as soon as the temperature drops below a critical threshold (of 770°C), magnetism reappears spontaneously; at that temperature, the individual microscopic magnets no longer jiggle haphazardly, but they all line up in a same and unique direction, producing a cooperative macroscopic magnetic field. Here again, the symmetry of the tiny magnets is broken by cold temperatures.

The same phenomenon also occurs in so-called "superconductors." These materials lose all electrical resistance when chilled below a certain temperature. In an ordinary conductor, the billions of electrons that contribute to an electrical current move independently of one another, following complicated and erratic trajectories. By contrast, in a chilled superconductor, these same electrons move in unison in a highly correlated and organized fashion. This type of a cooperative behavior enables electrons to move along without wasting any energy into heat. Economy of energy is, of course, very attractive to industrial engineers, who would love to take advantage of superconductors for designing computers and building telephone and power lines.

This type of collective movement on a macroscopic scale is also observed in "superfluids," such as liquid helium. When chilled to a very low temperature, it loses all viscosity and starts flowing without any friction or energy loss. Once again, cold temperatures break symmetry by enabling electrons (in superconductors) or helium atoms (in superfluids) to move cooperatively in the same direction.

THE UNIVERSE BREAKS ITS SYMMETRY

Just as in the case of water, magnets, superconductors, or superfluids, cold temperatures allowed the universe to break the prevailing symmetry and open the door to complexity. As I explained in detail in my book *The Secret Melody*, the universe was born out of an infinitesimally small void through the bias of a quantum fluctuation.[1] Our understanding of the first moments of the universe is very hazy when it comes to times earlier than 10^{-43} second and sizes smaller than 10^{-33} centimeter (these limits are known as Planck's time and Planck's length, respectively), because we lack a quantum theory of gravity, as we have already mentioned. Planck's barrier constitutes for now an impenetrable wall to our knowledge. Behind it hides a reality that is still inaccessible, perhaps inhabited by superstrings vibrating in a space-time with 10 dimensions. Only when the cosmic clock strikes 10^{-43} second do our familiar time and three-dimensional space make their entrance. At that time, the observable universe still has the incredible temperature of 10^{32} degrees. At the 10^{-43}-second mark, gravity seceded from the union and went on to live separately. The remaining three forces were still united into an electronuclear force, and the universe was governed by a duo. At 10^{-35} second, the universe had cooled by a factor of 100,000, down to 10^{27} degrees, and a second round of symmetry breaking occurred. It was now the turn of the strong nuclear force to strike out on its own. The triumvirate of the gravitational, strong, and electronuclear forces ruled the universe.

This second symmetry breaking was to have important consequences for the future of the universe. Just as water releases a burst of energy as it breaks its symmetry by turning to ice, so did the universe liberate the enormous vacuum energy. This energy caused the universe to embark on a wild expansion, called "inflation," lasting an infinitesimally short duration of 10^{-32} second. That was long enough for the observable universe to triple its size one hundred times in succession and grow to the dimension of an orange. The end of the inflationary phase was marked by another break in symmetry:

Matter and light, which had up to then been unified in the form of energy, went their separate ways. They came into existence in accordance with Einstein's relation between mass and energy, and the original vacuum devoid of any structure was suddenly filled with myriad particles. Quarks, leptons, photons, as well as their antiparticles, all burst on the scene. The universe continued expanding, but at a much more moderate pace. It grew increasingly cooler, and by the time the cosmic clock struck 10^{-12} second, the temperature was only (!) 10^{15} degrees. This cooling by a factor of a million million triggered yet another break in symmetry. The weak force split off from the electroweak force. From that point on, there would be four separate forces to fashion the complexity of the universe. Thus, each symmetry-breaking event created new potentialities for structures yet to come.

THE UNIVERSE IS BIASED TOWARD MATTER

While forces born out of a series of symmetry breakings have molded the world, we owe our existence to yet another broken symmetry—that between matter and antimatter. The matter-antimatter symmetry is one of the sacrosanct laws of Nature. Matter is a form of energy and can be created in particle accelerators, but, when it is being produced, it is always accompanied by an equal amount of antimatter. If vacuum energy created matter and its associated antimatter in the primordial universe, why is it that today we live in a universe made almost exclusively of matter? What happened to the antimatter? We are quite certain that the Milky Way does not contain a multitude of antistars, each with its court of antiplanets, on which an antiyou might be reading an antibook! Cosmic rays—those particles born during the explosive death of massive stars—reaching us from deep space tell the story unequivocally: They contain less than 0.01% of antimatter. Moreover, the profusion of X rays, expected from the annihilation of matter and antimatter if, from time to time, a galaxy were to run into an antigalaxy, is conspicuously missing. One might speculate that matter and antimatter somehow got segregated and live now in different corners of the universe without ever coming in contact. However, such an assumption is not very plausible, since they were intimately mixed in the primordial soup.

As already mentioned, perfect symmetry would lead to equal amounts of particles and antiparticles, in which case we would not be around discussing the issue. Matter and antimatter would have completely annihilated each other and the universe would contain nothing but photons. Continuously losing energy as a result of the universe's expansion and cooling, these photons would no longer be able to give birth to particles and antiparticles (remember that a photon's energy would have to be at least equal to the added masses of the particle and its antiparticle, multiplied by the square of the speed of light). All that would be left is a universe filled with light, unable to generate stars, galaxies, and human beings. Since we are in fact here, the universe must have found very early on a way to break the symmetry between matter and antimatter. It is now believed that this happened

10^{-32} second after the primordial explosion, at the tail end of the inflationary phase, when vacuum energy generated particles and antiparticles. It appears that Nature then showed a very slight preference for matter. For each billion antiquarks created out of vacuum, Nature managed to produce one billion *plus one* quarks. Minuscule as this difference is, its consequences were momentous. As the universe approached the first millionth of a second and had cooled down to 10,000 billion degrees—low enough for quarks to assemble in groups of three to form protons and neutrons, and for antiquarks to form matching antiparticles—the overwhelming majority of quarks and antiquarks annihilated each other and turned into light. However, for each billion particles and antiparticles that disappeared to produce a billion photons, one particle of matter was left over, without any corresponding antiparticle to destroy it. We exist and our bodies are made of matter because Nature preferred matter over antimatter by a ratio of 1 to 1 billion. That is also why there are one billion photons for every particle of matter in today's universe.

Why was the symmetry broken? The answer remains a mystery, despite the efforts of a number of physicists, particularly the late Andrei Sakharov (1921–1989), the well-known Soviet dissident and human rights activist. The physical processes likely to produce an asymmetry between matter and antimatter appear related to the same ones that are invoked to explain the death of protons. That is another reason why physicists are so eager to catch a proton in the act of decaying.

Nature thus had to introduce some degree of controlled imperfection into perfection to allow for our existence. To that end, she spontaneously broke the symmetry between forces as well as that between matter and antimatter. Because the early universe was but a primordial soup of elementary particles, particle physics has allowed us to understand, at least partially, how symmetry was broken. An approach based on reducing the behavior of the entire universe to that of subatomic particles has proven effective and has enabled us to make great strides in our understanding of the cosmos. The question is whether this "reductionistic" approach is sufficient to make further progress, or whether we will have to resort to other, more "holistic" strategies that consider the universe as a whole.

THE LIMESTONE ROCKS OF THE BAY OF HA-LONG

In my native country, which goes by the sweet name of Vietnam, less than sixty miles from the port city of Hai-phong, some sixty miles East of Hanoi, is one of the great natural wonders of the world—the Bay of Ha-Long (Figure 63). Sitting in a boat rocked gently by the iridescent waves, I never get tired of admiring the majestic scenery and letting myself be absorbed by the beauty and serenity of the site. In front of me, the ragged profiles of numerous limestone rocks jut right out the vast expanse of water, pointing straight toward the sky. These masses of limestone, which give the bay its well-deserved fame, were sculpted over the ages by the erosion caused by rain and waves, and are

Figure 63. *The Bay of Ha-Long*. The bay is located on the coast of Vietnam, northeast of the port city of Haiphong. It is famous for its limestone rocks, which jut out of the ocean. (Photo by Anne Claire Tran Thanh Binh Minh)

now carpeted with a thick and dense green vegetation. These sculptures are made of an ordinary chemical compound called calcium carbonate. If you were to examine a sample under a microscope, you would find that the limestone is composed of innumerable fossils of minuscule animals that long ago teemed throughout the China Sea. It took their tiny remains millions of years to deposit themselves at the bottom of the ocean in stratified and packed layers, eventually producing these imposing limestone structures, which tourists now come to admire from all over the world.

We are all familiar with chalk; it, too, is made of limestone. Teachers use it to write on blackboards and teach algebra and philosophy to their students. Chalk is white. Why? We know the answer. The white light of the sun, which illuminates all objects around us and enables us to perceive them, is a mixture of all colors of the rainbow, from red to blue, going through green and yellow. The color of an object depends on which components of sunlight are *not* absorbed by the object, and therefore get redirected toward our eyes. For instance, the turquoise stones adorning the neck of elegant women have a color somewhere between sky blue and blue-green because the phosphates of aluminum and copper of which they are made absorb red and orange, but reflect blue and green. Likewise, chalk is white because calcium carbonate reflects all components of the white light emitted by the Sun; it absorbs none of the colors of the rainbow (it does absorb infrared and ultra-violet sunlight, but these colors are of no concern to us here since our eyes cannot see them). If so, why are the limestone rocks in the bay of Ha-Long not white, the way chalk is? The reason is that they are not made exclusively of calcium carbonate: Microscopic dust grains have been deposited on their surface over geological times, which gives them a slightly dull, grayish color.

Why do some objects, such as turquoise stones, absorb certain colors of the rainbow, while others, such as chalk, do not? There are two reasons for this. The first reason has to do with the quantized nature of light: A beam of light contains innumerable particles without charge or mass (the photons), each with a well-defined energy. The second reason concerns the quantized nature of matter, which is composed of atoms and molecules. An atom can exist only in certain energy states. This energy cannot change arbitrarily, but only by specific amounts. An atom (or a molecule), which is generally in the lowest-energy state (also called ground state), can switch to a higher-energy state only by absorbing a photon with a particular energy. That energy has to be precisely equal to the energy difference between the final and initial states of the atom. It just so happens that atoms of copper, which are contained in turquoise stones, have an energy state that differs from their normal lowest-energy state by an energy exactly equal to that of a photon with a red-orange color. That is why the red-orange portion of the solar spectrum is absorbed, and why a turquoise stone looks blue-green. On the other hand, the molecules in limestone have no state whose energy exceeds that of the ground state by an amount corresponding to photons in the visible portion of the solar spectrum. These photons are, therefore, not absorbed and all the solar light ends up being reflected in our eyes, which is why we perceive chalk as white.

Why can atoms and molecules exist only in well-defined energy states? Why do these energies have particular values? Quantum theory, particularly Schrödinger's wave equation, provides the answer. Just as the air in a church organ can vibrate only at certain frequencies, so do particles in atoms exist only in specific energy states.

Why is matter composed of particles? Why does light exist? Like an inquisitive child who gets on your nerves with his endless string of "why?" and his insatiable curiosity, we have asked ourselves a long series of questions that eventually led us to elementary particles and the "standard model" describing their nature and behavior. This strategy attempts to reduce the richness and beauty of the world to nothing but particles, fields of forces, and interactions. It postulates that any physical system can be dissected into its elementary components, and that its global behavior can be understood and explained in terms of the behavior of the individual components viewed as fundamental. Such a reductionistic approach assumes that all the complexity of the world—the fragrance of lavender, the bark of a dog, the smile of a child—can ultimately be explained by the laws of physics.

REDUCTIONISM HAS A GLORIOUS PAST

Reductionism has exerted a profound influence on Western scientific thought. Because Nature is complex, reducing it to its simplest elements was the key to making progress. Physics as we know it would not have been possible without this process of simplification. We have had an opportunity to appreciate the power of the reductionistic method while musing over the color of rocks in the Bay of Ha-Long. We could have used the same approach for any of their other properties.

Take, for example, their chemical composition. Limestone is composed of three chemical elements—calcium, carbon, and oxygen—the weights of which are always in the same proportion of 40 percent, 12 percent, and 48 percent. Why this particular proportion, and why is it so remarkably constant? The answer is that the weights of the atoms of calcium, carbon, and oxygen are themselves in the ratios 40/12/16, and that one molecule of calcium carbonate contains one atom of calcium for one of carbon and three of oxygen. Why three times more atoms of oxygen than of calcium and carbon? The number of atoms of each chemical element within a molecule is determined by the number of electrons that element exchanges with its neighbors. In the present case, carbon has 6 electrons, oxygen 8, and calcium 20. According to quantum mechanics, four of the six electrons in a carbon atom are available for sharing, while that number is two out of a total of twenty in a calcium atom. An oxygen atom, on the other hand, has no electron to share, but, instead, can accept two of them from other atoms. That is why three oxygen atoms can grab a total of six electrons; they do so by combining with one carbon atom and one calcium atom, providers of those electrons. The electrical force exerted by the shared electrons is what keeps the atoms together in the molecule.

Why this precise ratio of 40/12/16 between atomic weights? That is because the atomic nucleus is made of protons and neutrons, of which the calcium nucleus contains a total of 40 (20 protons providing the 20 positive charges necessary to offset the 20 negative charges of the electrons—an atom being electrically neutral—plus 20 neutrons), carbon contains 12 (6 protons and 6 neutrons), and oxygen contains 16 (8 protons and 8 neutrons). What purpose do neutrons serve, since they have no electrical charge? The answer is that they keep the atomic nucleus together, as we have seen. Without neutrons, the electrostatic repulsion between protons would cause the nucleus to blow apart. Neutrons attract protons via the strong nuclear force so intensely that the nucleus is 100,000 times smaller than the entire atom, which, therefore, ends up being mostly void. Why does a neutron and a proton have almost the same mass, and why is a neutron electrically neutral, while a proton does have a charge? They have roughly the same mass because they are both made of the same number and types of quarks—"up" and "down" quarks. The different electrical charges are due to the fact that a proton is composed of two "up" quarks, each of charge $2/3$, and one "down" quark, of charge $-1/3$, while a neutron is made of two "down" quarks and one "up" quark.

Once again, our string of "why?" has brought us back to the standard model. The question of the chemical composition of the rocks in the Bay of Ha-Long has found once more an answer in particle physics. And yet again, reducing a complex system to its elementary components has enabled us to make progress and reach a higher level of understanding. The reductionistic method is effective; indeed, it has a glorious past. But is it the only one conceivable? Can we really reduce all phenomena in the world to particles and their interactions? What about the most complex phenomenon of all— life? After all, the rocks in the Bay of Ha-Long were themselves generated by life: They are composed of thousands of small fossilized animals that had absorbed the calcium salts and carbon dioxide dissolved in ancient oceans and used those materials to generate calcium carbonate shells around their fragile bodies to protect themselves against predators.

THE TWISTED DNA DOUBLE HELIX

There can be no denying that the reductionistic approach has racked up a remarkable record of success in the study of life. In that approach, living beings are nothing but a collection of atoms and molecules responding to the influence of the four traditional forces, in accordance with the familiar laws of physics. Explaining life would then boil down to understanding the behavior and interactions of the various molecular components that make up living beings. The only difference between animate and inert matter would be their degree of complexity. In other words, biological beings would simply amount to extremely complex machines. Such a "mechanistic" view of life asserts that biology is merely a branch of chemistry, which is itself but a branch of physics.

Studies of the molecular underpinnings of life have led to impressive breakthroughs, the most spectacular of which is without a doubt the deciphering of the DNA structure by the American James Watson (1928-) and the Englishman Francis Crick (1916-) in 1953. Their discovery of the twisted DNA double helix ranks among the greatest scientific adventures of the twentieth century. We now know that life on Earth results from the harmonious cooperation between two classes of very large molecules—nucleic acids and proteins. There are two nucleic acids (the name is a reminder that they reside in the central parts of living cells), which were given the unwieldy designations of ribonucleic and deoxyribonucleic; thankfully, they are more commonly known by their acronyms RNA and DNA.

Each nucleic acid has its own specific function. DNA molecules are long chains made of alternating sugar and phosphate molecules, onto which structures made of four types of so-called "base" molecules are attached. Two of these bases are relatively small molecules—thymine and cytosine— made of atoms of carbon, nitrogen, oxygen, and hydrogen, arranged in the shape of a hexagon. The remaining two bases—guanine and adenine—are larger molecules in which atoms are arranged in a hexagon sharing one side with a pentagon. These four bases constitute the entire alphabet of the genetic code, much as the twenty-six letters of the Roman alphabet are the foundation of the English language. Appropriately, they are usually referred to by their initials—T, C, G, and A. Bases hook up to each other in complementary pairs. Their respective structure and dimension are such that adenine always goes with thymine, and cytosine with guanine. A DNA molecule is configured like a double helix: Two chains of sugar and phosphate twist around themselves, cross-linked by complementary base pairs. Picture a spiral staircase whose side railings are made of the sugars and phosphates, and whose steps are formed by the bases. At each step, a large base is matched against its smaller counterpart (Figure 64a). The structure is such that one-half of the double helix contains the complete code; the other half is, in

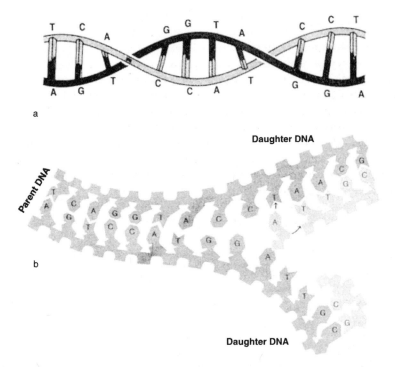

Figure 64. *(a) The twisted double helix of DNA.* The two helices are joined by four types of molecules called "bases," which form the alphabet of the genetic code. They are: adenine (A), cytosine (C), guanine (G), and thymine (T). Because of their structure and dimensions, bases can combine only in pairs and only one way: A with T, and C with G.

(b) The DNA molecule replicates itself. The two helices of a parent DNA molecule detach and separate from each other. Two daughter DNA molecules form when each of the two helices latches on to a complementary helix. Because the "bases" that link two complementary helices can fit in only one way (A with T, and C with G), the sequence of bases in the two daughter molecules is identical to that in the parent DNA. That is how the parent DNA replicates itself, and why children resemble their parents.

effect, superfluous. This redundancy is what makes it possible for the DNA molecule to replicate, by means of a process that is responsible for our children resembling us and is as essential to life as sex. When a cell divides, the two chains break off. Each base from one chain reassembles with a matching base attached to another chain, and the result is two identical DNA molecules (Figure 64b). While the molecular bonds between bases are fragile, those in a chain are more resilient, which enables the chains to remain intact during separation and reproduce without errors most of the time. Occasionally, though, replicating errors do occur as molecules get reshuffled, leading to a genetic mutation. Mutations can also result from spontaneous rearrangements of molecules.

Proteins are made of long chains of smaller units called "amino acids." There exist twenty different types of amino acids, which are found in every living species, be it humans, cats, flowers, or bacteria. Proteins are the jack-of-all-trades and building blocks of life. They do all the work at the molecular level and see to it that living cells function properly. To ensure a harmonious collaboration between DNA molecules and proteins, in order for them to communicate, a mechanism for translating the four-letter alphabet of DNA into the twenty-letter alphabet of proteins is necessary. The dictionary holding the key to this translation was worked out in the 1960s. The internal machinery of the cell reads the sequence of bases in groups of three at a time. Upon reading a set of three bases, a signal is activated directing the cell to manufacture a first amino acid. The next three bases are then read, triggering the manufacture of another amino acid, and so on. As the amino acids are being produced, they line up and combine into proteins, somewhat like on an assembly line in an automobile plant. Amino acids arrange themselves in accordance with instructions stored in DNA molecules. That is where the other type of nucleic acid—RNA—comes in. Sections of the base sequence containing the instructions get copied onto the RNA molecule, which consists of a single chain. The RNA chain acts as a messenger transmitting the information to molecules of great complexity called "ribosomes," which can be viewed as huge protein-manufacturing factories. The proteins ensure the proper execution of chemical reactions by either breaking or establishing molecular bonds, somewhat like factory workers drilling holes or welding parts.

ARISTOTLE'S FINAL CAUSE AND DEMOCRITUS'S BLIND FORCES

This brief foray into biology showed us that the molecular components of life are organized into a marvelous, well-tuned piece of machinery. Such a mechanistic description of life at the molecular level, with its lexicon of words such as "machine," "manufacture," "factory," and "assembly line," has enjoyed great success and has enabled us to unlock deep mysteries; it permitted us to understand the secrets of replication and gene transmission, and to comprehend why our children look like us. Yet a serious problem still persists when a living organism is considered in its entirety. If each molecule reacts blindly to forces acting on it at a given time and location, how does it manage to integrate and coordinate its individual behavior so as to form a coherent and harmonious whole? How can individual atoms—which move and react in accordance with the causal laws of physics, under the effect of localized forces produced by neighboring atoms—act in a concerted and cooperative fashion on the scale of an entire organism and over sizes much larger than intermolecular distances?

It is as though molecules all had a common purpose and carried within themselves a master plan not contained within the laws of physics, which govern only local behaviors. The notion that the components of living organisms develop and act in a "holistic" manner, as if they all strove toward one and the same end, is not new. Already some twenty-four centuries ago, the Greek philosopher Aristotle believed that the development of a living organism was predicated on a preexisting blueprint, guiding its behavior toward a predefined goal. This concept of "teleology" (from the Greek *teleos*, meaning "goal") meant that all living beings were driven toward a final cause. The universe itself was considered a giant organism evolving toward a predetermined destiny, according to a grand cosmic scheme. The idea of a holistic harmony ran diametrically counter to the "atomistic" vision championed by the Greek philosopher Democritus. Democritus believed that all structures and forms in the world were merely different arrangements of atoms, and that any change or evolution simply reflected a rearrangement of these atoms. He thought of the universe as an immense machine in which each atom responded blindly to the influence of its neighbors—and to nothing else. Any notion of final cause was anathema. The conflict between holism and reductionism was already taking shape.

THE DIVERSITY OF LIFE

It is impossible to see a flower bloom, listen to the soothing ruffle of leaves on a tree, watch a dog exult in happiness at seeing his master return, hear the birds sing, or contemplate the complex relationships between human beings, with all their joys and hurts, without coming to the conclusion that living organisms represent the ultimate form of organized and active matter. Life possesses an astounding degree of complexity and diversity. The American biologist Edward Wilson has estimated the number of living species on the planet at 1.4 million.[2] So many others remain to be discovered that the actual number could be as high as 10 or 100 million. The vast majority of these living organisms are insects (751,000 different catalogued species) and plants (248,000 species), while animals account for only 281,000 species. The remainder includes bacteria, viruses, algae, protozoas, and mushrooms, although here again these species have been so poorly studied that many are yet to be discovered.

This stunning variety does not apply just to the number of species. Even within a given species, the diversity of forms and characteristics is virtually boundless. For instance, human beings belong in different races and have distinct faces, sizes, nose and eye shapes, and hair and eye colors. Each human being has unique and specific characteristics that make his own individuality. This great specificity of living organisms distinguishes them from particles in the subatomic world, which all possess the same exact properties if they belong in the same class. When you have seen one electron, you have seen them all. A proton does not change one iota, whether it is in a nucleus of carbon or iron.

There is another property that sets living beings apart from subatomic particles. Identical twins do not exhibit exactly the same behavior, even though they carry precisely the same genes. This highlights the second crucial difference between the living and inanimate worlds. Subatomic particles fall squarely in the realm of the innate. They were born with a number of properties and retain them for the rest of their life. By contrast, living beings are a complex mix of characteristics that are innate and others that are acquired. The genetic properties they were endowed with at their birth are supplemented by additional attributes acquired as a result of the interactions of living organisms with their environment. The behaviors of identical twins are not exactly the same because their courses through life—friends, lovers, teachers, role models, etc.—were themselves different. Life on Earth is possible only to the extent that it is integrated into a complex network, the "ecosphere," made of a multitude of mutually interdependent living organisms that coexist in a state of dynamic equilibrium. Contrary to inanimate systems, which can be "closed"—that is to say, completely cut off from their surroundings—living organisms are "open" systems that constantly interact with their environment through endless exchanges, notably of energy and matter. Such exchanges can take place because a living organism is not in thermodynamic equilibrium with its environment. For instance, its temperature is typically not the same as that of ambient air. Such a nonequilibrium state is essential to life. It enables Nature to construct structures with higher degrees of organization and climb the ladder of complexity. It allows her to create order out of chaos, and diversity out of simplicity.

THE CHEMICAL METRONOME

Equilibrium is synonymous with structureless uniformity and sterility, while lack of equilibrium implies organization and creativeness. We can see it every day when we boil water in a pan to prepare our morning cup of tea. As the water at the bottom of the pan heats up, it becomes less dense and tends to move toward the top. As long as the system is very nearly at equilibrium— that is, when the temperature difference between bottom and top is small— nothing much happens. The viscosity of the water curtails any movement. But as soon as the temperature difference exceeds a critical threshold, structures begin to appear. The liquid gives up its passivity, and convection cells pop up, organizing themselves in an orderly and stable flow. The mass of water, originally homogeneous and structureless, organizes itself as it gets driven out of equilibrium by the supply of heat. Convection cells persist only as long as the temperature is not too high. If the temperature increases some more, the orderly flow eventually becomes chaotic.

This self-organization of a system out of equilibrium is not unique to boiling water. A similar phenomenon has been observed by Ilya Prigogine (1917-), a Russian-born Belgian chemist, in a chemical reaction known as the Beluzov-Zhabotinsky reaction (after the two Russian scientists who first studied it). The reaction takes place in a mixture of cerium sulfate, malonic acid, and potassium bromate dissolved in sulfuric acid. In order to make it easier to monitor the reaction, color indicators are added for the purpose of identifying the different types of ions (ions are atoms that have lost one or more electrons). Blue points to acidic ions, while red indicates basic ions ("bases" in chemistry are substances that neutralize acids by combining with them). A pump is used to inject the various chemical elements into the mixture. As long as the rate of injection is slow, the system remains near its equilibrium state and nothing unusual happens; all the colors are mixed homogeneously. However, when the rate of injection exceeds a critical value, the chemical system goes into a nonequilibrium state, and an amazing series of events occurs. First, the entire mixture suddenly turns red. Two minutes later, it switches to blue. Another two minutes later, red returns, and so on. The change in color is periodic and occurs with the regularity of a metronome. It is no accident that such a system has been dubbed a "chemical clock."

This rhythmic pulsation is all the more extraordinary since the chemical mixture contains some 100,000 billion billion (10^{23}) particles. Their concerted action is analogous to the following situation. We have all watched a Lotto drawing live on television. Numbered balls are placed in a rotating drum to mix them. The hostess retrieves one numbered ball at random from the drum, fixing the first digit of the winning number. The remaining balls are mixed again, and a new ball is picked to give the second digit, and so forth. Imagine now that we replace the numbered balls with 50,000 billion billion black balls, and as many white balls, and mix them all in the rotating drum. Most of the time, we will see a grayish color since, on average, as many black balls as white ones will come into view. True, from time to time, there might be a few more white balls than black ones, or vice versa, but such statistical fluctuations are small and completely random. The equivalent of the blue-red rhythmic pulsations of the chemical mixture described above would be to suddenly see nothing but black balls one moment, then white balls two minutes later, followed by nothing but black balls again after another two minutes, and so on. We would be flabbergasted and cry foul. Yet that is exactly what happens with open chemical systems in a nonequilibrium state: They have the ability to self-organize and create order out of chaos—something that closed chemical systems in equilibrium cannot do.

To try to explain the rhythmic behavior of such a chemical clock, cyclical changes at a certain frequency determined by the respective concentrations of the various components in the mixture have been invoked. Such periodic changes would be caused by a phenomenon called "autocatalysis." Catalysts are substances that accelerate chemical reactions; autocatalysis occurs when the presence of a particular substance generates even more of that same substance. But that raises a new question: The forces between individual molecules act over distances of one-tenth of a millionth of a centimeter, while the rhythmic pulsation extends over a range of several centimeters. Once again, we are confronted with the fact that the billions and billions of atoms present in a chemical mixture exhibit a holistic behavior and follow a global plan. The chemical clock constitutes one more example of Nature spontaneously breaking symmetry to go further up the ladder of complexity. The initial state, being devoid of organization, was symmetrical with respect to time; that symmetry was broken the moment the oscillations began, since the successive states of the mixture are no longer time-independent.

The organization of the Beluzov-Zhabotinsky chemical reaction takes place not only in time but also in space (Figure 65). Two kinds of extraordinary spatial structures appear: either blue-colored, circular, concentric waves emanating from a central source and propagating on a red background toward the periphery, or blue spiral structures spinning like a wagon wheel around its center, also on a red background. As was the case for boiling water, the chemical mixture has the ability to self-organize into a highly structured configuration when driven beyond a critical threshold and out of its equilibrium state. Ilya Prigogine called this kind of chemical mixture a "dissipative system," because it must dissipate energy in order to maintain its organized state. The system can also bifurcate to a completely chaotic state. In particular, if the injection rate keeps increasing, the periodic behavior of the chemical clock turns to more and more complex oscillations that eventually lead to complete chaos—a phenomenon we have already discussed at length. Under these circumstances, we can no longer determine ahead of time which bifurcation branch the system will take. It is as though inanimate matter had a mind of its own.

THE MYSTERY OF FORMS

While our understanding of the holistic character of inanimate matter and of its ability to self-organize in an open, nonequilibrium system remains very sketchy, the challenge is even more formidable in the case of living matter. The problem can be summarized as follows: How can a collection of molecules, which a priori experience the action of their nearest neighbors only, know how to organize itself to form a living organism that is coherent over distances much larger than the separation between molecules? The problem is particularly thorny when it comes to living beings. Proteins are long chains that must curl up in a very complicated fashion in order to adopt the three-dimensional shape required for them to perform their intended purpose. How do completely distinct parts of the molecule know how to coordinate their behavior so as to produce the desired shape? How can an embryo

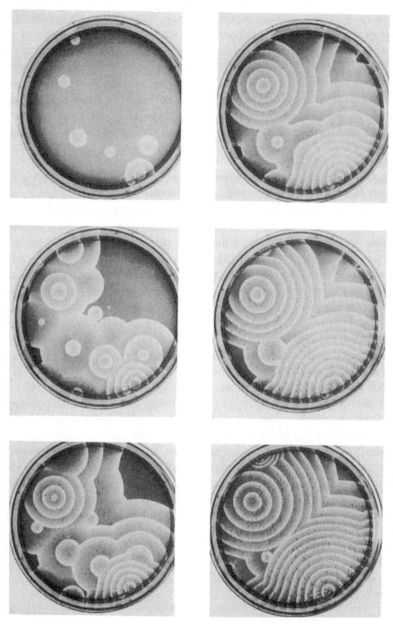

Figure 65. *A self-organizing chemical system*. When the chemical system known as "Beluzov-Zhabotinsky reaction" is driven out of equilibrium by injecting chemical elements at a high rate, it becomes chaotic, but can also spontaneously organize itself into spatial structures such as those shown here. Concentric circular waves propagate out of central points toward the outer edges. The implication is that nonequilibrium can create order out of disorder.

develop from a single fertilized cell into a living being of tremendous complexity, each part of which is assigned a specific function? That is the whole problem of cellular differentiation. How do particular cells in the embryo know that they are to become blood cells, while others are to develop into bone cells? There is also the problem of position in space: How does a particular cell know the location it is to occupy in relation to the other parts of an organism and move accordingly? How do ear cells know that they must proceed to the face, rather than to the stomach? How can such exquisite and precise coordination take place in both time and space? We are grappling here with the great mystery of the elaboration of forms (or morphogenesis) out of the embryo.

Here again, we cannot help but invoke some "grand blueprint" guiding the behavior of individual cells toward a final destiny as the embryo evolves into a fully developed organism. The French biochemist Jacques Monod (1910–1976), who is widely known for his passionate defense of reductionism and who certainly cannot be accused of being partial to holism, had this to say: "One of the fundamental characteristics common to all living beings without exception [is] that of being endowed with a purpose or project, which at the same time they show in their structure and execute through their performances. . . . Rather than reject this notion (which some biologists have tried to do), it must be recognized as essential to the very definition of living beings. We shall maintain that the latter are distinct from all other structures or systems present in the universe by this characteristic property, which we shall call teleonomy."[3]

This impression of purpose and finality is further reinforced when we realize the extraordinary ability of certain living systems to form completely, even if the embryo becomes mutilated during its formation stage. A process called "regulation" replaces the missing cells with new ones, or allows misplaced cells to regain their "correct" places. Even more amazing is the fact that certain living organisms, already fully developed, can reconstitute themselves when mutilated. An astonishing sight is that of an earthworm cut in two that goes on living as two separate worms. Salamanders have the ability to regenerate a whole new limb to replace one that has been severed. The most extraordinary case may be that of a small freshwater animal called the hydra, endowed with six to ten tentacles; it can divide spontaneously into two or three separate pieces, each of which generates a complete animal. Better yet, if you cut it into small pieces, it will reassemble itself whole.

THE "ÉLAN VITAL"

How can we explain that living beings seem to be "invested with a purpose," that they possess a teleological quality, to use Aristotle's phrase, or a "teleonomic" component, to paraphrase Monod? To speak of "final cause" or "grand project" is anathema in the view of many scientists—not without some justification, considering that modern science was born and flourished by systematically rejecting the explanation of natural phenomena in such terms. To say, as did Bernardin de Saint-Pierre (1737–1814), that Providence "divided the pumpkin in slices because it was made to be shared by the whole family" does little to heighten our knowledge in matters of pumpkin! Yet, as far as living beings are concerned, the evidence is persuasive and cannot easily be dismissed. Monod put it this way: "Objectivity nevertheless forces us to acknowledge the teleonomic character of living beings, to accept that, in their structures and actions, they pursue and realize a purpose. Therein lies, at least by all appearances, a profound epistemological contradiction. The central problem of biology is this very contradiction, whether the avowed goal is to resolve it if it is only apparent, or to prove it fundamentally insoluble if it truly is so."[4]

In an attempt to resolve this "profound epistemological contradiction," some have taken the view that living beings constitute a separate class, that they are subject to laws other than those that govern inert matter. These laws possess an extra ingredient that is missing from conventional laws. In the so-called vitalist theories, this extra ingredient takes on the attributes of what the French philosopher Henri Bergson (1859–1941) referred to as the "élan vital," or life energy.[5] It confers on living matter special properties that enable it to organize itself and evolve in a harmonious and creative manner.[6] A "flow of life" battles with inanimate matter and forces it into organizing itself. Today, such vitalist ideas are no longer fashionable. They have fallen into disfavor, for lack of experimental proof.

HOW CAN THE LOCAL INFLUENCE THE GLOBAL?

Other ideas have been proposed to explain the mystery of morphogenesis. One of them advances that, if indeed there is a "blueprint," it must be stored somewhere within the DNA molecules of the fertilized egg. However, given the fact that the DNA is the same in every part of a living organism, how would each DNA molecule choose its own particular role in the overall plan and execute it? One would have to invoke some sort of "metaplan" containing the relevant instructions. If so, where is the metaplan to be found? This line of reasoning would quickly lead us to an endless regress. The currently most promising theory of morphogenesis is one based on "activated" genes. It postulates that there exist special genes whose function is to regulate the behavior of other genes. They are normally dormant, but awaken and spring into action at the appropriate time to transmit instructions to the other genes. These special genes have been studied in flies, although they exist in other species as well, including man. While the discovery of these genes may explain the mechanism of morphogenesis, it still sheds no light on how genes with molecular sizes can coordinate their action on the scale of a macroscopic organism—in other words, how local actions can exert an influence on a global scale.

On several occasions in our exploration of the world of inert matter, we have already encountered situations where local interactions can give rise to global organization. Recall the examples of boiling water or of the chemical mixture that goes into rhythmic pulsations in a clockwork fashion. As we have seen, an open system out of equilibrium can bifurcate from a structureless state to one that is highly organized. Could it be that a system of living cells in the embryo also goes through a series of bifurcations allowing it to achieve organization and complexity? The answer is very likely no, because there are some fundamental differences between living cells and, for instance, inanimate convection cells in water. The cells in boiling water organize spontaneously without adhering to a global purpose. They have no need for instructions of the type contained in the genetic code. Furthermore, the shape and nature of convection cells in boiling water are completely unpredictable and changing, whereas the shape of living organisms is stable and invariant. Starting from a given configuration of a DNA molecule, the final form of the resulting organism is quite precisely predetermined.

Another idea already encountered in the world of particles for the purpose of linking the local with the global is that of "fields." As we have seen, electric fields are not localized just where particles reside, but extend much farther out. Could there be a "morphogenetic field" in living organisms, just as there is an electromagnetic field in molecules? Such a field would extend over the entire organism and transmit instructions to individual cells to guide their evolution. Unfortunately, there is a problem with this notion: Since the genetic code is supposedly localized within DNA molecules, how can these molecules impart to the field the information they carry, and how do they coordinate their local behavior on the scale of a global field? This is not even the only problem: How do DNA molecules, which are identical throughout a given organism, communicate different instructions to the field, depending on where they happen to be? The instructions must vary from location to location to allow for cell differentiation. If we assume that DNA molecules tell the field what form to adopt, and that the latter in turn tells the molecules how to behave, we find ourselves trapped in a circular reasoning and we have not really resolved anything.

One way to break out of this vicious circle is to postulate that the field already contains within itself all the instructions required for morphogenesis without having to go through the intermediary of DNA molecules. In this picture, DNA molecules would merely be receiving instructions, without needing to provide them. Such an idea has been advocated by the British biologist Rupert Sheldrake. He has theorized that the field already harbors the relevant information concerning the form an embryo is to take, and that it uses it to guide the embryo's development. However, he further added the notion of "morphic resonance," which rests on the hypothesis that the field possesses a memory. Once Nature has learned how to develop a particular form, she remembers it to guide—by a process of "resonance"—the growth of other organisms. For instance, when a dog learns how to accomplish a new task, the memory retained by the morphogenetic field would help other dogs learn the same task more easily. The same would apply to inanimate matter: Sheldrake cites the case of substances that, although never before seen in crystal form, suddenly decide to crystallize at the same time in different geographical locations. This hypothetical morphogenetic field would not act causally in space and time in accordance with the usual laws of physics. Because it involves new physical laws yet to be discovered and thus far devoid of any convincing experimental proof, Sheldrake's theory has met with considerable skepticism and is not widely accepted.

QUANTUM MECHANICS AND LIFE

Quantum mechanics is another area we have explored where we were forced to abandon the notion of local reality. The "EPR experiment" made it clear to us that all particles in the universe have a mysterious connection, that they are all part of the same global reality. Can the nonlocality of quantum mechanics help us resolve the mystery of morphogenesis? Even though living beings are macroscopic entities, they are made of DNA molecules that are a priori subject to the laws of quantum mechanics. The physicist Erwin Schrödinger (of wave function fame) actually demonstrated that quantum mechanics is essential for understanding the stability of the genetic code in DNA molecules.[7]

But there is a world of difference between using quantum mechanics to study the behavior of elementary particles and doing the same with living organisms. As Niels Bohr correctly stressed, it is impossible to determine the quantum state of a biological system without killing it. The reason is that quantum theory ascribes a primordial role to the observer. It is he who creates reality by doing the observing. In order to examine a living cell, one has to interact with it, which will inevitably perturb it and interfere with the molecular mechanisms that are essential to maintaining life. Moreover, quantum mechanics has a statistical character ill-suited to living beings. It describes reality in terms of probabilities, which implies that it can be verified only by observing the behavior of many identical systems. Suppose you wish to determine the probability that a coin tossed in the air will land on heads or tails. You would be unable to predict the outcome on any given toss. All you can say is that it has a 50 percent chance of landing on heads and an equal probability of landing on tails. But to verify that, you would have to toss the coin a great many times. Likewise, quantum mechanics requires the observation of many entities that do not differ by one iota. That poses no problem for elementary particles of a given type, since they are all identical. But it is quite a different story when it comes to living beings, which, as we have seen, are characterized by their specificity. No one living being is totally like another. That makes it quite unlikely that quantum mechanics can be applied without modification to the phenomenon of life.

It would probably be necessary to invoke a new set of principles, specific to biology, that would complement the laws of quantum mechanics without contradicting them. This point of view was expressed by the main creators of quantum theory. Niels Bohr, for one, asserted that, just as describing electrons as particles is complementary to describing them as waves, so are the laws of biology complementary to the laws of physics when it comes to describing Nature. Likewise, Schrödinger did not exclude the possibility of new laws in biology: "From all we have learned about the structure of living matter, we must be prepared to find it working in a manner that cannot be reduced to the ordinary laws of physics." More recently, the physicist Walter Elsasser has proposed the idea of "biotonic laws," which, according to him, would act in a holistic fashion on living organisms, and would possess a logical structure completely different from what we are accustomed to. They would augment the familiar laws of physics but could not be deduced from them. All these ideas remain to be proven.

So the question we asked ourselves at the beginning of our excursion into the living world has failed to find an answer. We do not understand how living cells develop from shapeless embryos into the complexity, richness, and diversity of the living forms that surround us. The "profound epistemological contradiction" that Jacques Monod was talking about continues to dog us. One conclusion, however, emerges: It does not appear that living beings can be explained in reductionistic terms as an ensemble of particles interacting locally. An organizing principle with a holistic character, acting on a global scale over the whole organism, seems to be a prerequisite.

THE SPARK OF LIFE

While the processes guiding the development of biological forms continue to retain their secrets, the greatest mystery of biology without a doubt has to do with the origin of life. How did inanimate matter give rise to living matter? How did the spark of life spring forth from a collection of inert atoms and molecules? Once life got started, it is possible to imagine various mechanisms enabling it to develop and proliferate; but how did the first living cell arise?

We have seen that life results from harmonious cooperation between nucleic acids (namely, DNA) and proteins. Because they go hand in hand and possess complementary roles, these two agents of living matter must have entered the stage simultaneously in order for life to manifest itself. The drama of life could not have unfolded without either one of these two actors. As discussed earlier, nucleic acids carry the genetic code, but they are so chemically inactive that they cannot accomplish anything on their own. It is the proteins that do all the chemical work, thanks to their remarkable catalytic power. But proteins are themselves assembled according to the instructions contained in nucleic acids. We are faced with the old chicken-and-egg problem: Which comes first? If it is the nucleic acids, their chemical impotence makes it impossible for life to get started; if it is the proteins, how can they assemble themselves without the genetic code in DNA? Scientists who have studied the problem are divided into two camps.

The first camp maintains that nucleic acids were the first to evolve. A structure capable of storing the genetic instructions and replicating itself would have appeared first. This "primordial" gene would not necessarily have been DNA; it could have taken the form of RNA, which, in contrast to DNA, would not be chemically inept and could replicate without the help of proteins acting as catalytic enzymes. DNA would have appeared only later in the history of biological evolution. In order to test this scenario, the American biologist Leslie Orgel has conducted laboratory experiments to induce RNA to replicate without the assistance of proteins, but so far without success. The German biologist Manfred Eigen, on the other hand, believes that RNA can arise spontaneously from an assemblage of complex organic chemical substances (that is to say, containing carbon), through a hierarchy of interdependent and mutually reinforcing chemical cycles, which he called "hypercycles." These ideas are supported by experiments with viral RNA (viruses are the most primitive living organisms and possess one type of nucleic acid only), and are further buttressed by highly complex mathematical models.

The second camp is convinced that proteins came first. This view is supported by experiments performed by the American biologist Sidney Fox, who showed that a mixture of amino acids heated to high temperature can give rise to molecules bearing a striking resemblance to proteins, which he named "protenoids." Even though these molecules are not found in living beings, they can form tiny spheres that are strangely reminiscent of the shape of living cells. The British biologist Graham Cairns-Smith, for his part, believes that life arose not from carbon-containing organic compounds but from clay substances. The idea came to him because certain clay crystals can replicate in a crude way. In time, the first clay organisms would have become more complex and begun to react chemically with organic substances. These would have gradually taken over the function of replication, causing the clay origin of life to be lost.

THE PRIMORDIAL TERRESTRIAL SOUP

We do not know whether it is the egg or the chicken that came first. But can we re-create in the laboratory the physical conditions that prevailed some 3.5 billion years ago, when the first living organisms were appearing on Earth, and test whether we can trigger the spark of life out of inanimate atoms? In a now-classic 1952 experiment, the American chemists Stanley Miller (1930–) and Harold Urey (1893–1981) tried to duplicate in their test tubes Earth's primitive atmosphere by mixing hydrogen, methane, and ammonia (as we have seen, these gases constitute today the major part of the atmosphere of giant planets such as Jupiter) with boiling water. The mixture was then subjected to electrical discharges to simulate the effect of the numerous electrical storms raging on the young Earth in these early times. A week later, a reddish-brown liquid had formed, which contained organic substances essential to life, notably several amino acids (Figure 66). Scientists now believe it more likely that the early atmosphere on Earth was made of a mixture of carbon dioxide, nitrogen, and water vapor spewing out

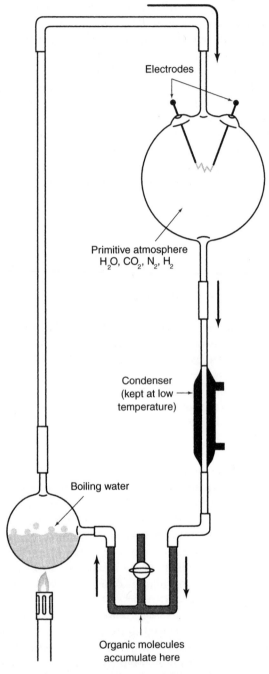

Electrodes

Primitive atmosphere
H_2O, CO_2, N_2, H_2

Condenser
(kept at low →
temperature)

Boiling water

Organic molecules
accumulate here

Figure 66. *The experiment of Miller and Urey*. Modern versions of this classic experiment demonstrate that the organic molecules necessary to life can be synthesized from gases present in the primitive atmosphere of Earth, such as carbon dioxide, nitrogen, water vapor, and hydrogen, subjected (for a week) to electrical discharges simulating lightning strokes from storms raging all across the atmosphere of the young Earth.

of the many volcanoes erupting then, in addition to hydrogen. The Miller-Urey experiment was repeated with this new mixture and, once again, a variety of organic compounds appeared.

Miller, Urey, and their successors were on the right track to decipher the mystery of life. But the goal remains distant, because there is a long way indeed from simple organic substances and amino acids to the twisted helix of DNA, and its ability to replicate life. The probability of a virus forming spontaneously, purely by chance, after a billion years (the time elapsed between Earth's birth and the appearance of the first living cell), solely through the interplay of molecules randomly associating and dissociating in the primordial terrestrial soup, is equal to the unbelievably small number of $10^{-2,000,000}$ (the digit 1 comes after 2 million zeroes)! It is less likely than the chances of a coin tossed in the air landing on tails 6 million times in a row.[8] The probability of such an event is so incredibly small because the number of possible chemical combinations in the primordial soup is itself unimaginably large.

EMERGENT PRINCIPLES

The chances of life appearing spontaneously solely through the play of random encounters of atoms and molecules in the primordial soup are thus practically nil, even though the complexity of a DNA molecule is not as great as that of a virus. What to do in the face of such evidence? We can always claim miracle, but that is not a very scientific tack. Some, like the Swedish physicist Svante Arrhenius (1859–1927), or, more recently, the British biologist Francis Crick and the British astrophysicist Fred Hoyle (1915–), have argued that, since life has not had enough time to develop spontaneously on Earth, it had to come from space, perhaps in the form of microorganisms delivered to Earth through the loyal and trustworthy services of asteroids and comets. Arrhenius called this the "panspermia" hypothesis. If microorganisms came from elsewhere, they would have had more time to develop than the one billion years or so available on Earth. The hypothesis of life originating elsewhere has recently taken a new turn with the discovery of pieces of rock that were ejected from Mars by asteroid impacts and landed in Antarctica. If Mars supported living organisms in its past, could it be that life made it to Earth by way of Mars (Figure 67)? Yet, even if life did come from outer space, the question of its beginnings would merely be shifted to an earlier date, a different place in the universe, perhaps under different physical conditions.

Since there are not sufficient experimental proofs that life came from elsewhere, we will restrict ourselves to examining the hypothesis that it began right here on Earth. If it cannot arise spontaneously by chance, we have to postulate the existence of one or more organizing principles able to guide molecules in the primordial soup along their evolutionary path from simple molecules to twisted DNA helices. We have to assume that "laws of complexity" exist that allow Nature to progress. These organizing principles

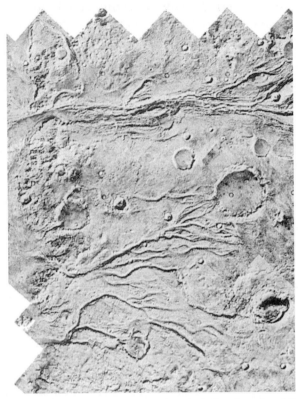

Figure 67. *Life in Mars's past?* The presence of dried-out riverbeds on the surface of Mars, such as those shown here, photographed in 1976 by the Viking space probe, suggests that water once flowed there in abundance under what must have been a much denser atmosphere. The atmosphere on Mars volatilized away and is today 100 times less dense than that of Earth. Any water exposed to the extremely tenuous Martian atmosphere would boil away and vaporize, or freeze instantly. The presence of water in Mars's past suggests that life may have existed some 3 billion years ago in the form of microorganisms buried in the Martian soil. There is a disputed claim that a piece of rock from Mars that fell into the Antarctica may contain trace evidence of these ancient microorganisms. (Photo courtesy of NASA)

do not necessarily require the intervention of mysterious new forces in addition to the four known forces, or the existence of yet-to-be-discovered interactions. They do not hinge on any extra ingredient of the kind required by vitalist theories. True, the possibility that new forces acting exclusively on living matter may someday be uncovered cannot be ruled out: The hypothesis that living matter and inert matter are both subject to the same physical forces is far from being verified with the desired strict scientific rigor. Nonetheless, the organizing principles considered here do not demand the existence of such new forces, even though such a possibility would in no way be incompatible with the known forces. For example, the strong nuclear force would continue to perform its function of holding protons and neutrons together in atomic nuclei, and electromagnetic forces would go on binding together atoms and molecules. Organizing principles would not contravene the traditional forces because they would operate on different levels. Whereas the familiar forces exert their influence on a local scale, organizing principles act on a global scale, guiding and channeling the collective behavior of atoms and molecules in a holistic manner.

These organizing principles and the traditional laws of physics can coexist at different levels because the latter have lost the rigid determinism in which Newton and Laplace believed so unconditionally. We have already shown how quantum fuzziness swept aside any notion of strict determinism, and how an open system out of equilibrium can bifurcate toward an organized (or chaotic) state without our being able to predict it. Organizing principles do not exist at the simplest level of elementary particles, but they "emerge" as soon as the level of complexity exceeds a critical threshold. Indeed, they are referred to as "emergent principles." These principles cannot exist at the same conceptual level as that of elementary particles, just as concepts of "teleonomy" or "design" are irrelevant at the level of a proton or electron. We can thus assume that biological systems possess a hierarchy of organizational levels, and that at each level of this hierarchy new concepts and behaviors emerge that cannot be deduced from an analysis of entities at a lower level. The behavior of a complex and organized ensemble, composed of a multitude of particles, cannot be extrapolated from that of just one or even a few of these particles.

MACHINE AND PROGRAM

The chemical properties of sugars and alcohols cannot be deduced directly from those of atoms of carbon, hydrogen, and oxygen which these substances are made of, even though there may exist causal relationships between the two categories. Likewise, the origin of life cannot be deduced by studying inanimate particles. To make use of computer science terminology, elementary particles, atoms, and molecules constitute the hardware, while the organizing principles are the software. The hardware serves as material support for the software. The sequence of instructions, called "program," is stored in the memory of the computer, but operates at a higher level than the material one. The program acts as an organizing principle, guiding the behavior of the machine. It tells it how to calculate, manipulate statistics, even play chess. The electrons racing through the complex circuits of the computer obey rigorously the traditional laws of physics, but it would be preposterous to assert that the rules of chess can be deduced from the laws of electromagnetism and quantum mechanics that govern the behavior of electrons. Electrons and chess programs belong in different levels.

What form could such organizing principles take? We have already run into one of them. We have seen that in open systems that are out of equilibrium, such as the chemical clock, matter does not behave haphazardly, but can organize itself under certain conditions. Whereas randomness is associated with equilibrium, nonequilibrium can generate complexity and organization. It is entirely conceivable that the system of particles in the primordial soup on Earth might have gone through a series of abrupt bifurcations that led it, level by level, step by step, to increasingly complex stages of self-organization, culminating in the emergence of structures capable of self-replication, and ultimately life itself.

Granted, all of this is mere conjecture. On the other hand, what is quite probable is that nucleic acids and proteins did not emerge at the same time, in a single phase, and that life requires emergent organizing principles that cannot be deduced from the behavior of elementary particles.

THE ASCENT OF LIFE

It was the extraordinary diversity of living species he observed on the globe, notably in the Galapagos archipelago, while cruising around the world on the H.M.S. *Beagle*, that gave the English naturalist Charles Darwin (1809–1882) the inspiration for his theory of the evolution of species.

This theory first appeared in Darwin's 1859 masterpiece *On the Origin of Species by Means of Natural Selection*. It injected a time dimension into the study of the living world. Living species were no longer thought of as having assumed their final form from the outset. Paleontological studies tell us so: Forms have changed, evolved, and gone through differentiation during the course of geological times so as to adapt to changing environments. Darwin's work had enormous psychological repercussions because it demoted man from his privileged station in the realm of the living. Darwin proclaimed that man, rather than being God's favorite child descending from Adam and Eve, actually had far less noble origins; going back in time, his genealogy included primates, mammals, birds, reptiles, fish and insects, and ultimately primitive cells. The shock was all the more severe because the Polish canon Nicolas Copernicus (1473–1543) had already struck a serious blow to man's perception of himself by dislodging him from his central place in the solar system. With his new theory, Darwin discarded once and for all the last vestiges of Aristotelian thought concerning the evolution of living beings. A teleological explanation would no longer do. The evolution of life on Earth would no longer unfold according to a "grand design"; nor would it tend toward a final cause. On the contrary, it developed at the whim of random mutations and was driven by natural selection.

The extraordinary progress accomplished in the areas of paleontology, biology, and genetics since the publication of Darwin's work has consistently supported his main idea: Living beings have evolved over time. The study of fossils indicates that the first living organisms established a foothold on Earth some 3.5 billion years ago in the form of extremely rudimentary single-cell organisms, the bacteria and blue algae that were teeming in the primitive oceans. Since that time, living forms have never stopped evolving and undergoing mutations. Advances in molecular biology allow us to understand why: Mutations occur when groups of molecules called genes are accidentally rearranged in the DNA of a living organism, either spontaneously or through errors in replicating the genetic code.

While the idea of the evolution of species in Darwin's theory and in its modern form—Neo-Darwinism—is almost universally accepted today, the same cannot be said of the mechanisms responsible for this evolution as proposed by Darwin and his successors: random genetic mutations and natural selection. Some biologists have raised objections against the excessive role attributed to chance as the primary cause of genetic mutations. How could a living organism as extraordinarily complex as an animal or a human being, endowed with highly effective organs such the nose and the ears, all functioning with exquisite harmony (should one of these organs fail, we become ill), be the result of chance or of a series of accidents? How could an organ as perfect as the eye, with its hundreds of millions of cone-shaped cells transmitting light messages from the retina to the brain, have arisen fortuitously? How could totally disordered events lead to organisms capable of dealing with and adapting to environmental changes as catastrophic as the ones once caused by a huge asteroid smashing into the Earth, triggering a global winter because of the massive amount of dust ejected into the atmosphere and screening out the Sun? More important, how can completely random genetic mutations be responsible for totally new and highly complex structures such as our brain, this network of some 100 billion neurons able to generate thought, experience love, and appreciate beauty?

These problems are further magnified when we consider the evolution of our biosphere in its entirety. The history of life has oftentimes been described as an evolution from bottom to top, a progression from very crude and simple forms toward organisms of unimaginable complexity and sophistication. The evolution of life is usually compared to ascending a ladder of organization or a pyramid of complexity, with man perched at the very top. The notion that we sit at the summit of the evolutionary pyramid is probably far too anthropocentric. There is no reason to believe that we should be the final destination of the journey of life. Nevertheless, it is undeniable—many paleontological studies have borne it out—that, during the last 3.5 billion years in the history of Earth, life has evolved from the simple to the complex, that it has progressed from the most rudimentary organization to the very highest levels of sophistication. How could chance alone be responsible for this relentless march toward complexity?

CHANCE ALMOST ALWAYS LEADS TO DISORDER

We all feel intuitively—and the laws of thermodynamics confirm it—that, left to its own device, chance tends to dismantle rather than build, to sow disorder rather than promote order. The more complex a system, the more prone it is to degradation and malfunction. A grain of dust in your walkway does not require any care, but it takes all the skill and know-how of an expert gardener to ensure that the flowers bordering that walkway maintain their vivid colors and delight your eyes. A rock lying on a path has no need for anyone to take care of it; but the magnificent Gothic cathedrals built of cut stone would fall into ruins without constant maintenance and restoration work. Left solely to the blind action of chance, a complex system goes haywire much more readily than does a simple structure. Should an engineer make a benign mistake in designing a wheelbarrow, that would not have overly drastic results. The wheelbarrow will still function and you will still be able to use it for you gardening chores. But if errors creep into the design of a car, the repercussions could be fatal. The slightest flaw in the design of an aircraft or a space shuttle can have even more catastrophic consequences.

There is a good reason why complex systems are more prone to failure. A complex system can exist in many more different states than a simple one and, among all the possibilities, disordered states far outnumber ordered ones. You can easily appreciate this by experimenting with a deck of cards. If you take a deck in which the cards are arranged in a certain order (for instance, in numerical order) and shuffle it, the probability of finding the cards in the same original order is practically zero, because the combinations in which the cards are disordered are far more numerous than those in which they are ordered. That is precisely the reason why casino croupiers always carefully shuffle a deck before dealing out the cards. Shuffling the cards almost always leads from order to disorder, or at least from one type of disorder to another, very rarely from disorder to order.

Similarly, if chance alone dictated the rearrangement of genes responsible for genetic mutations, we would expect an increase in disorder, which would degrade the complexity of living organisms, rather than a higher level of order, making them ever more structured and efficient. The study of mutations in a particular species of flies, called "drosophila," illustrates the point well: Most mutations have injurious, rather than beneficial, effects, and the living organism tends to regress, rather than show improvements.

We may also formulate the problem in terms of information content. The information needed to construct a living organism is contained in its genes. An organism with increased complexity requires more information. The more sophisticated the organism becomes, the more it is necessary for the DNA molecule to acquire extra information. But where does it manage to find this additional information? Information theory—as well as our intuition—tells us that random perturbations occurring haphazardly (what we call "noise") invariably reduce, rather than increase, the information content. Anyone who has ever tried to speak over a phone line drowned in static, or to listen to the weak signal of a distant radio station, can attest to this plain fact: A large portion of the words, hence of the information, is lost in the parasitic noise.

NATURAL SELECTION

When confronted with the objection that chance cannot be a reliable source of order and organization, and that random mutations cannot lead the marvelous diversity of living beings all around us, the Neo-Darwinists' usual reply is that, while haphazard changes do indeed cause degradation most of the time, they can from time to time produce improvement and ameliora-tion in a living organism. This amelioration will then be picked out, pre-served, and amplified through the filter of what Darwin called "natural selection," to the point that the altered organism becomes the norm. Nat-ural selection is that process through which living species poorly adapted to their environment lose out in their constant struggle for limited resources and are eliminated. The better-adapted species are more likely to survive, reproduce, and proliferate.

There is no disputing that natural selection operates in Nature. Exam-ples abound. Picture a wooded area inhabited by a single animal species, with a light-colored coat contrasting sharply with the shrubs. One day, genetic mutations produce the birth of a few animals with a dark coat, blending in completely with the vegetation. It is easy to guess how the two populations are going to evolve: In time, the light-colored species will be decimated by predators and hunters because they are much easier to spot. The other species, on the contrary, will escape extinction, thrive, and multi-ply. Eventually, all their descendants will have a dark-color coat. The species will thus have evolved from a light- to a dark-colored coat in order to adapt to the color of the woods.

Gene filtering through natural selection, which enables living species to become better adapted to their environment, certainly does take place. The central idea of the theory of natural selection, based on the notion that "those fittest for survival will prevail," has become almost a tautology and is not in question. What is far more difficult to understand, however, is the sys-tematic character of the filtering process of natural selection. Repeated genetic mutations relentlessly push evolution forward and drive it consis-tently from simplicity to complexity. Human beings are without question infinitely more complex than the single-cell organisms that appeared on Earth some 3.5 billion years ago. Yet it is by no means evident that this sys-tematic progression from the simple to the complex follows directly from Darwin's theory.

ARE BACTERIA BETTER ADAPTED THAN GIRAFFES?

A common perception is that "lower" organisms have a lesser ability to adapt than "higher" ones. But who is to say that single-cell organisms have been less successful than we in the struggle of life? After all, they have survived for more than 3 billion years (there are now approximately 27,000 different known species of algae), and no one can guarantee that the human race will last this long, given its propensity for playing with nuclear weapons and for upsetting and polluting the ecosphere. As our fight against AIDS demon-strates, we are all too often on the losing side against the threat of viruses (there are roughly a thousand known species of these very simple organ-isms, compared to 1.4 million known animal, plant, and microbial species). Giraffes and rhinoceroses may be more interesting creatures to study than bacteria and microbes, but are the former truly better adapted to their environment than the latter? If one measures the ability of a living species to adapt to its environment by the number of its offspring, there can hardly be any doubt that bacteria have been far more successful than tigers and lions.

That said, why did Nature bother to "tinker"—as the French biologist François Jacob (1920–) put it—with increasingly organized and complex organisms? Why did she not stop at the stage of single-cell organisms? Why is Earth not populated solely by cells that spend all their time dividing furiously? Why did it embrace biodiversity?

Darwinism provides no explanation for this systematic progression toward complexity. All it can say is that the first living organisms were extremely simple and that evolution had no choice but to make them more complex. When one is at the bottom of the ladder, one cannot go any lower. The only choices are either to stay put or go up. If so, why did Nature not choose to stay where she was?

While studying the mutations of fruit flies (drosophila) in the laboratory, scientists have tried to quantify the occurrence of harmful mutations relative to beneficial ones. But this runs into several difficulties. To begin with, what appears harmful or beneficial to us is not necessarily so to flies. Next, how does one quantify, from the vantage point of adaptation to the environment, the advantage provided by a long neck (in the case of the giraffe), many legs (in the case of the millipede), or a large volume (in the case of the elephant)? How can one estimate the number of additional offspring an animal produces because its neck is a few centimeters longer? To make matters worse, we can never have a complete and detailed knowledge of the environmental conditions (climate, geography, etc.) that have existed during the past 3.5 billion years, nor of the exact morphology and behavior of the living organisms evolving in those environments of yesteryear.

Realizing that chance is incapable of driving evolution toward complexity, that it cannot be responsible for the richness and diversity of the biosphere, a number of scientists and philosophers have gone back to the idea that there exists a principle guiding the evolution of living species on Earth toward ever-higher levels of organization. The idea is not new: Aristotle already spoke of evolution as driving toward a final cause. He believed that the components of living organisms behave in such a way as to form a coherent whole, because a perfect "idea" of an organism exists even before that organism begins to develop. As discussed before, the idea of a guiding principle is also at the root of vitalist theories. According to Henri Bergson, an "élan vital," or life energy, incites biological systems to organize themselves and develop in a creative and performing fashion. Other thinkers have added a religious dimension to this organizing principle. For instance, the French Jesuit Pierre Teilhard de Chardin (1881–1955), a noted theologian and paleontologist, was of the opinion that the evolution of living species tends toward a final stage he called the "omega point," which in his view epitomized communion with God.

A SELF-ORGANIZING BIOSPHERE

Recent developments pertaining to the laws of self-organization and complexity suggest, however, that it may not be absolutely necessary to invoke mystical or transcendent principles to explain evolution. We have seen that open systems (those that interact with their environment)—whether they be physical (as boiling water) or chemical (as the chemical clock)—can go through "bifurcation points" that suddenly push them into new, more highly organized states. In that light, it is not unreasonable to assume that biological evolution might have proceeded in the same way. Instead of being driven by random genetic mutations and natural selection, evolution perhaps progressed by means of a series of bifurcations, of self-organization coming on top of self-organization, so as to climb the ladder of complexity.

There are several reasons why this scenario is not implausible. First, living organisms are open systems par excellence. Life never exists in isolation. It continuously exchanges energy with its environment, whether to absorb food or to excrete wastes. Furthermore, there are inevitably agents of change at play—both internal and external—that upset the equilibrium of the biosphere and drive it out of equilibrium. Changes can be gradual or sudden. The progressive oxygen enrichment of Earth's atmosphere is one example of a gradual change. In a process called "photosynthesis," vegetal species absorb soil water through their roots and carbon dioxide through their leaves, using solar energy to convert these elements into sugars and expelling oxygen into the atmosphere. Earth is the only planet in the solar system to have oxygen in its atmosphere because it is the only one to support plants and flowers. Other agents of change can be quite sudden, as in the case of solar flares that send high-energy particles toward Earth, or cosmic rays originating in supernovae and raining down on the planet, capable of inducing genetic mutations in living organisms. In our discussion of contingent events fashioning reality, we have already mentioned the enormous and brutal change caused by the collision of a giant asteroid with Earth 65 million years ago: It was responsible for the extinction of the dinosaurs as well as many other animal and vegetal species, thereby permitting the emergence and development of mammals and, ultimately, of our ancestors.

Whether they are gradual or sudden, catastrophic or benign, such changes thrust the biosphere out of its dynamical equilibrium and drive it into states of higher organization. Assuming that principles of self-organization apply equally well to living and inert matter, evolutionary changes must have occurred not gradually, but in fits and spurts, much as physical and chemical systems change abruptly when a certain critical threshold is crossed. Modern biological studies seem to support this view. The idea so dear to Darwin that evolution must proceed one small step at a time, imperceptibly and continuously, has been challenged since 1972 by the American biologists Stephen Jay Gould and Niles Eldredge. Drawing on paleontological evidence, they have shown that if evolution had proceeded continuously, we should be able to find many more fossils corresponding to all intermediate forms bridging the main species of living creatures. Darwin had earlier rejected this argument, maintaining that the geological record is very spotty because we have dug at too few places around the planet, and not all ancient species end up preserved as fossils. According to Gould and Eldredge, the reason why so many links are missing is that they simply do not exist. They take the view that biological evolution proceeds in successive stages of "punctuated equilibrium." Living species would remain unchanged for extremely long stretches of time, and then undergo profound changes in relatively short periods. To borrow a term from the quantum theory of atoms, evolution would occur in "quantum jumps." It is very likely that the spark of life appeared during the first such "jump."

And so, we are once again led to the conclusion that a global organizing principle of a holistic nature, one that cannot be reduced to the level of elementary particles but does not violate the laws that govern them, presides over the evolution of the biosphere. Living species do not result from a series of accidents. Molecules performing their metabolic dance in cells, cells assembling themselves to form living organisms, organisms integrating themselves in turn in an ecosphere of dazzling beauty and astonishing variety and complexity—none of this stems solely from random "tinkering." Natural selection certainly has an important role to play, but it is not the exclusive agent fueling evolution in its march toward increasing complexity. The credit goes to a principle of self-organization, which emerges naturally and spontaneously from a biosphere driven out of its dynamical equilibrium by phenomena that are either gradual or sudden, contingent or predetermined.

THE ARROW OF TIME AND THE DEATH OF THE UNIVERSE

Biological evolution confers a well-defined direction to time. Like an arrow propelled by the string of a bow toward a target, evolution moves forward, from the simple to the complex, from lack of structure to elaborate organization, from disorder to order. This state of affair seems to flatly contradict what is known as the "second law of thermodynamics" (thermodynamics is the branch of science that studies the behavior of heat). The law states that, in a closed and isolated system, disorder (as measured by a quantity physicists call "entropy") must increase (or at best remain constant) as time progresses. We can see manifestations of this law virtually every day, as it causes disorder to take over things left to themselves: A castle falls into ruins for lack of maintenance, a bed of flowers is invaded by brambles and weeds when a gardener no longer takes care of it, and so on. Disorder almost invariably wins over order inasmuch as there exist a thousand more ways to create disorder than to institute order, as we have pointed out above. If you shuffle a deck of cards, do not expect to find the cards arranged in numerical order. If you toss the pieces of a puzzle on a table, it is unlikely that they will arrange themselves spontaneously so as to reconstruct the original picture. There are many more ways to produce a disordered deck of cards or puzzle than an ordered one. Disorder is so much more probable than order that it almost always prevails.

The German physicist Hermann von Helmholtz (1821–1894) uttered this cry of despair in 1854: "The world is running to its death!" He was foreseeing a universe relentlessly invaded by disorder. The increase in entropy that inevitably accompanies all natural processes would lead—or so he feared—to the end of any creative activity within the universe. Every second ticking away, every passing hour, the universe would squander its energy reserve and dissipate it into heat; because it is a degraded form of energy, heat could never again be put to use as a source of energy and would become worthless as far as the running of the universe is concerned. Ultimately, the entire supply of energy would be exhausted and the universe would lapse into a sterile state of thermodynamic equilibrium from which any temperature difference would be excluded, any creativity banished, and where decay would take over. The universe would die suffocated in a vast bath of degraded energy that would be of no use whatsoever. The entire cosmic machinery (planets, stars, galaxies, etc.) and all the creations of man's genius (Beethoven's symphonies, Van Gogh's *Sunflowers*, and everything else) would be buried under the debris of a universe irremediably in ruins. The Sun shining by day and the stars twinkling at night are perfect examples of how the universe's energy degrades and becomes worthless. They will someday exhaust their supply of nuclear fuel and die. Myriad stellar remnants, black dwarfs, neutron stars, and black holes will then litter the galactic expanses. The universe will be plunged into a dark and glacial night and will no longer be able to sustain life.[9]

STARS CREATE DISORDER

If indeed the second law of thermodynamics leads inexorably to the decay and death of the universe, if it replaces order with disorder and offers no hope, how do we account for the organization and harmony of the cosmos? How to explain that we do not live in a totally chaotic universe? Astronomical, geological, paleontological, and biological studies all show that, on the contrary, the universe progresses—at least in some places—from disorder to order, from the simple to the complex, from the disorganized to the organized. Out of an energy-filled vacuum, the universe has manufactured elementary particles, galaxies, stars and planets, and, on one of these planets, life. It knew how to foster ever more complex and sophisticated states of energy and matter. Instead of following an arrow leading to sterility and despair, it was able, here and there, to follow another one guiding it toward creativity and hope. Does this mean that, in those particular places, the second law of thermodynamics is being violated? Not in the least! The law in question in no way precludes pockets of order and organization from appearing in some places, as long as their order is offset by more disorder somewhere else. The stars are the agents creating the disorder necessary to compensate for the order required to maintain the cosmic organization.

Much as warm water cools off when it comes in contact with cold air and imparts the disorder of its molecules to those of air, thereby increasing the disorder of the universe, so do stars disgorge their warm light into their colder environment, enhancing the overall disorder of the universe.

The total amount of disorder generated by stars more than makes up for the shortfall (what physicists call "negentropy") created by the organization of matter in increasingly complex structures. The net disorder of the universe increases with time, and the second law of thermodynamics is respected overall. Likewise, life could appear on Earth because the Sun, by radiating its fire, generates enough disorder to offset the order resulting from the emergence of terrestrial life. Elementary particles, atoms and molecules, stars and galaxies, and, most of all, life were able to organize themselves and escape the disorder and decay so feared by Helmholtz because they are open systems capable of constantly exporting disorder into their environment, which allows them to fight against the permanent threat of degradation embedded in the second law of thermodynamics. The universe, by contrast, is a closed system. Being—by definition—the largest possible entity, it has no environment to export disorder to and spare itself from ultimate degradation. Yet the universe did not slavishly conform to the dictates of thermodynamics. It showed itself infinitely more inventive by creating islands of order where open systems in nonequilibrium were allowed to self-organize. To the arrow of despair, it managed to oppose an arrow of hope. For void and barrenness, it was able to substitute a marvelous cosmic architecture in which life and conscience could emerge.

« 7 »

The Unreasonable
Effectiveness
of Thought

LIVING BEINGS TEND TOWARD A GOAL

The brain is unquestionably the most complex and highly organized system ever created by Nature. It has the ability to shape our behavior, make us feel joys and pains, fill us with wonder at the birth of a baby, but also to overwhelm us with grief at the passing of a loved one. It can generate thought, and produce sublime musical pieces and immortal literary works. It is the seat of free volition and sparks questions about the universe that begat it. It soothes our sleep with pleasant dreams, but can also disturb it with horrible nightmares. The question arises once again: Can such an elaborate structure be explained solely on the basis of the laws of physics and chemistry? Can human behavior be reduced to a matter of neuronal networks and elementary particles, or do we have to invoke principles of self-organization and complexity operating on a level different from that of the laws of particle physics?

We need only observe the behavior of living beings around us to realize that their actions are directed toward a goal, that they conduct themselves with the purpose of accomplishing a specific task: a bird seeking food and bringing it back to its nest perched high atop a tree to feed its progeny, a dog playing with a bone, a cat dozing in the warmth of the Sun, and particularly, human beings going about their daily routines.

Take, for instance, a blue jay feeding its young. Its entire organism operates in a harmonious fashion toward a singular purpose—finding food. All the different functions of the bird's body participate in an integrated and coordinated fashion in a common strategy. The eyes seek and locate food, the wings carry the bird to where an earthworm has been spotted, the beak clasps the worm, the wings spread out again to carry the bird back to the nest where the chicks are waiting. It all takes an extraordinary coordination of extremely complex and interdependent functions. It is difficult to escape the conclusion that the bird's behavior has a teleological character, that it is focused on a goal. It is impossible not to think that the bird has in its brain a clear mental picture of the objective it is trying to reach—to feed its young.

It is extremely hard to account for such a decidedly teleological behavior on the basis of pure reductionism. From that perspective, the atoms and molecules in the bird's body are supposedly influenced by their nearest neighbors only, through the action of local forces obeying the laws of conventional physics. If that is true, how do they coordinate their actions on the scale of the entire body of the blue jay in order to fulfill its "will"? We are once again confronted with the "profound epistemological contradiction" that Monod pointed out.

INSTINCTIVE BEHAVIOR

Faced with this contradiction, reductionists are wont to retort that all behaviors in living beings are not necessarily voluntary or conscious, that some fall in the domain of instinct. For instance, it is not obvious that a spider weaving its web has in mind a global strategy, a grand design it executes step by step. According to classical biology, it merely acts out of instinct. No one ever had to teach a spider how to weave a web. This knowledge is embedded in its genes. Likewise, a colony of ants building a nest is not "conscious" of the task it carries out. There is no master plan conceived by engineer ants and followed by worker ants. Rather, the plan is contained in their genes, and they just execute it blindly. Likewise, their social organization (a colony of ants is like a superorganism composed of worker ants toiling around the queen ant) is dictated by instinct.

Yet even instinctive behavior is not easy to explain in a reductionistic manner. How can a simple rearrangement of molecules in a spider or ant gene translate into an activity as structured and organized as weaving a web or building a nest? The problem is similar to the one we encountered in our discussion of morphogenesis, except that it is now compounded. Whereas previously the issue was one of spatial organization of concrete forms, we are now dealing with an activity taking place on an abstract level. Could it be that the instructions are written sequentially in the genes and "read" by the spider or the ant the way a computer "reads" a program designed for playing chess?

Such an analogy is not entirely satisfactory, though, because the behavior of living beings, even when it is instinctive, shows a remarkable ability to adapt to its environment, something a programmed machine cannot do. For instance, when a spider runs into an obstacle while weaving its web, it will figure out a way to get around it so as to complete the task. Likewise, when an ant colony moving in a single file finds its way blocked, it will, of course, be momentarily confused, but will quickly alter its strategy and adjust to the new circumstances. By all evidence, the behavior of spiders and ants is not dictated blindly and mechanically by a series of predetermined instructions, as is the case for a computer. There must exist in their organisms some feedback mechanisms that enable them to modify their behavior so as to adapt to changing conditions in their environment that they detect through their sensory organs.

Unlike a computer, an ant cannot be considered a closed and completely isolated system. Its individual behavior is integrated in a holistic way into that of the entire colony. The survival instinct of the colony is far stronger than that of an individual member. That is why, when a wasp happens to wander into an ant's nest, it will face retribution not from a single ant but from the entire colony, united by the instinct to protect the queen and defend their territory. This sense of preservation of the colony and its social structure is so powerful that ants act like genuine kamikazes, ready to die to protect the nest against any intrusion or danger threatening their society. It is probably this protective instinct that explains why ants are so numerous on Earth: They alone account for over 10 percent of all the animal biomass in the Amazonian forest, and nearly half the biomass of all insects living on Earth. More than 20,000 species of ants have been inventoried across the globe.

THE STARRY SKY IN THE BIRDS' DNA

One of the most spectacular examples of instinctive animal behavior has to do with the migration of birds. Cold weather drives them toward more clement and hospitable skies, until they return to delight us once again with their songs as soon the mildness of spring reappears. Apparently helped by a knowledge of the sky and Earth's magnetic field, birds are capable of astonishing navigational feats. Experiments have shown that certain species can travel thousands of kilometers without straying from their predetermined route. This despite the fact that they have never before made the trip and nobody ever taught them how to do it. What is even more remarkable is that birds released hundreds, if not thousands, of kilometers from their habitat, in totally unfamiliar surroundings, fly in a nearly straight line back to their nests.

Die-hard reductionists would assert that this extraordinary capacity for orientation and navigation is programmed in the birds' DNA molecules. But even if that were true, we still would not have the slightest clue as to how a living organism manages to translate a particular arrangement of genes into so remarkable a behavior, capable of adapting to unpredicted and nonprogrammed events. Pushing the argument to the extreme, we might even claim that, if we knew how to decode the information stored in the bird's DNA, we could in principle reconstruct the entire map of the sky. Better yet, we might even be able to retrieve the entire time-lapse record of the sky as it changes over the course of a year. Indeed, birds must take into account the changing appearance of the sky as Earth completes its yearly revolution around the Sun; the summer constellations are not the same as those in the winter. A brilliant geneticist could thus learn all about astronomy without ever having to raise his eyes to the sky. Such an absurd conclusion can only lead us to think that new emergent principles must be involved in guiding the flight of birds. These principles are without significance at the lower level of elementary particles, but they manifest themselves at a higher level as soon as the system crosses a certain threshold of complexity. At each new level of complexity, new laws of self-organization emerge that cannot be deduced from the laws prevailing at the lower levels. Once again, a purely reductionistic and mechanistic explanation is doomed to failure.

"I THINK, THEREFORE I AM"

If instinctive behavior is difficult to explain in terms of particles, the problem is far more complex still when this behavior becomes conscious. Defining consciousness is no trivial task. The dictionary tells us that it is the "perception, the more or less clear knowledge, that each individual can have of his existence and that of the external world." Our existence is defined by thought, but also by memory, volition (what we might call the "mind"), as well as by all the feelings and emotions that come from interacting with the external world: love or hatred, hope or despair, attraction or repulsion, joy or grief, etc. It is all part of what we call the "human condition."

The quest for understanding the meaning, purpose, and workings of the mind is as old as civilization itself. The Greek philosopher Aristotle believed that the mind resides all over a living organism, forming a single entity with the body and infusing life in it. All forms of life—man as well as plants and animals—were endowed with a mind; Aristotelian thought was strongly "animist." Curiously, though, Aristotle did not connect the mind with the brain. In his view, the only function of the brain was to regulate body temperature. Instead, the organ with which he associated the mind was the heart. On the other hand, the Greek philosopher Plato (ca. 427–347 B.C.) did make a connection between mind and brain. Since the brain was more or less spherical in shape—the one closest to perfection—it was the ideal candidate to harbor the mind. To this day, our culture continues to echo these two points of view. We routinely associate—at least metaphorically—our emotions with the heart. A love story ending in tragedy "breaks our heart." Amorous feelings are often depicted by a heart pierced by an arrow shot by Cupid. By contrast, when we are dealing with mental or intellectual faculties, we readily attribute them to the brain. "He is a brain," we say with admiration of a student who graduates with top honors.

Neither Aristotle nor Plato made a distinction between the mind and the body. It would not be until the seventeenth century that the two became separate. The French philosopher René Descartes (1596–1650) was the first to explicitly formulate this division.

In 1629, he retreated to a Dutch inn to ponder the mysteries of the mind. He was intent on finding a principle that would bring certainty to the reality of the mind. Descartes had realized that the perception of reality by our senses can be quite deceptive. After all, our senses tell us that, since the Sun rises in the east and sets in the west, Earth must be standing still at the center of the solar system, and the Sun and all other celestial objects revolve around it. Yet Copernicus had demonstrated as early as 1543 that exactly the opposite was true: The Sun was in fact at the center of the solar system, and Earth revolved around it. The Cartesian system of thought is based on doubt: Everything must be challenged, for we cannot trust our senses, which are apt to lead us to errors. Objects appear to us as real in a dream as when we are wide awake. Who is to say that life itself is not but a dream?

That night, in his room at the inn, Descartes had a sudden revelation. If everything can be questioned—the room he was in, the chair he was sitting on, etc.—at least one thing was beyond dispute: That was the very fact that he had doubts. When he was doubting, he had to be thinking, and because he was thinking, he had to exist as a thinking being. Descartes expressed this conviction in his famous maxim: *Cogito, ergo sum,* "I think, therefore I am."

THE DUAL NATURE OF MAN

For Descartes, reality had two distinct forms: that of the mind (or thought), and that of the material world. Such is the essence of Cartesian dualism. The mind is pure consciousness, has no spatial extent, and cannot be subdivided. By contrast, matter is devoid of consciousness, extends through space, and can be subdivided. Thus, man has a dual nature: He thinks, but he is also endowed with a material extension that is his body.

Descartes believed that the body was a perfect machine, in the image of the automatons in the Royal Gardens at Saint Germain, in Paris, which had fascinated him. In his *Treatise of Man,* published in 1664, he described a visitor walking in the Royal Gardens, stepping onto flagstones to cause water from a reservoir to flow through pipes and activate the automatons. Completely distinct from the mind, the brain was for Descartes one component of this perfect machine. It would function in accordance with principles similar to those of automatons. For him, the sensory organs were stimulated by the environment just as the automatons were activated by the weight of the visitor pressing on the flagstones of the Royal Gardens. The brain had its own pipes and valves, as well as a reservoir of fluid whose flow was controlled by these stimuli. The thinking mind was stationed near the reservoir and, like an attentive chief engineer, looked after the proper operation of the mechanism, occasionally intervening directly to open or close the valves in the brain. Contact between the mind and the brain took place at a particular location in the brain that Descartes called the "pineal gland." Like a mathematical point, this gland had no spatial extent. Through it, the mind could respond to the passions and humors of the body, although it also had the power to distance itself from the body's "base" impulses, such as lust and hatred, and to operate completely independently from it.

Cartesian dualism was to exert a considerable influence on Western thought. To this day, the brain and the mind are regarded as two distinct entities in Western medicine. When we have a headache, we consult a neurologist; when we are depressed, we are told to see a psychiatrist. True, the chasm has been narrowing as a more holistic view of mind and body is gradually gaining ground, spurred on by a new kind of medicine that treats bodily and mental dysfunctions at the same time, more in the tradition of Oriental medicine.

IS THE WORLD MERELY MENTAL, AND IS MAN NOTHING BUT MATTER?

Cartesian dualism professes that mind and matter are distinct but coexist. Other schools of thought advocate far more extreme positions. At one end of the spectrum, only mental events exist; at the other, only matter has a reality.

The first position was promoted by the Irish bishop George Berkeley (1685-1753). He maintained that the material world has no real existence and that the things it contains are merely images in the mind, God being the source of perception of these mental constructs. This philosophy, called "idealism," was vigorously attacked by the Englishman Samuel Johnson (1709-1784), who took the view that material objects exist independently of the perception we have of them. He argued: "You need only kick a rock hard to realize that it really does exist!"

The other extreme position is known as "behaviorism." This movement, whose goal was to elevate psychology to the status of objective science, was born in the early part of the twentieth century in the United States with the American psychologist John Watson (1878-1958), and subsequently developed by, among others, the American psychologist Burrhus Skinner (1904-1990). Behaviorism promotes the study of living beings exclusively by studying their responses to external stimuli originating in their environment. Because they reject anything that cannot be observed directly, behaviorists deny the existence of the mind and of any mental act. In this spirit, all that is needed to determine the behavior of a living being is to know the stimuli that act on it. Man's behavior is nothing more than a series of reflexes, much as the dogs of the Russian physiologist Ivan Pavlov (1849-1936) salivated at the sight of a bone. All mental activities that do not translate into observable acts—such as reasoning, making a decision, or thinking—do not exist. Behaviorism is completely deterministic and reductive.

DO ANIMALS THINK?

Man occupies a special place in the Cartesian scheme. He alone is endowed with a mind. Descartes believed that animals did not possess one, that they were simply extremely complicated automatons. Other thinkers have rejected this point of view and proposed to endow all matter in the universe—living or inanimate—with a consciousness. This "panpsychism" has been promoted by, among others, Teilhard de Chardin and, more recently, by the British-American physicist Freeman Dyson (1923-), who holds that mind is present in every particle of matter.

What can we tell about the existence of mind in animals when we observe their behavior? They clearly experience feelings and emotions similar to our own. Anyone who has ever seen a lioness nurture her cubs cannot question her maternal love. Anyone one who has heard the strident shrills of a bird being pursued by a cat cannot doubt its fright. Anyone who has witnessed a dog jumping at his master upon his return cannot fail to appreciate that it is overcome with joy. Still, can animals think—that is to say, form mental images? They seem to be able to recognize abstract properties, such as shapes and colors. But it is far less certain that they possess an awareness of themselves and of their existence. No one has ever seen, or expects to see anytime soon, a chimpanzee—the closest animal to man genetically—paint the *Mona Lisa* or write *War and Peace*.

THE MIND EMERGES FROM THE COMPLEXITY OF THE BRAIN

In light of what we have said previously concerning principles of self-organization that emerge in systems as they cross a threshold of complexity, what insight can we acquire on the issue of mind and body? After all, the brain is the most complex system Nature ever produced. We should, therefore, expect emergent phenomena to manifest themselves at a higher level, with properties different from those of neuronal processes (such as the flow of electrons in neurons), which belong in a lower level. These emergent phenomena can be assimilated with what we call the "mind." Consciousness "emerges" from neuronal activity in the brain, but at a higher level. Once generated, mental states follow their own causal laws, which are different and cannot be deduced from those governing the workings of neurons at the lower level. Emergent mental states can in turn act on the neurons that produced them by means of their holistic and collective behavior.

As the American Marvin Minsky, an expert on artificial intelligence, put it:

> The atoms in the brain are subject to the same all-inclusive laws that govern every other form of matter. Then can we explain what our brains actually do entirely in terms of those same basic principles? The answer is no, simply because even if we understand how each of our billions of brain cells work separately, this will not tell us how the brain works as an agency. The "laws of thought" depend not only

upon the properties of those brain cells, but also on how they are connected.[1]

Thus, psychology cannot be reduced to physics and chemistry. Behaviorists arrived at their deterministic and reductive conception by copying the methods of the so-called "objective" sciences (even though, as we have seen, quantum mechanics has greatly diminished this objectivity by ascribing a primordial role to the observer in creating reality). That does not mean, however, that psychology should reject the laws of physics and chemistry. These operate at a different level. Psychology requires additional principles that function on a higher level of organization.

We can once again use the example of a computer playing chess: The program provides the computer with the instructions on how to move the pawns; but under no circumstance are the rules of chess, as programmed into the computer, in conflict with the laws of physics that control the workings of the electronic circuits inside the computer.

CAN MACHINES SURPASS US IN INTELLIGENCE?

Studying the laws of complexity and self-organization thus casts a new light on the old mind-body problem. In all probability, both exist in their own right. Extreme positions like idealism and behaviorism are no longer considered plausible. Matter is not just a manifestation of the mind, and mental states in living beings cannot be reduced to elementary particles and reflexes. Nor are mind and body considered separate entities any longer, contrary to the assertions of Cartesian dualism. The mind can emerge from the body. Our "self," together with its consciousness and free will, can emerge from the organization of matter in the brain. It operates at a different level from that of the material extent that is our body. The laws that govern this superior level are new and cannot be deduced from those that preside over the lower level of matter. By the same token, they do not contradict them.

One question arises: If the mind can emerge from the complexity of the brain, does it mean that, as technology advances, computers will in due time cross a certain threshold of complexity and begin to think? Will they become conscious of their existence, will they be transported with love and consumed by hatred, overcome with sorrow and pity? Will they exult in joy and moan in pain? Will they agonize over moral issues? Will they start creating literary works and composing music? Will they ponder over philosophical questions and get engulfed in theological controversies? Could they even surpass us in intelligence?

The British mathematician Alan Turing (1912–1954) proposed in 1950 a simple test to assess the degree of intelligence of a machine. Suppose, he argued, that you engage in a conversation with two hidden interlocutors, one of which is a human being and the other a computer. If, during the conversation, you find yourself unable to tell the difference between the two, you will be forced to conclude that the computer is as intelligent as the human being. He predicted that, by the year 2000, computers would be conversing with such sophistication that they would dupe an "average" questioner at least 30 percent of the time after about five minutes of dialogue. Perhaps that is true. But could they fool an "intelligent" questioner at the conclusion of a very extensive conversation?

I do not believe so. Our brain is so extremely complex that an understanding of how it works remains enveloped in a thick fog that is not about to lift. We have not the slightest clue about the processes that allow us to create, feel, and love. Despite the unquestionable progress of modern neurology, we still do not fully understand how memory functions, how information is stored, and how we retrieve it. As long as we do not have a clearer picture of how the brain works, we will be in no position to build a computer capable of rivaling the human mind, let alone surpassing it. Such an understanding of the brain will probably not happen until we have a deeper comprehension of the relationship between observer and reality in quantum mechanics. Why does an event not realize itself until after an observer has registered it in his brain, after he has become aware of it? The British mathematician Roger Penrose has postulated that reaching this level of comprehension will require new physical laws that go beyond quantum mechanics.

THE ABACUS AND THE COMPUTER

As long as computers amount to no more than complex circuits in which electrical currents flow in accordance with programmed instructions, they will remain machines incapable of thinking, feeling, loving, and hating. They will do no more than blindly manipulate binary digits—that is to say, strings of 1s and 0s, 1 being represented by the presence of an electrical impulse, and 0 by its absence. In other words, a computer is nothing but an extremely sophisticated version of an ancient instrument called the *abacus*, in which "1s" are represented by balls sliding along metal rods, while "0s" correspond to gaps along those same rods. Instead of electronic components passing or blocking electrical currents, depending on programmed instructions, the fingers on a hand move the balls and leave behind empty spots according to certain precise rules. Of course, the computational power of an abacus is severely limited by the small number of balls and rods, as well as by the time required for fingers to move the balls. Supercomputers can calculate infinitely faster than an abacus. Their computational speed makes it possible to solve mathematical problems that simply cannot be handled by hand on the time scale of a human life. For instance, no human mind could ever calculate the number pi to millions of decimal places the way a computer can. No individual could manipulate such huge numbers with such lightning-quick speed, precision, and flexibility. Be that as it may, a computer remains a giant abacus. Its mechanical manipulation of 0s and 1s cannot be compared to thought. To claim that the computer "thinks" would be tantamount to saying that the balls on the abacus "cogitate" as they are being moved to carry out an addition.

In 1997, the press made a big to-do when a supercomputer named Deep Blue defeated the grandmaster Garry Kasparov in a chess tournament. Some journalists interpreted the outcome as evidence that the supremacy of humankind is being threatened. I totally disagree with this view. The computer beat the human being because of its great ability to review all possible future moves consistent with a starting configuration of the chess pieces. A human chess player can anticipate only a few of those. He relies in large part on experience and intuition to discard those moves that are likely to prove disadvantageous. Deep Blue, on the other hand, could systematically review all possible future combinations for at least the next ten moves. As a matter of fact, it could examine 200 million positions per second. It was this phenomenal capacity that enabled it to win over Kasparov. Thus, the computer passed the intelligence test described by Turing. Had Kasparov not known the nature of his opponent, perhaps he could never have guessed that he was doing battle with a machine. Nevertheless, in no way does it mean that Deep Blue had acquired an intelligence comparable to that of a human being. Its victory over Kasparov was completely unconscious. Indeed, the computer was no more aware that it was playing chess than a car knows it is on its way to New York City. It could not have cared less whether it won or lost. It did nothing more than "stupidly" and strictly following the instructions programmed in its electronic circuits by humans. The will to win, the anxiety, the nervousness, the tension, the frustration at having made a poor move, or the pleasure at having devised a winning strategy—all of this was beyond Deep Blue's ability. Perhaps it is precisely because Kasparov experienced all these emotions that he lost the match. But it is a huge chasm from Deep Blue's electronic circuits to the brain of a bird capable of using the positions of the stars to orient itself, and an even greater one to the intelligence and emotions of a human being. As long as the workings of the human brain remain a mystery, the goal of making machines intelligent will be out of reach.

COLLECTIVE CONSCIOUSNESS

The mind and consciousness of an individual living being are not at the very top of Nature's pyramid of organization and complexity. There is an even higher level. Beyond individual consciousness comes collective consciousness, the kind that results from the experience of society as a whole. Collective consciousness is the origin of religion and culture; it molds artistic, literary, and scientific works, as well as social and political institutions. To use the terminology of the Austrian-British philosopher Karl Popper (1902–1994), these abstract entities issued from collective consciousness belong in World 3, while Worlds 1 and 2 are those of material objects and mental states, respectively.

Once again, the laws governing World 3 cannot be reduced to those of World 2, which themselves do not flow from those of World 1. They have their own structure and dynamics, and are independent of those that rule the other two worlds. For instance, the existence of a social organization (World 3) is not necessarily linked to a conscious state (World 2). We have seen that a society of ants is highly structured and organized, even though its behavior is rooted in instinct. The same is true of a colony of bees. In contrast to animal societies, human society is fashioned by conscious and deliberate decisions; the laws governing the dynamics of human society, the rules controlling our political, economic, and social systems, cannot be deduced from the mental state of a single individual, and even less from the laws of physics.

Whereas there is no causality toward the top—that is to say, the laws of World 3 cannot be deduced from those of Worlds 1 and 2—there can, however, exist a reverse causality toward the bottom, meaning that events in World 3 can influence those of Worlds 1 and 2. For instance, a change in government, accompanied by new political, economic, and social policies, or a collapse of the stock market—both events belonging squarely in World 3—can have important repercussions on the mental state of individual citizens, which is part of World 2. The material world around them (World 1) can also be affected. For instance, an economic crisis can prevent the construction of a superhighway or the renovation of a football stadium.

All through these pages, we have made the long climb up the pyramid of self-organization and complexity. We started with inanimate matter and made it all the way to socioeconomic and political systems, encountering along the way living beings endowed with consciousness. Does this conclude our ascent? Have we reached the summit of the pyramid with World 3?

Probably not. There are some who think that there exist still higher organization levels—for example, those who believe in destiny and postulate the existence of a greater law guiding the course of events toward a specific end. Believers in astrology hold that the destiny of each human being is controlled by a global cosmic principle that is reflected in the harmony of the heavens.

Nature herself, with her laws, principles, and exquisite fine-tuning, which allowed the emergence of life and consciousness, constitutes the most striking example of a superior organizing principle.

WHY IS THE UNIVERSE COMPREHENSIBLE?

In its relentless ascent toward complexity, the universe has given rise to man, endowed with consciousness and intelligence. Stardust coalesced to trigger the spark of life and cause the appearance of a living being capable of comprehending the cosmos. Poets have sung its beauty, artists have painted its harmony, but it fell to scientists to unveil its mysteries and reveal to us this truly miraculous fact: We live in a rational universe ruled by very precise laws that can be perceived and analyzed by human reasoning.

Why was man granted the gift of understanding the universe? After all, we could just as well have emerged in a universe of such complex organization that it would be beyond the reach of our comprehension; in such a universe, we would have been reduced to submitting to the laws of Nature without having the slightest idea what they might be. As Einstein declared: "What is the most incomprehensible is that the universe is comprehensible." Is the fact that we managed to explain the world by means of the scientific method merely the result of a fortunate coincidence, or was it somehow pre-programmed? Is our ability to grasp the universe a fluke, or was it inevitable that biological organisms emerging from the cosmic order should exalt this order by understanding it? Are our spectacular scientific advances merely accidents in the long, drawn-out history of the universe, or do they result from an intimate cosmic connection between man and the world?

In attempting to answer these questions, we must examine the essence of the immutable and intangible laws of Nature and how they apply to a universe in perpetual evolution.

PERMANENCE AND IMPERMANENCE

As it tries to understand the world, the human mind finds itself confronted with a profound dichotomy between the temporal and the intemporal, the becoming and the being, or, in Buddhist terms, impermanence and permanence. It is undeniable that there is an element of permanence in our lives—happily for our mental well-being! We recognize ourselves in the mirror every morning. There exists a reassuring constancy in the personality of people around us. The collectibles that decorate our living room, the pictures adorning the walls, the verdant trees sheltering the birds, the mountains in the distance, the full Moon returning every month, the spring flowers recurring every year, everything takes on an air of quasipermanence that confers a relative stability on our life and allows us to assign a distinct identity to each object and being. We refer to "Pierre" or "Carol," and we speak of the Moon or of a particular painting by Vermeer. Yet under the appearance of permanence hides a constant current of impermanence, an uninterrupted flow of changes, an endless festival of transformations. The people around us change not only physiologically—they break limbs, their skin becomes wrinkled, their hair turns gray—but also psychologically. Collectibles break, paintings fade, trees wither, and erosion gouges mountains; the Moon will no longer project her soft glow and spring will no longer blossom when the Sun runs out of nuclear fuel some 4.5 billion years from now. The universe itself is in constant evolution: It sprang into existence out of a vacuum some 15 billion years ago and will eventually die, either in an infernal brazier or in a slow glacial cooling. The present moves on and fades into the past, while the future remains to be built.

Plato was one of the first to reflect seriously on this fundamental paradox of existence—the dichotomy between the temporal and the intemporal. He believed that, because of this dichotomy, there existed two levels of reality. First was the reality of the world accessible to our senses, a world that is changing, ephemeral, and illusory. This sensible and temporal world would in fact be merely a pale reflection of the "true" world, that of eternal and immutable Ideas, in which mathematical relations and perfect geometrical structures reign supreme. According to Plato, all living beings in the sensible world are but imperfect copies of eternal forms residing in the world of Ideas. For instance, birds of a given species look alike because they are all material representations of the same perfect bird Idea. Birds in the changing and ephemeral world may well age, become sick, and die, but the bird Idea is eternal, immutable, and intangible. This makes it possible for a given species of birds to perpetuate itself through the centuries, since it can materialize over and over again in the image of the same immutable bird Idea.

THE SHADOWS IN THE CAVE

To illustrate the dichotomy between the sensible world and the world of Ideas, Plato introduced his famous allegory of the cave in his dialogue *The Republic*. Imagine, he said, a group of men imprisoned in a cave. They have their backs turned to the entrance of the cave and are facing the walls. Outside the cave is a world vibrant with light, colors, and shapes, but the men in the cave are totally unaware of it. All they can see are the shadows that the objects and living beings in the outside world project against the walls. For them, the world of shadows is the only reality that matters, since they do not know anything else. They cannot appreciate that the shadows are but pale reflections of the glorious reality existing outside the cave. They cannot realize that, if any of them could free himself from his chains and escape to the outside world, he would be blinded by the brightness and beauty of the beings and objects illuminated by the golden brilliance of the Sun. The sullen darkness and indistinct outline of the shadows would be replaced by the exuberance of colors and the sharpness of shapes in the "real" world. That world would be vastly more beautiful and perfect than the world of shadows.

In Plato's view, the world accessible to our senses is akin to the world of shadows experienced by the men in the cave. It is merely an imperfect manifestation of a perfect world—the world of Ideas, "illuminated by the Sun of intelligibility."

A TRAGIC MISTAKE

Plato's allegory of the shadows in the cave reminds me of a beautiful legend in my native country, on the far side of Earth where the Sun rises, which also has to do with mistaking the world of shadows for the real world, a mistake that can lead to tragic consequences. I leave it to the Vietnamese writer Pham Duy Khiêm to tell the story in his own moving words: "There once was a woman whose soldier husband had been sent to a border post, at the far end of 'the country where one goes by traveling upriver.' In those days, communications were very difficult and, during the three years he was absent, she received only sporadic news. One evening, she was sewing by the light of an oil lamp, next to her sleeping child, when a thunderstorm broke out. A gust of wind blew the lamp out, thunder began to rumble, and the child awoke. He became frightened. The mother relit the oil-soaked wick and, pointing to her shadow on the wall, she said to her child: 'Fear not, my darling; father is here to look after you.' The child looked up and stopped crying. The next day, at bedtime, he insisted on seeing his father again. The mother smiled contentedly and placed herself so that her shadow would be in clear view of her son. She taught him how to join his hands and bow before the shadow while saying: 'Good night, Father.' It quickly grew into a habit and, every evening, they would go through the same ritual. . . ."

Finally, the husband returned. His wife was moved to give offerings to the ancestors to thank them for having brought her husband back safe and sound. She went to the store to buy the offerings, leaving the child in her spouse's care. While she was out, the husband played with his son and tried to goad him into calling him 'Father.' Much to his surprise, the child refused, explaining that the man could not be his father since he said goodnight to the latter every evening before going to sleep. The man, concluding that his wife had been unfaithful, left the house without a word, his heart filled with pain. In despair, the distraught wife took her own life by throwing herself into the river. Upon learning of her death, the man began to have doubts and came back home. In the evening, he lit the lamp, which projected his shadow against the wall. He was stunned to see his son join hands and bow before the shadow. Only then did he understand his tragic mistake.[2]

AN IMMORTAL SOUL

Because there are two levels of reality, man also has a dual nature—or so Plato believed. He has a material body, which changes and ages with time, in which he evolves and through which he communicates with the imperfect and impermanent world of the senses. However, he also possesses an immortal soul endowed with reason, which can come into contact with the perfect and permanent world of Ideas. The soul preexists the body, although as soon as it moves into its habit of flesh, it loses any memory of the world of Ideas. As it discovers natural forms in the world of the senses—a cypress tree, a rose, a horse, a baby, etc.—a vague and distant memory of the world of Ideas is reawakened. Man then realizes that, when he sees a flower, what is present before his eyes is but an imperfect representation of the Idea of a perfect flower. This explains his constant nostalgia for Perfection and the soul's intense desire (what Plato calls *eros*) to return to the perfect world of Ideas, where it truly belongs. If, like the child in the Vietnamese legend, we confuse shadows for reality and look no further, we do violence to our deep longing for the perfection of our immortal soul.

The duality of the world entails a duality of the Supreme Being. The world of Ideas is ruled by the Good, an eternal and immutable being existing outside time and space. Meanwhile, in the world of senses, the Demiurge fashions matter according to blueprints from the world of Ideas; the demands for his attention are never-ending because the material world keeps wearing out and deteriorating, which requires constant interventions. By invoking two supreme beings, Plato does not attempt to resolve the profound dichotomy between the changing world of experiences and the unchanging world of eternal forms. He merely declares that the changing world of the senses is illusory and that the eternal world of Ideas is the only one that really matters.

TIME AND ETERNITY

Aristotle, a disciple of Plato's, did not try to resolve the paradox of time and eternity, either. He, too, decided in favor of a single world, but, contrary to Plato, he opted in favor of the temporal world, rejecting the Platonic world of Ideas. As we have seen, Aristotle proposed an animistic universe that develops toward a predetermined final goal, much like an embryo. Living beings are also endowed with souls guiding them toward the desired goal. However, unlike Plato, Aristotle believed that the soul did not transcend the body, but, rather, was immanent to it. The distinction occurred not at the level of the world of senses and that of Ideas, but between the heavens and the Earth. The Earth was the realm of imperfection, degradation, and death, whereas the heavens were the realm of perfection, eternity, and permanence. Celestial objects described circular and immutable orbits, the circle being the most perfect geometrical shape.

Saint Augustine, bishop of Hippo, abandoned the idea of a Demiurge, but seized upon the Platonic idea of a god operating outside the confines of time. This view raised a number of serious theological problems in the context of Christian doctrine. How can a god removed from time suffer and die on the cross, given that such events clearly belong in time? The debate was still raging in the thirteenth century, when Saint Thomas Aquinas, although endorsing the idea of a god residing in a Platonic sphere beyond time and space, tried to demonstrate logically, in the manner of a mathematical proof, that God had to possess a number of qualities such as perfection, simplicity, timelessness, omnipotence, and omniscience. That hardly resolved the problem of reconciling a god outside time with one who is merciful and attentive to our prayers set in time.

The next developments came in the sixteenth and seventeenth centuries. The science born in medieval Europe culminated in the work of the Italian Galileo Galilei (1564–1642), the father of the experimental method in physics, followed by that of the Englishman Isaac Newton (1642–1727), who authored a monumental theory of gravitation. The world of senses rose to preeminent status. The scientific method was rooted in precise observations and quantitative measurements of natural phenomena. Nature came to be thought of as rational, obeying eternal and immutable laws, which themselves reflected God. This was strongly reminiscent of the immutable and eternal god residing in Plato's world of Ideas, or of Thomas Aquinas's concept of a god living beyond time and space. The dichotomy between the temporal and the intemporal was making a powerful comeback. God was intemporal, but manifested himself in the form of temporal natural phenomena. Newtonian physics described a changing world, one that evolved in time—an apple falling to the ground in an orchard, the Moon orbiting Earth, the rising and falling tides. But the laws governing these phenomena were immutable and invariant. Such is the problem of the "arrow of time" in the physical sciences, which continues to be the focus of much attention today: How can intemporal laws describe temporal events?

To come up with even partial answers, it is helpful to examine the very nature of physical laws.

THE LAWS OF NATURE

The dictionary states that a law is "a general proposition acknowledging necessary and constant relationships between scientific facts." The phrase "necessary and constant relationships" implies "regularity." Our distant ancestors, a few tens of thousands of years ago, were already quite aware of such regularities. The dolmens and menhirs at Stonehenge, in England, aligned with sunsets and sunrises at specific times of the year, attest to it. Our forebears also realized intuitively that they could hit and kill precious game by throwing a stone in a particular way. But many natural phenomena remained unpredictable and mysterious. Cavemen ascribed spirits to them. They came to live in a magical universe in which the Sun spirit would light the Earth spirit during the day, to be replaced by the Moon spirit at night. The rain spirit would provoke downpours feeding the river spirit. The world of spirits mirrored the world of humans and was moved by the same desires and impulses. It was all simple and familiar, with a human dimension.

This familiarity and innocence disappeared as knowledge accumulated until, about 10,000 years ago, the human magical universe was supplanted by a superhuman mythical universe. Spirits deserted trees, flowers, and rivers, and were replaced by gods with superior powers. All natural phenomena, including the creation of the world, came to be viewed as consequences of the actions, loves, unions, hatreds, and conflicts of these gods. Yet, in the midst of this mythical universe, around the sixth century B.C., the Greek miracle took place, sowing the seeds of the scientific universe we know today. The Greeks had the revolutionary idea that natural events were not the exclusive province of gods, but that human reasoning could also apprehend them. They cast their curious and inquisitive gaze on subjects as varied and diverse as the structure of matter, biological, geological, and meteorological phenomena, the nature of time, geometry, and mathematics. Still, the concept of "natural law" as we understand it today had not taken root yet.

Aristotle explained the behavior of a natural system not by means of laws but in terms of a final cause. For him, natural systems behave so as to realize a goal, much as living beings do. They have a "teleological" behavior. Aristotle built an elaborate system of causality by distinguishing four kinds of causes—material cause, formal cause, efficient cause, and final cause. If

asked "Why does it rain?", the Greek philosopher would not answer simply
that rain falls because air, when cooling, causes water vapor in the atmosphere to condense into water droplets that fall to Earth's surface by virtue of
gravity, as a modern meteorologist would. Rather, he would make a distinction between the material cause, which is the water droplets, the efficient
cause, which prompts water vapor to condense into droplets, and the formal
cause which makes the droplets fall to Earth's surface. Most important,
instead of appealing to gravity to explain why the droplets fall, Aristotle
invoked a final cause: The drops fall to Earth because plants and animals
need water to live and grow.

The Aristotelian causality system can also be illustrated with the example
of a house: The material used to build it (bricks, tiles, cement, etc.) constitute the material cause. The efficient cause is the mason who assembles
those materials to erect the structure. The formal cause is the shape and
arrangement of the house. And the final cause is the architect's plan, which
the mason follows while building the walls. Each object in Nature would
behave so as to realize a predetermined goal. Mandarins and oranges ripen
so that men can eat them. It rains so that rivers do not run dry. The same
principle applies to the movements of material objects: A Stone falls to the
ground because Earth is the natural place of all heavy objects. Smoke rises
to the sky because its natural place is in the ethereal regions beyond the
heavens. Aristotle's "laws" boil down to elaborate descriptions of how the
final cause is realized. Every physical object has its own characteristics, and
the complexity of the world manifests itself in a boundless variety and diversity of these characteristics.

The notion of law familiar to us appeared with the emergence of
monotheistic religions such as Christianity and Islam. God became independent, divorced from his creation. Laws were no longer inherent to physical
systems, but came to be imposed from the outside by a supreme Being.
Nature became a domain ruled by divine decrees. The cause was no longer
contained within the object itself, but was related to the effect through physical laws originating with God.

The concept of laws imposed from the outside by a sovereign god met
with considerable resistance. As late as the thirteenth century, Saint Thomas
Aquinas was still defending the Aristotelian notion of innate tendencies in
physical systems, which were assigned by a providential god. However,
Aquinas's views were firmly condemned by the archbishop of Paris in 1277,
and the idea of a god creating and promulgating laws gained the upper
hand.

As modern science emerged in medieval Europe, most early scientists
were still working with the conviction that the order and regularity in Nature
reflected a grand divine plan, and that deciphering the plan would exalt
God's glory.

It was in this spirit that the German astronomer and great mystic
Johannes Kepler sought God's perfection in the movements of the planets.
For the English physicist Isaac Newton, the universe was a vast machine reg-

ulated with extreme precision by a rational Engineer-God. The universe supposedly functioned like a clock powered by a spring. However, opinions diverged about God's role once the clock had been wound: Did the mechanism keep running on its own, with God retreating into the distance since there no longer was anything for him to do? Or was the initial fillip that got the universe started not sufficient, in which case constant subsequent interventions would be required? Newton leaned toward the second alternative. He believed—wrongly, as we now know—that, without constant divine intervention, gravity, which attracts all things, would cause the universe to collapse on itself, which is clearly not happening. The German philosopher and mathematician Gottfried Leibniz (1646–1716), a contemporary of Newton, was of the opposite view. He stressed the absurdity of the notion that God would not know how to build a piece of machinery able to keep going by itself. For Leibniz, as well as for the French philosopher and mathematician René Descartes, God was the originator and the guarantor of Nature's rationality, and this rationality allowed human reason—itself a gift from God—to comprehend the cosmos.

Because, once wound, the machine ran by itself, God became increasingly distant. Reason reigned supreme, relegating faith to the background, until God was eventually no longer necessary. At the end of the eighteenth century, as Napoleon Bonaparte chided the marquis Pierre Simon de Laplace for not having mentioned even once the Great Architect in his work *La Mécanique céleste* (Celestial Mechanics), Laplace replied: "Sire, I have no need for this hypothesis!" Ever since, science and religion have continued to drift apart. Today, most scientists study the regularities of Nature, which they refer to as "laws," without inquiring—at least openly—about their origin.

WHY WAS SCIENCE BORN IN THE WEST?

Infused with the Christian concept of a God manifesting himself in Nature's rationality, permeated by the notion of civil law in society, medieval Europe was a particularly fertile breeding ground for the idea of natural laws—in other words, science.

One might wonder why science was not born in China, which enjoyed a sophisticated and complex thousand-year-old culture and was technologically ahead of the West in many areas (the Chinese had invented, for example, gunpowder and the compass). I believe that the answer lies in the way the Chinese perceived Nature. For them, the natural world resulted not from the action of a god creating and decreeing laws, but, rather, from the reciprocal and dynamical action of two opposite forces—the yin and the yang. Because the notion of the laws of Nature was alien to them, the Chinese had no incentive to look for them.

Furthermore, the Chinese had a holistic conception of Nature, every part interacting with every other part, forming a harmonious whole that added up to more than the sum of the individual parts. Such a holistic view did not promote the idea that Nature can, in a first step, be broken down into its parts, and each part can be studied in isolation, independently of the others. This concept is at the basis of the reductionistic method and is responsible for much of Western science, which would not have been possible absent the ability to understand a small fraction of the universe without comprehending the whole.

It is evident that a purely reductionistic approach cannot be the final word. We have seen that systems considered in their entirety possess "emergent" properties that cannot be deduced from the properties of individual components. For instance, it is impossible to explain life on the basis of individual inanimate elementary particles. Nonetheless, holism is not exclusive of reductionism: The two approaches are complementary and can both be helpful in unlocking Nature's secrets.

One intriguing question remains, however: How is it at all possible to understand a very small part of Nature without understanding the whole?

WHY DOES THE REDUCTIONISTIC METHOD WORK?

We could very well have emerged in a universe in which every physical phenomenon in a given place would be so intimately connected to the rest of the universe that it would be impossible to study it without understanding the entire universe. Everything would be so intricately interwoven that no simple law could ever be deduced. Our understanding of the universe would then be a matter of all or nothing. Instead, science made it possible for us to catch bits of information without knowing the entire plot, to hear a few notes of music without grasping the whole melody. The mystery is all the more puzzling since the available evidence suggests that the universe forms a highly interconnected whole. The fall of an apple in an orchard is determined not only by Earth's gravity, but also by that of the Moon, the Sun, and all the other planets. The Einstein-Podolsky-Rosen experiment showed us that the reality of elementary particles is not local but global. Two particles interact and fly apart. If one of them is perturbed, the other one knows it immediately, even if it is on the other side of the universe.

Yet, despite the interconnectedness of the universe, despite the strong conviction of many physicists that the principles of physics will someday culminate in a Theory of Everything, the reductionistic method works. It enables us to progress step by step, to assemble the puzzle one piece at a time without ever having seen the completed picture. Happily for me, I do not have to solve all the problems of the universe at the same time to do astrophysics.

Why does reductionism work? The answer is contained in two properties of certain physical systems. The first is called "linearity." A physical system is said to be "linear" when the whole is exactly equal to the sum of its parts—nothing more, and nothing less. In such a system, the sum of the causes produces precisely the sum of the effects, and it suffices to study the behavior of each component separately and simply add them up to deduce the behavior of the whole.

To understand what linear behavior means, consider a dry sponge placed under a faucet (Figure 68). At first, the weight of the sponge increases proportionally to the number of water drops it soaks up. The sponge doubles its weight when it has absorbed twice the number of drops, and triples it with three times as many drops. The relationship between the weight and the number of drops is said to be linear (the word comes from the fact that if you were to plot the number of absorbed water drops horizontally and the resulting weight of the sponge vertically on a graph, the points would all fall along a straight line). However, the behavior of the sponge changes by the time it has soaked up a lot of water. There comes a point when its ability to absorb extra water diminishes, until it becomes zero. When that happens, the weight of the sponge no longer increases in proportion to the number of drops falling on it. The behavior switches from linear to nonlinear. Finally, the weight of the sponge reaches a steady level, as it cannot absorb any more water. It has reached saturation.

We go through the same stages when we visit a museum full of works of art. At first, our aesthetical pleasure increases in proportion to the number of paintings we get to admire. We are in the linear regime. After a while, though, the pleasure wears thin; we gradually grow numbed by too many masterpieces, and we enter a nonlinear regime. Eventually, we reach a point where we cannot stand it any longer: We have become saturated with paintings.

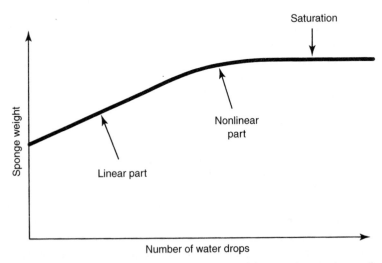

Figure 68. *The sponge and water drops.* Initially, the weight of the sponge increases in proportion to the number of water drops it absorbs. The relation is said to be linear because it is represented by a straight line in the above graph. When the sponge approaches saturation, it can no longer absorb all the water dripping on it, and the dependence becomes nonlinear. Systems with nonlinear properties can exhibit a complex and unpredictable behavior, and become chaotic.

AN INTELLIGIBLE UNIVERSE

The spectacular successes scored in physical sciences by the reductionistic method over the last three centuries is attributable to the fact that physicists were able to isolate physical systems with a linear behavior. As long as they remain in that regime, such systems can be "analyzed"—that is to say, understood in their totality by studying their individual components. For instance, electric and magnetic fields behave linearly. The electric field produced by a system of two electrons is the sum of the electric fields associated with each electron.

So impressive was the success of reductionistic and linear physics that it gave the false impression that the world is made primarily of linear systems. In reality, physical systems almost invariably become nonlinear beyond a certain threshold, much like a sponge saturated with water or a museum visitor exposed to too many paintings. At that point, these systems can no longer be analyzed, because the whole becomes more than the sum of its parts. The effects cease to be proportional to the causes. Chaotic systems are nonlinear par excellence. As we have seen, they are extremely sensitive to the slightest perturbation. The effect becomes incommensurate with the cause and can no longer be predicted: A molehill can become a mountain; a butterfly beating its wings in Australia can trigger a thunderstorm in New York.

THE SEPARABILITY OF THE WORLD

No physical system is ever perfectly linear. Even the orbits of planets, once believed to be stable and immutable, can go awry. But, in that case, why isn't the world completely chaotic? Why, despite a profusion of nonlinear phenomena, can we still manage to isolate a vast array of physical systems whose behavior is linear and predictable? Why can we do science?

The reason is due not only to the linear properties of certain systems in which effects are proportional to causes, but also to a second important property they exhibit, known as "locality." A physical system is said to be "local" if its behavior depends only on forces and influences restricted to its immediate surroundings. Global interactions, such as those involved in the Einstein-Podolsky-Rosen experiment, are excluded. The criterion is met when forces and interactions are either short-range or very weak, or both. The range of the fundamental forces in Nature is generally short. The strong nuclear force acts only on the scale of an atomic nucleus (10^{-13} centimeter), while the range of the weak nuclear force is about 10 times smaller (10^{-14} centimeter). Electromagnetic and gravitational forces have much longer ranges, but their intensity decreases as the square of the distance between two electrical charges or masses. Increase the separation by a factor of 10, and the corresponding force decreases 100-fold. This property ensures the locality of forces. The influence of forces at a given location comes entirely from the immediate surroundings, not from the entire universe. That is why the gravitational influence of the Moon, the Sun, and other celestial objects on the fall of an apple in an orchard is negligible when compared to that of Earth.

The intensities of the forces in Nature depend on about fifteen numbers called "physical constants." For instance, the intensity of the force of gravitation depends on what is known as the gravitational constant, which happens to be quite small (it is equal to 6.67×10^{-8} in the *cgs* system of units, in which lengths are counted in centimeters, weights in grams, and time in seconds). We know how to measure these constants with extreme precision in the laboratory, but have no theory thus far to explain why they have the particular values we measure. These constants were "handed down" to us when the universe was born. They were tuned with phenomenal precision to allow the universe to harbor life and consciousness (as we have seen, this delicate tuning is called the "anthropic principle," from the Greek *anthropos*, meaning "man"). Furthermore, this fine-tuning made it possible to separate the universe into distinct entities that could be studied individually with the help of the reductionistic method.

One question lingers on, though: How can the macroscopic universe escape the quantum fuzziness and global interconnectedness characteristic of the subatomic world? After all, the infinitely large arose out of the infinitesimally small. Early in the universe's history, when it was infinitesimally small, it had to obey the laws of quantum mechanics, with their inherent fuzziness. If so, why isn't the orbit of the planet Jupiter just as fuzzy as that of an electron around an atomic nucleus? Why do macroscopic systems with a local reality exist? Why has quantum fuzziness given way to the definiteness of everyday life? Why did the global reality of elementary particles turn into the local reality of macroscopic objects?

According to the American physicists James Hartle (1939–) and Murray Gell-Mann (the discoverer of quarks), this transformation could take place only because the universe's initial conditions (such as its density, temperature, and initial rate of expansion) were tuned to extreme precision, much like the physical constants mentioned above. Hartle and Gell-Mann's calculations show that a subatomic universe with arbitrary initial conditions could not, in most cases, result in a macroscopic world containing material objects sharply localized in space with a well-defined time. In such a universe, the separability of the world into distinct objects in time and space would be impossible, and the world would not be intelligible.

Thus, both the linearity and locality of certain physical systems in the universe—hence its intelligibility—are a direct consequence of the initial conditions and physical constants of the early quantum universe. The fact that we were able to progressively discover Nature's laws and gradually perfect our understanding of the world—indeed, that the scientific method works at all—was predetermined with extraordinary precision from the beginning. The unreasonable capacity of human thought to unravel the world is a consequence of the unreasonably precise tuning of the early universe.

THE SPIRIT OF LAWS

Natural laws are characterized by a set of generally accepted properties, which are strangely reminiscent of those usually attributed to God.

First, the laws of Nature are universal. They apply everywhere in space and time. The universality of physical laws has been verified many times over by astronomical observations. Because light takes time to reach us, seeing far is equivalent to seeing early. The light reaching us just now from a galaxy located 12 billion light-years away brings us information about a very young universe, when it was only 3 billion years old. As a result, our telescopes enable us to travel back in time. Not once in our many journeys into the past have we ever discovered physical laws different from those governing our small corner of Earth. No matter where we point our telescopes, we keep seeing the same physical phenomena.

Second, the laws of Nature are absolute. They depend neither on the individual who studies them nor on the state of the system observed. A physicist from Vietnam will deduce the same laws as one from America.

Even though natural laws connect distinct states of a given system at different times, they themselves do not vary with time. That is the third important property of these laws: They are eternal and intemporal, exactly as Plato's world of Ideas was.

Fourth, they are omnipotent. There is absolutely nothing in the universe that can escape their dictates, from the smallest atom to the largest super-cluster of galaxies.

Finally, they are omniscient, in the sense that material objects in the universe do not have to broadcast "information" on their particular states in order for laws to act on them. The laws "know" ahead of time.

ARE PHYSICAL LAWS DISCOVERED OR INVENTED?

Although there is a certain consensus about the properties of natural laws, disagreement rules when it comes to their status: Do these laws reveal genuine patterns in Nature, or are they merely products of man's imagination? Do Newton's law of gravitation and Maxwell's laws of electromagnetism translate fundamental and objective relations in Nature, or are they simply brilliant inventions that Newton and Maxwell used to describe regularities they perceived only subjectively?

Two separate camps have formed around this question. On one side is the "realist" camp, which holds that laws exist independently of ourselves and are just waiting to be discovered. Following in Plato's footsteps, a realist insists that the laws of Nature reside in the world of Ideas and possess a reality of their own, distinct from sensible reality. On the other side is the "constructivist" camp, for which natural laws arise only in the fertile imagination of physicists and exist nowhere but in the neurons and synapses of human beings.

Between these two diametrically opposite points of view, I come down squarely on the side of the realists. I am convinced that the regularities I perceive in Nature through my telescope are not a creation of my mind. The young stars emerging from stellar nurseries, the delicate arms gracing spiral galaxies, have a reality of their own. A constructivist is likely to reply that the human mind has a tendency to see regularities and patterns where there are none. He would point to the regularities the ancients thought they saw in the constellations in the sky; they exist only in man's mind, since the forms attributed to these constellations—a bear, a swan, or a lyre, et cetera—have

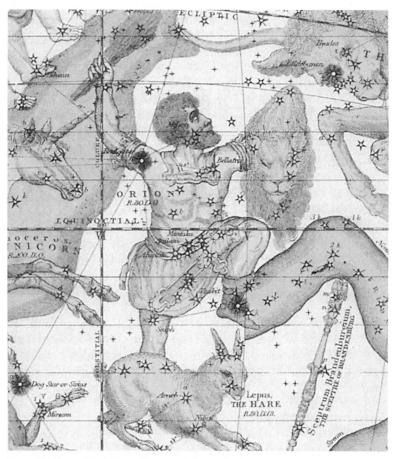

Figure 69. *Man and the heavens.* The constellations in the sky change during the yearly journey of Earth around the Sun. Here we see a depiction in an 1835 astronomical atlas of the constellation Orion, which is visible in the Northern Hemisphere in the wintertime. Throughout the ages and cultures, man has often projected onto the heavens his aspirations and dreams. Many constellations were named after mythical characters of antiquity. In Greek mythology, Orion was a giant-size hunter slain by Artemis, the goddess of hunting, whom he had offended. As punishment, Orion was changed into a constellation. While the forms men thought they saw in the constellations exist only in their minds, the stars in the celestial vault have a very real existence, and their reality is independent of the human mind.

varied according to the cultures and the times (Figure 69). Despite this pro-
clivity of the human mind to imagine patterns, I believe that constructivists
are wrong when they question regularities at a deeper level, the ones we
refer to as the "laws of Nature." These laws reflect very real regularities that
are not figments of our imagination but exist independently of ourselves.

There are several reasons for my conviction. To begin with, scientific
pursuits would have no meaning if regularities were imaginary. Second, laws
deduced from regularities that are already perceived bring novelty by revealing additional and unexpected patterns in Nature. Indeed, the great theories in physics do more than just provide a simple description of regularities
already uncovered. They lead us onto unknown paths, uncharted areas, and
they uncover unsuspected harmonies. Newton built his great theory of gravity for the purpose of accounting for the movement of planets; but it also
allowed him to understand the to-and-fro motion of ocean tides. The English astronomer Edmund Halley (1656–1742) used it to predict that the
famous comet now named after him would return to visit mankind every sev-
enty-six years. When Maxwell worked out the laws of electromagnetism, he
never anticipated that his equations would also reveal to him that electric
and magnetic waves were none other than light. When Einstein developed
his theory of special relativity, he had no idea that it would lead him to discover that matter can transform into energy, that this matter-energy equivalence is what makes the Sun shine, and that it would be responsible for the
devastation of Hiroshima and Nagasaki. He could not have predicted that
his theory of general relativity would lead to objects as strange and exotic as
neutron stars and black holes. Because laws reveal unsuspected connections,
regularities that never crossed our mind, they could not possibly be a pure
product of our imagination. To use a metaphor from the computer world,
the laws are Nature's software, while physical states are the hardware. The
software exists independently of the computer's electronic circuits.

Finally, there is a third reason for me to believe that natural laws are not
pure intellectual constructs. All laws can be expressed in a common language, which is that of mathematics. As it turns out, there are very good reasons for assuming that mathematics is not a human invention, but, rather,
inhabits a Platonic world of Ideas that is completely independent of the sensible world.

THE LANDSCAPE OF MATHEMATICS

Here again, opinions diverge concerning the nature of mathematics. As
they did concerning natural laws, the constructivists believe that mathematical entities do not really exist, except in the mind of mathematicians. They
agree with the Scottish empiricist philosopher David Hume (1711–1776)
that "all our ideas are merely copies of our impressions," and that only experiences based on the senses have any reality. For instance, geometrical
shapes would have a reality only to the extent that they are found in Nature.

On the other hand, for the realists, on whose side I am, mathematical entities constitute a vast landscape that we can explore and discover with our reason, just as one can explore the Amazonian forest and the Sahara desert. Integers are not a product of our mind; they simply are, whether human beings are conscious of them or not. Their reality is distinct from sensible reality. René Descartes, a fervent realist, wrote the following remark about geometrical figures in his *Meditations on First Philosophy*:

> When I imagine a triangle, even though such a figure may exist nowhere in the world except in my thought, indeed may never have existed, there is nonetheless a certain nature or form, or particular essence, of this figure that is immutable and eternal, which I did not invent, and which in no way depends on my mind.[3]

More recently, the British mathematician Roger Penrose (1931-) had this to say:

> There often does appear to be some profound reality about mathematical concepts, going quite beyond the mental deliberations of any particular mathematician. It is as though human thought is, instead, being guided towards some external truth—a truth which has a reality of its own, and which is revealed only partially to any one of us.[4]

A scientist investigating the mathematical landscape in the space of the mind is every bit as much an explorer as the American astronaut Neil Armstrong (1930-), hopping on the arid surface of the Moon, or the French oceanographer Jacques-Yves Cousteau (1910-1997), discovering the wonders of the underwater world. Novelty emerges from it as it does from any exploration. The knowledge growing out of studying certain mathematical entities is far richer than that initially available to the mathematicians who set out to construct these entities. The German physicist Heinrich Hertz (1857-1894) put it this way: "We cannot help but think that mathematical formulas have a life of their own, that they know more than their discoverers do, and that they return more to us than we have invested in them."

MATHEMATICAL INTUITION

The way mathematicians work strongly supports the notion of a Platonic sphere of mathematical forms. We defer once again to Roger Penrose:

> I imagine that whenever the mind perceives a mathematical idea, it makes contact with Plato's world of mathematical concepts. . . . When mathematicians communicate, this is made possible by each one having a *direct route to truth*. . . . Each is directly in contact with the *same* externally existing Platonic world. . . . In their greatest works,

they are revealing eternal truths which have some kind of prior ethe-real existence.[5]

This magical contact with the world of Ideas manifests itself in sudden and unexpected ways in the form of mathematical intuitions. Examples abound. Here is how the German mathematician Carl Gauss (1777-1855) described a sudden inspiration after years of fruitless research on an arithmetical theorem:

> Finally, two days ago, I succeeded, not because of my painstaking efforts, but through the grace of God. As if in an instant flash, the enigma was suddenly resolved. I myself could not speculate on the nature of the thread connecting what I knew before and what made my breakthrough possible.

Also revealing is the account by the French mathematician Henri Poincaré, the precursor of chaos theory, on how the solution to a mathematical problem that had stumped him for weeks suddenly appeared to him as clear as day, completely unexpectedly and out of the blue, during a geological field trip:

> Just at this time, I left Caen, where I was living, to go on a geologic excursion under the auspices of the School of Mines. The incidents of the travel made me forget my mathematical work. Having reached Coutances, we entered an omnibus to go some place or other. At the moment when I put my foot on the step, the idea came to me, without anything in my former thoughts seeming to have paved the way for it. . . . I did not verify the idea; I should not have time, as, upon taking my seat in the omnibus, I went on with a conversation already commenced, but I felt perfect certainty. On my return to Caen, for conscience' sake, I verified the result at my leisure.[6]

Unpredictability (the solution can leap out in the most unexpected places, as evidenced by Poincaré stepping onto the bus, or Archimedes [ca. 287-212 B.C.] shouting "Eureka!" in his bathtub), suddenness, brevity, and immediate certainty are the hallmarks of mathematical illumination. This fleeting contact with the world of Ideas occurs after a long gestation period of subconscious processes. Poincaré described it thus:

> The privileged phenomena—those that are apt to become conscious—are those that, either directly or indirectly, have the most profound impact on our sensibility. It may surprise some to invoke sensibility in matters of mathematical proofs, which would seem to be of relevance only to intelligence. But that would ignore the sense of mathematical beauty, of the harmony of numbers and shapes, and of

the eloquence of geometry. It is a genuine aesthetical sense that all true mathematicians experience. And that falls squarely in the realm of sensibility.

Here again is the theme of beauty and harmony guiding right intuition.

THE UNIVERSALITY OF MATHEMATICS

The aesthetic pleasure a mathematician feels when he does mathematics is remarkably akin to that experienced by an artist creating a work of art. It reflects the same exalting feeling to have come close for a fleeting instant to the divine and to have briefly glimpsed a small part of eternal truth. Yet there is an important difference between artistic creation and mathematical work. As is true of the laws of Nature, mathematical discoveries have a universal character. They do not clearly carry the imprint of their author, whereas a work of art is distinctly marked by the style and individuality of the artist. That is why mathematicians from different cultures and traditions, working in diverse societies and using sometimes quite dissimilar methods of proofs, get the same results and arrive at the same theorems. While the *Ninth Symphony* is indissolubly linked with Ludwig van Beethoven, and *Madame Bovary* could not have flowed from the pen of anyone other than Gustave Flaubert, things are quite different in the world of mathematics.

That is yet another reason to conclude that mathematics exists independently of our mind. It possesses an objective character that is dissociated from the scientist making discoveries, unlike artistic or literary works, which are subjective and reflect the personalities of their authors. The objective nature of mathematics explains why it lends itself to teamwork, whereas it is generally harder for two or several people to collaborate on a painting or a novel.

RAMANUJAN, A MATHEMATICIAN
WITH A PRODIGIOUS INTUITION

The feeling that mathematics exists outside our brain, in the Platonic world of Ideas, is magnified and reinforced when we consider the case of unusually gifted mathematicians, whose genius not only produces spectacular proofs of theorems but can also formulate with astounding ease new and surprising results, based not on any rational proof but solely on intuition. It often takes several long years of arduous work to give such intuitive results a more rigorous logical foundation. It is as if these brilliant mathematicians have had direct access to the world of Ideas, enabling them to bypass the laborious and tortuous path of rational thought. There are many such examples.

In his last testament letter, written on the eve of the duel that would take his life, the French mathematician Évariste Galois (1811–1832) mentioned a theorem on integrals that would not be proven until twenty-five years after his death.

Another French mathematician, Pierre de Fermat (1601–1665), scribbled in the margin of a piece of paper, almost as an afterthought, a theorem whose first complete proof would not be completed until 1994 and required hundreds of pages of calculations.[7]

Perhaps the most famous and striking illustration of mathematical intuition is the case of the Indian mathematician Srinivasa Ramanujan (1887–1920). Born in a poor family in the city of Madras, in southern India, Ramanujan received a very limited formal education. Largely self-taught, isolated from other mathematicians, he rediscovered all by himself and with his own methods many famous mathematical results, blazing his own trail in the process. With his unconventional approach to mathematics, he also discovered a large number of completely new theorems, in an intuitive manner and without rigorous proof. The British mathematician Godfrey Hardy (1877–1947) was astonished when he became aware of Ramanujan's work: "I have never seen anything like it. These theorems could only have come from the mind of a mathematician of the highest caliber." Hardy had to muster all his considerable talent to prove just a few of these theorems. As for the rest, he had to concede defeat, surmising that "they must be true, for no one could have enough imagination to make them up." He even added: "At the game whose rules he knows, Ramanujan would beat hands down any other mathematician."

In 1914, Hardy invited the young genius to come to England and work at the University of Cambridge. Unfortunately, cultural shock, great difficulties in adjusting to a different social environment, and illness all took a heavy toll on Ramanujan. He still managed to dazzle his English mentor with his mathematical prowess. The following anecdote is typical:

One day, Hardy came to visit Ramanujan, who was already ill. He mentioned that he had arrived in a taxi bearing the tag 1729, "quite a dull number." "Not at all," replied Ramanujan. "It is actually a very interesting number: It is the smallest integer that can be decomposed two different ways into a sum of two cubes. Indeed, 1729 is equal to $12^3 + 1^3$ as well as $10^3 + 9^3$. . . ."

Tuberculosis claimed Ramanujan at age thirty-two. He left behind a treasure trove of mathematical conjectures. Even today, no one knows how he got them. Hardy described his working style this way: "His ideas about what constituted a mathematical proof were extremely hazy. All his results, whether new or old, true or false, rest on a mixture of reasoning, intuition, and induction."

By a strange coincidence, the problems Ramanujan tackled in such an original and intuitive manner were generally the very same ones that preoccupied the more traditional mathematicians of the day. This consonance constitutes more evidence supporting the contention that mathematics has an objective existence of its own. Here we have someone raised in a radically different cultural and social environment, who was not trained in the traditional academic mold, and yet was coming up with the same mathematical ideas as his more conventional counterparts. Apparently, Ramanujan drew his inspiration from the same Platonic world of mathematical forms as did his colleagues. Another instance in which we cannot help but invoke a communication with the world of Ideas is when we hear about someone capable of prodigious mental calculations, flawlessly multiplying numbers with hundreds of digits without the slightest inkling about the mental processes that led to the result; or when we come across "autistic savants," psychologically impaired individuals able to solve mathematical problems on which most of us would promptly give up.

NATURE IS SUBTLE, BUT SHE IS NOT MALICIOUS

The astonishing successes of science, reported almost daily on the front pages of newspapers and on evening television news broadcasts, tend to make us forget that it is almost a miracle that science works at all. We have already stressed that one reason for this is that Nature exhibits regularities, which we call "laws." But a second, equally important factor is the fact that our brain has the ability to perceive these regularities. Indeed, Nature sends musical notes our way, but she does not provide the complete melody. It is up to us to reconstruct it. Whereas we are fairly proficient at noticing regularities, their meaning is not always immediately obvious. The mauve and orange tints of sunrises and sunsets are for sure a delight to our eyes. But it was not until the French physicist Léon Foucault (1819–1868) hung his pendulum from the ceiling of the Pantheon, in Paris, and observed its oscillation plane rotate with respect to the ground as the hours went by, that we had incontrovertible proof that the Sun rises and sets because the Earth spins on itself once every twenty-four hours. It took all the genius of Johannes Kepler and Isaac Newton to wrest the secret of the motions of planets and to understand that they move along elliptical orbits, with the Sun at one of the foci.

Nature's message comes to us in a coded form, and it falls on us to decode it. The amazing fact is that our brain is capable of deciphering, at least partially, the cosmic code, and that we can thus progress toward an ever more complete understanding of the world. We might have emerged in a universe in which regularities are so well hidden, patterns so obscured, and the melody so secret that deciphering the cosmic code would have required incomparably more mental capabilities than the human brain possesses. Or we might have inhabited a universe in which the regularities are so evident and transparent that they would jump out at us and that no mental effort would be required to comprehend their meaning.

There is a first interesting coincidence: We live in neither of these extreme situations, but, rather, in an intermediate universe where the difficulty of the cosmic code seems to have been mysteriously adjusted to match the ability of the human brain to understand it. The task is certainly not easy, but it is by no means insurmountable. It is sufficiently difficult to present us with a serious challenge, but not so complex as to make us feel discouraged and give up. "Subtle is the Lord, but malicious He is not," said Einstein. Subtle, because the secret melody is not handed to us on a silver platter. And not malicious, for if we apply ourselves sufficiently, we can decipher the code. We are not reduced to the status of larvae subjected to natural laws without the possibility of ever understanding them.

A second, no less remarkable, coincidence is the ability of our brain to absorb concrete facts and abstract concepts, its capacity for learning physics, chemistry, biology, mathematics, computer science, and so on, which enable us to do science. Following high school graduation, it generally takes about six years of higher education for a student to gain a sufficient mastery of these disciplines and acquire the required scientific maturity to begin contributing to research. By then, an individual is normally in his mid-twenties or early thirties. Coincidentally, experience shows that, in the physical sciences, the most creative period of a researcher happens also to be roughly between the ages of twenty-five and thirty-five. It is typically in this time window that most major advances and discoveries occur. Newton discovered the law of gravitation when he was twenty-four years old. Einstein was twenty-six when he published, simultaneously, his revolutionary work on special relativity, statistical mechanics, and the corpuscular nature of light. He was thirty-six by the time he published his work on general relativity. The time required for a scientific education and the period during which the human brain performs at its peak dovetail surprising well, providing man with an ideal opportunity to pursue science under optimum conditions. Yet such a concordance is not a priori necessary. We could easily imagine a universe so complex that we would have to spend an entire lifetime learning the necessary facts and concepts before being able to practice science. But by then, our brains would be in an advanced state of deterioration, or death would have caught up with us.

INTELLECTUAL KNOWLEDGE IS NOT REQUIRED FOR SURVIVAL

What can we conclude in the face of these amazing coincidences between the properties of the brain and those of Nature, which allow us to decipher the secret of the melody? To the ardent Darwinist, there is nothing mysterious about such coincidences: They are simply the outcome of natural selection, which molded and shaped man's brain so as to give him the ability to understand Nature, better adapt, and enhance his chances for survival.

I do not consider this a persuasive argument. Darwinian evolution was certainly a key contributor to the development of our mental abilities to cope with changes in our environment. However, these abilities are of a concrete and practical nature. For instance, man learned to hide from his predators, protect himself against inclement weather, seek shelter from the cold and heat, ferret out food, and dodge objects falling on his head. But it is highly doubtful that the struggle for survival requires a grasp of the laws of gravity or of the propagation of light, an understanding of the formation of galaxies and black holes, or an advanced comprehension of the structure of atoms. Such purely intellectual knowledge offers a priori no biological advantage. It is hard to see how mathematics might improve our chances for survival. Understanding how the universe formed and what it is made of does not seem terribly relevant to the competition for the planet's limited natural resources.

An avowed Darwinist might counter that knowing the mysteries of the atom makes it possible to build a nuclear arsenal, which might prove useful in gaining access to these limited resources. However, science is generally not done for utilitarian and practical purposes. When Einstein discovered, with his theory of special relativity, that matter is a form of energy, he had no vision whatsoever of building an atomic bomb. Nor was he motivated by issues of survival and competition. Rather, he was totally engrossed in the relativity of time and space. It is no less true that a deeper knowledge of Nature, even when originally motivated by purely aesthetic and intellectual considerations, generally ends up translating into technological applications, a good part of which contributes to improving society's well-being. History is replete with such examples: Understanding the laws of gravity led to airplanes capable of transporting us to the opposite side of the globe in less than a day, and made it possible to send men to the Moon; Maxwell's laws of electromagnetism allowed images from anywhere in the world to make their way almost instantly onto small screens in our living rooms, and to converse at will with almost anyone, anywhere on the planet. Still, as in Einstein's case, Newton and Maxwell were driven by an intellectual desire for knowledge, not by the need to survive.

A CONSCIOUSNESS OF THE UNIVERSE

Knowledge of the world comes to us in two distinct ways: First, there is perceptual knowledge, which is direct and instinctive; and second, there is intellectual knowledge, less immediate and more abstract. When we watch an apple fall in an orchard, we may be quite content with a type of knowledge that relies exclusively on our senses. We may admire the gorgeous deep red color of the apple's skin, follow its fall from the branch to the ground, listen to the sharp sound it produces as it hits the grass, and leave it at that. Or we may consider the apple's fall on a much more abstract and deeper level. With a knowledge of Newton's law and of mathematics, we can calculate with great accuracy its trajectory during its fall, the time it takes to reach the ground, its velocity at the moment of impact, and verify it all with precise measurements. There is a priori no evident connection between these two modes of knowledge. They are separate from each other: Intellectual knowledge is not just a more sophisticated extension of perceptual knowledge.

Perceptual knowledge fills a biological need. The fact that we are aware of the apple's fall enables us to avoid its hitting us on the head and hurting us. On the other hand, intellectual knowledge has nothing to do with survival instinct. As a matter of fact, some concepts in certain scientific disciplines, particularly in physics and astrophysics, can at times be so bizarre that they fly in the face of common sense and are of no use whatsoever in everyday life. For instance, the notion that an electron manifests itself as a wave when no one looks at it, only to transform into a particle the moment one observes it, hardly belongs on the list of requirements for survival. To know that space folds back on itself and time comes to a standstill near a black hole does not confer on us any obvious selective advantage. Surviving in "life's jungle" does not require an intellectual knowledge of Nature, only an awareness of those of its manifestations that directly touch our senses. The survival of species depends not on the search for a hidden order or a concealed code, but on a direct appreciation of the world. When we run to dodge a projectile, or when we jump over a puddle to avoid getting wet, we clearly do not analyze the situation in terms of the laws of mechanics and gravitation. When some object speeds straight toward us, we do not think (there is no time for that), we simply act instinctively to duck it. This reflex-based behavior is analogous to what takes place in animals. When a dog leaps in the air to catch a bone you just threw at him, he certainly does not invoke Newton's laws. A bird in flight flaps its wings instinctively, without any knowledge of the laws of aerodynamics. Such instinctive knowledge is inscribed in the genes of the dog or the bird, just as human instincts, molded by the experiences of our ancestors and our own, are stored in our genes.

If, indeed, only a perceptual knowledge of the world is required for survival, if it alone fills a biological need, why, then, can we also achieve an intellectual knowledge of our environment? My own view is that our ability to understand the universe is not the result of propitious chance. I believe that it was "programmed" in advance, much as the universe itself was tuned with extreme precision, from the outset, for life to appear. We still do not know how we think and create, but I would not be at all surprised if, once we do understand the mechanisms of thought, it turned out that the human brain was itself engineered with meticulous precision, just for thought to emerge. The existence of the universe has a meaning only if it contains a consciousness capable of appreciating its organization, beauty, and harmony. It was inevitable that this consciousness arisen from the cosmic order should exalt this order by comprehending it. The ability of our brain to understand the laws of Nature is no mere accident, but reflects an intimate connection between man and the cosmos.

THE UNREASONABLE EFFECTIVENESS OF MATHEMATICS

The conviction that the underlying regularity of Nature can be expressed in mathematical terms is at the very basis of the scientific method. Some scientists have even decreed too dogmatically that any discipline that cannot be articulated in a mathematical language does not qualify as "science." The idea that the physical world is merely a reflection of a mathematical order arose, as did so many others, in ancient Greece with the mathematician Pythagoras (sixth century B.C.): "Numbers are the principle and source of all things," he proclaimed. The essence of Nature resided in the harmony of numbers. Numbers that are even were associated with femininity and mother Earth, while odd numbers represented masculinity and the heavens. Every number had its own meaning. The number 4 symbolized justice, and 5 was associated with marriage. The notion of a mathematical order of the world reached its zenith in Renaissance Europe with the work of Galileo, Newton, and Descartes. They all expressed the regularity of Nature through mathematical laws. Galileo affirmed: "The book of Nature is written in the language of mathematics."

The surprising success of mathematics in describing reality constitutes a most profound mystery. The fact that mathematics "works" so well for the purpose of describing the physical world demands an explanation, for it is not at all obvious that such should be the case. The Hungarian-American physicist Eugene Wigner (1902–1995) expressed his own amazement when he marveled at the "unreasonable effectiveness of mathematics" in describing Nature. Why this synergy between the world and mathematics? Why do abstract entities, forged in the minds of mathematicians, and generally of no use in ordinary life, turn out to be so in concordance with natural phenomena? Why is pure thought in symbiosis with the concrete?

Some have postulated that the success of mathematics in describing the world is merely a cultural phenomenon: It is simply because man likes mathematics that he uses it to describe Nature. In other words, the mathematical character of the world is not inherent to it, but was imposed on it by man. Darwinian evolution shaped man's brain so he would have a penchant for mathematics, and it would only be natural for him to look exclusively for those aspects of Nature that can be described in this language. Some extraterrestrials, with a totally different biological evolution and brain physiology, might not think that Nature is mathematical at all.

It is true that scientists prefer to use mathematics to study Nature and tend to choose among all possible problems the ones that are more amenable to mathematical solutions. Those phenomena in Nature that are not likely to be tractable mathematically are often relegated to the background and ignored, as was the case for nonlinear chaos until very recently. It is no less true that some researchers are too quick to label as "fundamental" only those aspects of Nature that can be treated mathematically, which leads them to the tautological conclusion that "the fundamental aspects of Nature are mathematical." Nonetheless, I do not believe that the mathematical character of the world is a purely cultural phenomenon and that it stems solely from man's predilection for mathematics. The reason is the following: A large part of mathematics was developed strictly on an abstract level, without any concern for practical applications to the natural world. Researchers have studied numbers or geometrical figures for their mathematical interest, not as tools useful for describing Nature. Physicists have always been the first to be surprised when, drawn into uncharted territory by a new physical phenomenon, they have almost invariably found out that mathematicians had already been there, guided not by Nature but, rather, by pure thought.

This surprising and troubling consonance between mathematics and reality is further evidence of the independent existence of mathematics in the Platonic world of Ideas. If mathematics were a pure invention of our minds, there would be no reason for this consonance to exist.

Examples of the "unreasonable effectiveness of mathematics" are many in the history of physics. After Einstein discovered in the 1920s that gravity warps space, he could no longer use Euclidean geometry, which describes only a space that was flat. He was pleased to find out that the German mathematician Bernhard Riemann (1826–1866) had already discovered and studied non-Euclidean geometries in the nineteenth century. Certainly, Einstein cannot be accused of wanting to impose those new geometries on Nature since he had not previously been aware of their existence. Similarly, the young Évariste Galois developed group theory long before it found applications in the physics of elementary particles during the twentieth century.[8] So many times has pure thought been ahead of concrete problems. So much so that, when it fails to happen—as in the case of "superstring" theory, which treats elementary particles as vibrations of pieces of string, and where the mathematical tools have not yet been developed—physicists are taken aback.

The mathematics-based approach is not the only one that has been proposed for describing Nature. Aristotle thought of Nature as a living organism. For others, it was a machine. But in the end, the mathematical approach won the day.

ARITHMETIC AND NATURE

One possible explanation for this remarkable consonance between mathematics and Nature is the existence of physical systems in the natural world that lend themselves to arithmetic operations such as addition, subtraction, and multiplication. Counting, adding, subtracting, seem to us such common and ordinary operations that we could not imagine a world in which they would be impossible. Yet, in a universe where the laws of physics would be completely different and where material objects would not be separable into distinct entities, none of these operations would make sense. In ancient times, arithmetic was used for such highly practical purposes as counting sheep in a herd or coins in a purse. These operations would not have been possible if sheep and coins did not form distinct entities and discrete units, but, instead, constituted nonseparable and continuous objects, somewhat like a river emptying into the sea. The fact that we can say with confidence that "9 sheep + 5 sheep = 14 sheep," and assign an abstract number to a collection of objects is quite remarkable.

Indeed, the compatibility of the physical world with arithmetic operations has one very important consequence: It means that it is calculable. It makes it possible for computers to crunch models of the universe, galaxies, and stars. It allows us to simulate the collisions of galaxies and track the formation of stars from the collapse of an interstellar gas cloud. If the universe were not calculable, we would know considerably less about it.

HILBERT'S CHALLENGE

Because there is congruence between mathematics and reality, because numbers are the essence of Nature, mathematics has allowed us to forge ahead in deciphering the cosmic code. At the same time, mathematics has also revealed to us that there is a limitation to rational thought, that the mystery at the end of the trail will never be resolved by reason alone, that the secret of the melody will never be unlocked solely by logic. This limitation to rational thought came to light thanks to a brilliant tour de force achieved by a gifted Austrian mathematician, Kurt Gödel (1906–1978), in response to a challenge issued by the German mathematician David Hilbert (1862–1943).

In a now-famous address to the 1900 International Congress of Mathematics in Paris, Hilbert listed what he perceived as the twenty-three most important problems in mathematics on the eve of the new century. His goal was to "lift the veil obscuring the future, so as to predict mathematical advances in the centuries to come." His list of problems demonstrated an uncanny prescience and a clear and penetrating vision of the mathematical landscape of the time. Those problems that have since been solved have opened up numerous novel and fruitful vistas, while those that have not continue to be the focus of entire teams of researchers.

The problem that caught Gödel's attention was the second one on Hilbert's list. It was phrased as follows: "Prove the consistency of the axioms of arithmetic." The challenge Hilbert issued to his colleagues was to devise a general procedure aimed at determining whether any arithmetic proposition is true or false, thereby placing arithmetic (and, later, the whole of mathematics) on a consistent logical basis. What Hilbert had in mind was to "formalize" mathematics—in other words, to translate all of its propositions into a set of symbols. Once that was achieved, a proof would then merely consist in a series of manipulations of symbols. If a general procedure could be worked out, it would suffice to mechanically follow—a machine could do it—a series of predetermined steps to arrive at the result, without ever having to worry about the actual meaning of the symbols. One could then prove a mathematical theorem much the same way as one follows a recipe to prepare a gourmet dish. This was a reductionistic program in the most literal sense of the term: Mathematics would be reduced to a collection of symbols, while ideas and applications giving meaning to these symbols would be unnecessary and disposed of. Mathematics would amount to no more than a formal symbol-shuffling game. The objective was consistency, not meaning. Hilbert thought that, with a procedure of this kind in place, all mathematical problems could be solved. He often repeated loud and clear: *"Wir müssen wissen. Wir werden wissen"* ("We must know. We will know"), a motto he would later have inscribed as an epitaph on his tombstone.

Hilbert's desire for "formalization" is understandable. After all, most of us practice mathematics in such a formal way during the normal conduct of our affairs. When we do additions and subtractions to balance our checkbooks, we follow very strict simple rules to work out the figures without any concern about their deep significance. When we write 7 x 4 = 28, there is no compelling need for us to understand what the numbers 7 and 4, or the symbol x, really mean. All we need is to recognize the numbers and symbols, and blindly follow the rules of multiplication tables. As children, we did associate numbers with objects (for instance, the fingers on our hand), but the need for such props diminishes as we grow older and become more comfortable with abstraction. By the time we reach high school, we usually no longer cringe at the thought of manipulating abstract quantities, symbolized by x or y, as we learn algebra. Those of us who pursue a higher education in the sciences learn to manipulate mathematical entities that are totally disconnected from physical reality, such as "complex" numbers or elements of a group. Hilbert's dream was, therefore, reasonable. But was it achievable?

GÖDEL AND THE LIMITATIONS OF THOUGHT

The young Kurt Gödel (Figure 70) met Hilbert's challenge brilliantly in the year 1931, although perhaps not quite in the way Hilbert had hoped for. Instead of developing a general procedure to prove the truth of any mathematical proposition, Gödel showed that such a procedure actually could not exist. He arrived at a conclusion that is perhaps the most extraordinary and mysterious result in all of mathematics: A consistent and noncontradictory

Figure 70. *Kurt Gödel (1906–1978).* He is the author of a famous mathematical theorem named after him, which is widely regarded as the most important logic discovery in the twentieth century. Gödel's theorem states that any arithmetic system contains undecidable propositions, whose truth cannot be decided on the basis of the axioms contained within that system alone. Those propositions can be proved only by adding axioms external to the system. In other words, total truth cannot be contained within a finite system. Any finite system is incomplete. This "incompleteness" theorem implies that rational thought has inherent limitations and cannot attain absolute truth.

arithmetical system invariably contains "undecidable" propositions—that is, mathematical statements of which we can never say whether they are true or false; furthermore, we cannot prove that a system is consistent and noncontradictory solely on the basis of the axioms contained within this system; this can be done only by stepping out of the system and imposing one or several additional axioms that are external to it. In this sense, the system is incomplete by itself. This explains why Gödel's theorem is often called the "incompleteness theorem."

The consequences of this veritable thunderclap in the serene skies of mathematics were enormous. The existence of undecidable propositions seemed to undermine the very logical foundation of the field. Gödel's theorem shattered Hilbert's reductionistic dream of formalizing all of mathematics. In order to ascertain the truth of a mathematical proposition, it is not enough to reduce it to a series of symbols stripped of meaning and to manipulate these symbols. One also has to take into account the meaning of the entire proposition in the larger context of mathematical ideas.

Hilbert never accepted that Gödel's theorem spelled the demise of his entire program. He clung tenaciously to his position up until his death, in 1943. Meanwhile, Gödel's theorem had repercussions far beyond the field of mathematics. The shock waves it created continue to reverberate to this day in areas of thought as diverse as philosophy and information theory.

PARADOXES THAT UNDERMINE LOGIC

In truth, all was not well in the skies of mathematics prior to Gödel's thunderbolt. There were early signs of an impending storm and dark clouds had been gathering on the horizon. In particular, it was known that logic could fail when dealing with self-referencing propositions—that is, those that refer back to themselves. Consider, for instance, the statement "the present sentence is false." If the sentence is true, it is false; and if it is false, it is true. Logic is confounded and made ineffectual. This kind of logical paradox was already known to the ancients. Saint Paul (ca. 10–ca. 63) had cited one example as he spoke of a philosopher from Crete who is said to have declared: "All Cretans are liars." If the philosopher was telling the truth, he was lying, since he himself was Cretan. On the other hand, if he was lying, he was telling the truth.

The English mathematician and philosopher Bertrand Russell (1872–1970) showed that this kind of logical contradiction is not due to semantic difficulties associated with any particular language, but, rather, arises in all symbolic or mathematical representations. He himself gave many examples, such as the one concerning the barber of Seville: "A man of Seville is shaved by the Barber of Seville if and only if the man does not shave himself. Does the Barber of Seville shave himself?" If he does shave himself, he cannot be getting a shave from the barber of Seville, and, therefore, he does not shave himself; but if he does not shave himself, he gets a shave from the barber of Seville, and, therefore, he shaves himself.

Russell, who also dreamed of giving all of mathematics a solid logical foundation, felt that such logical contradictions could jeopardize his grand project. He expressed his deep despondency in the following words:

> At first, I supposed that I should be able to overcome the contradiction quite easily. . . . Gradually, however, it became clear that this was not the case. . . . Every morning, I would sit down before a blank sheet of paper. . . . Often when evening came it was still empty. . . . The two summers of 1903 and 1904 remain in my mind as a period of complete intellectual deadlock. It was clear to me that I could not get on without solving the contradictions, and I was determined that no difficulty should turn me aside from the completion of *Principia Mathematica*,[9] but it seemed quite likely that the whole of the rest of my life might be consumed in looking at that blank sheet of paper. What made it the more annoying was that my time was spent in considering matters that seemed unworthy of serious attention.

It was partly to try to get rid of these frustrating logical contradictions that Hilbert set out on his program of formalizing mathematics. He believed (in spite of Russell's work) that these were semantic difficulties that would disappear if mathematical propositions could be stripped of their meaning and converted to a string of symbols. As we have seen, Gödel demonstrated that Hilbert and Russell's dream of putting mathematics on a firm logical basis belonged in the realm of utopia. By examining the relationship between the description of mathematics and mathematics itself, he discovered that the logical contradictions plaguing self-referencing propositions were present in mathematical propositions as well. Just as one cannot decide whether all Cretans are liars or whether the Barber of Seville shaves himself, neither can one decide whether certain mathematical propositions are true or not. Because there exist undecidable propositions, a system will never be complete within itself.

A LOGICAL FLAW IN THE UNITED STATES CONSTITUTION

Gödel emigrated to the United States in 1938. Daily life can sometimes be difficult to cope with for a man with such a totally logical mind. To wit the following anecdote, which relates what happened as Gödel was becoming a naturalized American citizen in 1948:

To obtain his citizenship, he had to study the U.S. Constitution and demonstrate to a local judge that he was generally familiar with it. He also had to have two witnesses at his side, who would vouch for his moral character. He could not have selected more prestigious people, since one of them was none other than Einstein, his colleague and friend at the Institute of Advanced Study, in Princeton, New Jersey. The other was the Austrian-American economist Oskar Morgenstern (1902–1977). The day before the interview with the judge, Morgenstern got a phone call from a very excited

Gödel announcing that he had just discovered a logical flaw in the Constitution of the United States, a flaw that could allow a dictatorial regime to be installed in America. Morgenstern did his best to calm his friend, assuring him that it was absurd, and begged him not to mention his alleged "discovery" to the judge at any point during his interview.

The next day, the judge opened the interview by remarking: "You are a German national." "No, Austrian," Gödel corrected him. Not in the least taken aback, the judge went on: "At any rate, you used to live under a dictatorship. Fortunately, this could never happen here in America!" The word "dictatorship" had been uttered, and Gödel could not restrain himself: "On the contrary, I can prove that it could indeed happen!" It took all the combined skills of Einstein, Morgenstern, and the judge to calm Gödel down and prevent him from launching into a lengthy proof of the logical flaw in the Constitution. The story did have a happy ending. The judge was not unduly offended, and Gödel was granted American citizenship.

THE LIBERATION OF MATTER

We are about to conclude our explorations, and it is time to collect our thoughts. During our peregrinations through the epic of science, we have witnessed the emergence of a new vision of the world. This vision is radically different from the one that emerged with Newton in the seventeenth century and dominated the world for three hundred years.

In the Newtonian world, matter was simply an inert substance, blindly and slavishly submitting to external forces, and completely deprived of creativity. The universe was a well-oiled machine, an exquisitely tuned clockwork, which, once wound, kept running on its own according to strictly deterministic rules. Freedom and imagination were banned. The behavior of the smallest atom was determined in advance, prompting Laplace to proclaim that if the present state of the universe were known, its entire future would be determined. There was no place for time, since the future was already embedded in the present and the past. The Great Book was already written, as it were, and, in the words of the Belgian chemist Ilya Prigogine, "God's role was reduced to that of simple archivist turning the pages of the cosmic book." Moreover, effects were always in proportion to causes, and the whole was equal to the sum of its parts, neither more, nor less. As a result, Nature could be studied by decomposing it into its simplest components. Such a mechanistic, materialistic, and reductionistic view spilled over to other domains. In biology, living beings became "genetic machines," collections of particles subject to blind forces. Biological and mental phenomena were viewed merely as physical processes explainable in terms of matter or energy.

No one can deny that this reductionistic and materialistic approach has contributed importantly to our understanding of the world by allowing us to isolate morsels of reality and study them individually without understanding the whole. At the same time, it clearly also contributed to breaking the old alliance between man and the universe, and to alienating him from the world he inhabits. Demoralized, depersonalized, devalued, and disoriented, man came to feel lost in an implacable and inexorable vast machinery over which he had no control. The reason such a materialistic science gradually distanced and dissociated itself from the rest of culture is that the gloomy and depressing vision of man as an automaton devoid of will and creativity proved unbearable.

I have tried to show throughout these pages that materialism is dead. The twentieth century saw the advent of quantum mechanics, which completely changed our conception of matter. Newton's deterministic machine was replaced at the subatomic level by a phantasmagorical world of waves and particles governed not by the rigid laws of causality but, rather, by the emancipating rules of chance. Matter came to lose its very substance: Empty space teems with virtual particles with an ephemeral and impalpable existence. The theory of superstrings, which seeks to unify the fundamental forces of Nature, stipulates that particles of matter are nothing but vibrations of infinitesimally small "bits of strings" in a universe with 10 dimensions. Newton and Laplace's determinism is swept aside. Heisenberg's uncertainty principle tells us that we can never know ahead of time the behavior of an electron. This unpredictability does not stem from our ignorance, nor is it due to our inability to build sufficiently sophisticated measuring instruments; rather, it is an intrinsic property of Nature at the atomic level.

Yet, in some sense, remnants of the determinism of old lingered on. Whereas individual quantum events were undetermined, the relative probabilities of a set of possibilities were perfectly determined and could be predicted by the laws of statistics. We may not be able to calculate the exact trajectory of an electron, but we know how to calculate the probability that it will be found at a specific location. Indeed, this vestige of determinism is what enables your portable computer and stereo system to work. Even though the fate of an individual electron in the electronic circuits constituting these devices is unpredictable, the behavior of a collection of electrons is not haphazard but is perfectly well determined by the laws of probability. Quantum mechanics brings a liberating breath of fresh air to each electron individually, but it still constrains ensembles of particles to obey statistics.

This vestige of determinism was itself thrown overboard with the advent of chaos. Chaos is at work whenever a minute change in the initial causes results in an inordinate change in the effects. The effect is no longer proportional to the cause. A butterfly beating its wings on the island of Hawaii can trigger a thunderstorm over New York. This extreme sensitivity to initial conditions makes the behavior of chaotic systems fundamentally unpredictable, inasmuch as the initial conditions can never be known with unlimited precision. As was the case for quantum indeterminism, this impossibility is not due to human limitations but is inherent in Nature. Chaos frees matter from the straitjacket of determinism. It empowers it to display its inventiveness and creativity in fashioning complexity. The richness and beauty of the world result from a subtle mix of phenomena, some chaotic and some not. Even though we cannot predict the weather a month from now, we have every reason to expect that, for a long time to come, the Sun will rise every morning and spring will return every year to reawaken Nature and signal trees to bloom. Chaos is thus a powerful ally of quantum indeterminism in liberating matter from its deterministic shackles. One operates at the macroscopic level, while the other exerts its influence in the subatomic world.

One may wonder whether chaos might also reinforce quantum indeterminism to further magnify the unpredictability of the universe at the subatomic level. Is there such a thing as quantum chaos? Some believe that quantum effects would actually tend to moderate the influence of chaos. Instead of amplifying it, they would attenuate it. The definitive answer to this question remains unknown.

TIME REINSTATED

Chaos frees matter from its inertia. It allows Nature to engage in her creative games with abandon, to produce novelty that was not implicitly included in her previous states. Her destiny is "open," her future being no longer determined by the present or the past. No longer is the melody composed once and for all. Rather, it develops as it goes along. Instead of performing classical music, where each note has its place and cannot be changed or deleted without destroying the delicate equilibrium of the entire composition, it is as though Nature were playing a piece of jazz.[10] As a jazz musician improvises and ad-libs on a general theme to produce novel sounds at the whim of his inspiration and the audience's reactions, so does Nature show her creative and playful side as she experiments with natural laws to create novelty. She does not hold back. Just consider the astonishing richness and variety of animal and plant species all around us, even in the most unimaginable and inhospitable places. From underwater locales so deep that sunlight cannot reach them, to the most hostile deserts, from scalding-hot geysers to the most frigid glaciers, Nature explodes with creativeness. To get around problems, she invents not one solution but a thousand. The British-American physicist Freeman Dyson expressed this creative outpouring in a principle he named "maximal diversity": The physical laws and initial conditions of the universe were such as to produce as interesting and diverse a world as possible.[11] Because the future is no longer contained in the present and past, time regains its rightful place. The Great Cosmic Book is yet to be written, and God is no longer just an archivist turning the pages of an already completed text.

Matter has lost its preeminent role. Instead, center stage has been appropriated by principles that organize matter and allow it to rise to higher levels of complexity. In certain systems where the whole is greater than the sum of the parts, emergent principles step to the fore. Even the vocabulary has changed. Metaphors like "machine" or "clock" have been displaced by terms more familiar in biology than physics—words such as "adaptation," "information," and "organization." This new vision of the world has spread to other domains of human activity as well. It found its way, for example, into economics, yet another field where matter has lost its dominance. From the time of the industrial revolution, matter (iron, coal, and other raw materials) has been considered the primary source of a nation's wealth. Most people now think that in the twenty-first century the prosperity of nations will no longer come from mining natural resources, but rather from mastering information-transfer technologies (a perfect example is the Internet) and organizational strategies. The material world of inert particles is rapidly losing ground to a vibrant world of innovative creations issued from the mind.

A CONTINGENT WORLD AND A NECESSARY GOD

The creativity with which matter has been newly credited, together with its liberation from the yoke of determinism, allows us to take a fresh look at the problem of the dichotomy between the changing world of human experience and the unchanging world of eternal Forms; between an immutable and eternal god, outside time and space, and his manifestation in a changing and contingent Nature. As we have seen, Plato was well aware of this dichotomy. To account for it, he postulated two figures of God. One, which he called the "Good," was eternal and immutable, and resided in the world of Ideas; the other, which he designated the "Demiurge," fashioned matter in the contingent and changing world in accordance with the plans of the world of Ideas. Yet Plato never tried to reconcile the two, being content to declare that only the Good was true, the Demiurge being nothing more than a pale and illusive representation of it.

Christian doctrine attempted to resolve the problem by introducing the idea that the world was created *ex nihilo* by an act of free will on the part of a God outside time and space. Because God had a choice between creating the universe or not, the world was not a divine emanation; unlike God, it was not "necessary," but contingent. However, the intelligibility of Nature demanded a rational god who made a reasoned choice. Nevertheless, great conceptual difficulties persisted, which entire generations of philosophers and theologians failed to resolve: What motivated God's choice? Why did He choose this universe rather than another? Did He choose according to His own nature? What can we say about God's nature? That it is necessary? But in that case, He had no choice about his own existence and qualities. If so, was He able to make choices concerning the universe He created? The answer is yes only if His choice was not rational but totally arbitrary. This would imply that the universe would itself be totally arbitrary. We are once again faced with a dilemma: Either God is the creator and cause of the contingent world, and He must himself be contingent and temporal; or He is necessary, in which case His creation must also be necessary and, therefore, intemporal. A God who is necessary cannot create a world that is contingent.

We have seen that the statistical nature of atomic events and the instability of chaotic systems under the influence of minute changes ensure an undetermined future for the universe by giving free rein to its creativity. This freedom for the universe to experiment with the untried and produce novelty may provide a compromise between chance and necessity. God, creator of intemporal and necessary laws, would be responsible for the order of the world, not by direct action but by offering potentialities that the universe is free to actualize or not. There is chance in necessity, freedom in determinism, and unpredictability in the predictable. Because the universe has the liberty to choose from among a large array of possibilities, because contingency is an integral part of it, there is no need for God to be in time, His nature and designs (embodied in laws of organization and complexity that are outside time) being immutable and invariant.

A UNIVERSE CONSCIOUS OF ITSELF

Modern cosmology has rediscovered the ancient covenant between man and the cosmos. Man is the child of stars, the brother of wild animals, and the cousin of plants and flowers; we are all star dust. Astrophysics teaches us that the emergence of life and consciousness from the primordial soup depended on an extremely delicate adjustment of the laws of Nature and the initial conditions of the universe. A minute change in the intensity of the fundamental forces, and we would not be around to talk about it. The stars would not have formed and started their marvelous nuclear alchemy. None of the heavy elements that constitute the basis of life would have seen the light of day.

In fact, the laws of physics are special from an even more subtle point of view. Not only did they permit man to step onto the stage, but they also conferred on him the ability to understand the world in which he lives. The fact that man does not simply and blindly endure the laws of Nature without understanding them is highly significant. Darwinian selection certainly played a role in fashioning our brain to help us cope with the many challenges of life, but understanding the mathematical laws governing the universe seems to have come as a bonus. Our ability to do science and decipher the cosmic code suggests an intimate connection between the world of the mind and that of Platonic forms. The universe has produced human beings capable of understanding it. The loop is now closed. I believe that it did not happen by accident. We are endowed with the gift to understand because the universe is not just a collection of particles of inanimate matter. It is the manifestation of an infinitely more subtle and elegant principle. The universe does have a meaning, and it is man who, by understanding it, bestows that meaning on it.

Does this mean that man has reclaimed his central place in the universe? Hardly. The physical and chemical processes that unfolded on Earth and led to life and consciousness are probably not unique to our planet. An extraterrestrial intelligence endowed with scientific and mathematical knowledge would be just as suitable to give the universe a meaning.

THE UNIVERSE IS NO LONGER ALIEN TO US

Chance or necessity? Science cannot settle the issue. Both options are possible, as I have shown in my book *The Secret Melody*: Either man emerged by chance in an indifferent universe totally devoid of meaning, or his ascent was preordained from the outset so he could give meaning to the universe by understanding it.

If chance is the right answer, then the very precise tuning of the laws of physics and the initial conditions to allow consciousness to arise could be explained by the existence of a multitude of parallel universes. These parallel universes would contain all possible combinations of physical laws and initial conditions. Virtually all these universes would be barren and incapable of harboring life and consciousness—except ours, which, by pure coincidence, would hold the winning combination, with us as the grand prize. Quantum mechanics, as we have seen, is entirely compatible with the existence of these parallel universes; every time a choice or decision must be made, the universe would split: The electron would go east in one universe and west in the other. The observer himself would divide, even though, as the British mathematician Roger Penrose noted, it is not at all obvious that our mind and body can undergo such splittings without our being aware of it. Another scenario involving parallel universes has been suggested by the Russian physicist Andrei Linde, who proposed a big bang model in which our universe would be only one small bubble among a multitude of others within a meta-universe.

I personally reject the hypothesis of multiple universes and the chance that would result from it. I do not like the hypothesis because it runs counter to the principle of economy.[12] Why create an infinitude of infertile universes just to get one that is conscious of itself? Furthermore, postulating an infinite number of parallel universes, none of which is accessible to measurements and, as such, cannot be verified, is not consistent with the scientific method. Science is rooted in experimentation and observation. Short of this foundation, it quickly gets mired in metaphysics.

I reject the hypothesis of chance because, aside from the meaninglessness and despair it would entail, I cannot conceive that the harmony, symmetry, unity, and beauty we perceive not only in the world, from the delicate pattern of a flower to the majestic architecture of galaxies, but also—as I have tried to show all through this book—in a much more subtle and elegant way, in the laws of Nature, are the mere results of chance. If we accept the hypothesis of a single universe—our own—we must also postulate a Primary Cause that from the outset tweaked the laws of physics and initial conditions in order for the universe to become conscious of itself.

Science will never be able to decide between these two possibilities. It will never be capable of reaching the ultimate destination. Gödel's magic result demonstrated to us the limits of reason. We will have to rely on other modes of knowledge, such as mystical or religious intuition, informed and enlightened by the discoveries of modern science. Be that as it may, one thing remains certain: The universe is no longer distant and alien to us, but intimate and familiar.

« NOTES »

CHAPTER 2

1 On a human scale, the electromagnetic force always comes out on top. That is fortunate for us, for a world dominated by gravity would be exceedingly dull and depressing. The sphere would be the only shape allowed. The delicate figure of a rose petal, the perfect contour of a statue by Rodin, and the intricate iron lace pattern of the Eiffel tower would all be impossible.

2 See Jacques Laskar, "La Lune et l'origine de l'homme" (The Moon and Man's Origins), *Pour la Science*, August 1993, p. 14.

3 Pieces of the Moon and Mars ejected during such violent collisions have actually reached Earth. It is in fact in one of those asteroids, which originated from Mars and landed in Antarctica, that some scientists have recently claimed to have found evidence of Martian life in the form of of microorganisms. This claim is highly disputed.

4 See note 2 to this chapter.

5 The time interval is actually not exactly 12 hours, but 12.5 hours, because the moon also moves some distance along its orbit while the earth rotates.

6 Actually, the portion of the Moon visible from Earth totals 59 percent of the lunar surface, because the Moon's motion is not constant along its orbit. It speeds up and slows down along its path, so that synchronicity between the rotation of the Moon and its revolution around Earth is not quite perfect. In addition, the Moon's spin axis is not precisely perpendicular to our line of sight, which exposes some regions near the poles that would otherwise remain invisible.

7 This scenario is often referred to as "nuclear winter," because the same sequence of events would occur if an extremely powerful atom bomb were exploded anywhere on Earth.

8 Our aesthetic appreciation of comets is relatively recent. Before comets were understood to be just large dirty snowballs with a rocky core, most people believed that comets brought bad tidings. Humanity used to tremble in fear when comets came too close to Earth and worried that their long, sweeping tails could contain noxious gases that would contaminate Earth.

9 Yet, when you visit a museum of natural history, chances are you will get to see not a stony meteorite but an iron specimen. The reason is quite simple. Despite being more common, it is much more difficult to find and identify stony meteorites. They are easily confused with ordinary stones of terrestrial origin, while iron meteorites are not hard to spot with metal detectors.

10 As stated earlier, the Earth lost its primordial atmosphere, which was composed mostly of hydrogen (75 percent) and helium (23 percent), just as the solar nebula. When heated up by the young sun, hydrogen and helium atoms started moving furiously. Being too light to be trapped by Earth's gravity, they escaped into space and were lost.

CHAPTER 3

1 Thomas Kuhn, *The Structure of Scientific Revolutions* (Chicago: University of Chicago Press, 1970).

2 Jupiter is the planet that spins the fastest in the solar system. It rotates 27 times more rapidly than Earth.

3 The equation applicable to the Malthusian model (unbridled growth) reads: (new population) = (growth rate) x (old population). The simplest way to include a slowing down of the population growth due to wars, diseases, famines, and so on, is to multiply the right-hand-side of the previous equation by a factor (1 - old population). This extra factor has the effect of slowing growth since, when the population increases, the value of (1 - population) decreases, which reduces the entire right-handside. The final equation thus reads: (new population) = (growth rate) x (old population) x (1 - old population). The latter equation is then iterated many times by computer. One starts with an initial population that is used to calculate the new population. That result is then fed into the right-hand-side of the equation to calculate an updated value of the new population, which, in turn, is substituted again into the right-hand side, and so on iteratively.

4 See note 3 to this chapter.

CHAPTER 4

1 The example to follow is adapted from the one given by Paul Davies in *About Time* (New York: Simon & Schuster, 1995).

2 Instead of the time changing, it is the frequency of the sound that is being altered. The Doppler effect causes an ambulance siren to have a higher pitch when it is approaching, and a lower one when it is receding.

3 As we have seen, that same "ether" was also supposed to transmit electrical and magnetic forces.

4 The principle that expresses the resistance of electrons (or neutrons) to crowd too closely together is due to the Austrian physicist Wolfgang Pauli. It is known as Pauli's exclusion principle.

5 See preceding note.

6 See Kip S. Thorne, *Black Holes and Time Warps* (New York: W. W. Norton & Company, 1994).

7 What is described here is the fate of an open universe, one that does not contain enough dark mass for gravity to stop the receding motion of galaxies. In such a universe, expansion would continue forever. A closed universe, on the other hand, would have a finite existence and would come to an end in a "big crunch," as described in my book *The Secret Melody* (New York: Oxford University Press, 1995).

8 Thorne, *Black Holes and Time Warps*.

CHAPTER 5

1 Sadly, Lavoisier's brilliant contributions to chemistry did not spare him from the guillotine, as he had the misfortune of also being a deputy in the National Assembly during the turbulent French Revolution.

CHAPTER 6

1 Trinh Xuan Thuan, *The Secret Melody* (New York: Oxford University Press, 1995), pp. 120–124.

2 Edward O. Wilson, *The Diversity of Life* (Cambridge, Mass.: Harvard University Press, 1992), p. 134.

3 Jacques Monod, *Chance and Necessity: An Essay on the Natural Philosophy of Modern Biology* (New York: Knopf, 1971).

4 Ibid.

5 Henri Bergson, *Creative Evolution*, trans. Arthur Mitchell (Westport, Conn.: Greenwood Press, 1975).

6 It should be emphasized that Bergson rejected the idea of a final cause, or predetermined blueprint; in his view, such a notion was just another form of determinism, but inverted in time.

7 Erwin Schrödinger, *What Is Life?* (Cambridge, UK: Cambridge University Press, 1967).

8 Paul Davies, *The Cosmic Blueprint* (New York: Simon & Schuster, 1988), p. 118.

9 These chronicles of the foretold death of the universe can be found in my book *The Secret Melody, op. cit.*

CHAPTER 7

1 Marvin L. Minsky, *The Society of Mind* (New York: Simon & Schuster, 1986).

2 Pham Duy Khiêm, *Légendes des Terres sereines* (Legends from Serene Lands) (Paris: Mercure de France, 1989).

3 René Descartes, *Meditations on First Philosophy*, trans. John Cottingham (New York: Cambridge University Press, 1986).

4 Roger Penrose, *The Emperor's New Mind: Concerning Computers, Minds, and the Laws of Physics* (New York: Oxford University Press, 1989), p. 95.

5 Penrose, *The Emperor's New Mind*, pp. 97, 428.

6 Cited in Jacques Hadamard, *The Psychology of Invention in the Mathematical Field* (Princeton, N.J.: Princeton University Press, 1945), p. 13.

7 We are referring to Fermat's last theorem, which states that there exists no integer n greater than 2 for which the relation $A^n + B^n = C^n$ holds, where A, B, and C are themselves integers. Note that in the case of $n = 2$, the relation is verified for A = 3, B = 4, and C = 5, since $3^2 + 4^2 = 5^2$.

8 "Groups" are mathematical entities that make it possible to express the concept of "symmetry," which, as we have seen, is one of Nature's favorite principles in building complexity. Groups allow us to unify entire families of particles.

9 Alfred North Whitehead and Bertrand Russell, *Principia Mathematica* (Cambridge, UK: Cambridge University Press, 1910–1913). In this three-volume work, written in collaboration with the English mathematician Alfred Whitehead (1861–1947), Russell attempted to put mathematical concepts and propositions on a logical basis.

10 Hubert Reeves, *Atoms of Silence: An Exploration of Cosmic Evolution*, trans. Ruth A. Lewis and John S. Lewis (Cambridge, Mass.: MIT Press, 1984).

11 This principle rests largely on intuition and is far from being proven, since we do not know how to quantify "diversity."

12 The principle of economy, also referred to as "Occam's razor," after the theologian and philosopher William of Occam (end of thirteenth century–ca. 1350), states: "What can be done with fewer assumptions is done in vain with more"— that is to say, a simple explanation for a phenomenon is more likely to be true than a more complicated one (see Glossary).

« GLOSSARY »

Accelerator—A machine using electric fields to accelerate electrically charged particles (such as electrons, protons, or their antiparticles) to give them a high energy. Because linear accelerators would require impractical lengths to reach high energies, most accelerators are circular. They use magnets to bend the path of particles, which pick up more energy at each new turn around the loop.

Accretion disk—Gaseous disk surrounding a massive object (such as a black hole), whose matter spirals into the central object.

Amino acids—Proteins, which are the basis of life, are constituted of a sequence of amino acids.

Animism—Philosophy attributing a soul to natural phenomena and objects.

Anthropic principle—The notion that the universe was tuned with great precision to allow for life and consciousness to emerge. The word comes from the Greek word *anthropos*, which means "man."

Antimatter—Matter composed of antiparticles, such as antiprotons, antielectrons (or positrons), and antineutrons. Antiparticles have exactly the same properties as their corresponding particles, except for the electrical charge, which is of opposite sign.

Antiparticle—A constituent of matter with the same properties as its matching particle, with the exception of the electrical charge, which is reversed.

Asteroid belt—Region of the solar system, between the orbits of Mars and Jupiter, stretching from 2.2 to 3.3 times the distance between Earth and the Sun, where most asteroids reside.

Asteroid—An irregularly shaped rocky celestial body, devoid of atmosphere, orbiting around the Sun. The largest asteroid is Ceres, with a diameter of about 1,000 kilometers. The number of asteroids increases rapidly as their size decreases. There are approximately one thousand asteroids larger than 1 kilometer.

Atom—The smallest particle of an element that still retains the properties of that element.

Atomic nucleus—The most massive part of an atom, composed of protons and neutrons, around which electrons orbit; its size is 10^{-13} cm. The nucleus is 100,000 times smaller than an entire atom; therefore, matter is made almost entirely of void.

Background radiation—Radio-wave radiation field that is isotropic (the same in all directions) and homogeneous (the same at all points), bathing the entire universe. It dates back from the time the universe was only 300,000 years old, and is characterized by a temperature of 2.7°Kelvin (or - 270.3°C).

Baryon—Collective name designating all elementary particles composed of three quarks; a baryon has a half-integer spin ($\frac{1}{2}$, $\frac{3}{2}$, . . .) and is sensitive to the strong nuclear force.

Behaviorism—Psychological doctrine emphasizing behavior as object of study and observation as method; it excludes anything that is not directly observable, such as thought.

Bifurcation—Sudden change in the behavior of a dynamical system when a particular parameter of the system exceeds a critical value. The system then alternates between two different states.

Big bang—Cosmological theory according to which an extremely hot and dense universe would have been created in a huge explosion occurring everywhere in space about 15 billion years ago.

Big crunch—The opposite of the big bang. The hypothetical final stage of the universe collapsing in on itself under the influence of its own gravity. No one knows if the universe contains enough matter for gravity to eventually stop and reverse the current expansion.

Binary system—A pair of stars or galaxies bound by gravity and orbiting each other.

Black hole—Celestial object collapsed on itself; its gravity is so strong that the velocity required to break free of its grip is higher than the speed of light. As a result, neither matter nor light can escape.

Boson—Collective term designating particles whose spin is an integer $(0, 1, \ldots)$. The properties of bosons are markedly different from those of fermions, which are particles with half-integer spins $(\frac{1}{2}, \frac{3}{2}, \ldots)$. Photons, gluons, and W and Z particles are all bosons transmitting forces between quarks and leptons.

Butterfly effect—A phenomenon such that a very small change in the initial state of a dynamical system can dramatically alter its subsequent evolution.

Causality—The notion that the effect must come after the cause.

Celestial mechanics—A branch of astronomy that studies the movements and gravitational interactions of celestial bodies in the solar system.

Ceres—The largest known asteroid, and the first to be discovered.

Chaos—Property characterizing a dynamical system whose behavior in phase space depends extremely critically on the initial conditions.

Comet—Small body made of ice and dust, with a rocky core, orbiting around the Sun. When a comet approaches the Sun, the ice vaporizes, forming the head and tail of the comet.

Complementarity principle—Principle stated by the Danish physicist Niels Bohr, which asserts that matter and radiation can behave as both waves and particles, these two descriptions of Nature being complementary.

Complex number—Number of the form $a + ib$, where a and b are ordinary (real) numbers. It includes two parts—the "real" part a, and the "imaginary" part ib, where i is defined as the square root of -1.

Constellation—One of 88 regions into which astronomers have divided the sky. A given region is often named in accordance with the arrangement of its brightest stars, which may evoke an object, animal, person, or hero in ancient Greek mythology.

Converging series—An infinite series of numbers whose sum is finite.

Cosmic rays—Atomic nuclei (primarily protons) accelerated to extremely high velocities in supernovae. They penetrate Earth's atmosphere with very high energies.

Crater—Circular depression on the surface of a solid body (planet, moon, asteroid), generally resulting from an asteroid impact; from the Greek word for "vase."

Cygnus X-1—A massive object in the Milky Way that is probably a black hole. Hot gas originating from the atmosphere of a neighboring star is being drawn toward the object and emits copious amounts of X rays.

Demiurge—Supreme Being who, according to Plato, exists in space and time and fashions the material world after plans in the world of Ideas ruled by the Good, which is an eternal and immutable Being existing outside time and space.

Determinism—Philosophical doctrine according to which there exist cause-and-effect relationships between physical phenomena, making it possible to predict their behavior if one knows the initial conditions.

Deterministic system—Dynamical system whose subsequent evolution is completely preordained by specific laws.

Differential equation—Mathematical equation describing how a local (as opposed to global) property of a system changes as a function of time.

DNA molecules—Fundamental constituents of the nucleus of living cells. They are shaped like a twisted double helix and carry the genetic code.

Ecliptic plane—Plane of Earth's orbit around the Sun. The term is derived from the word "eclipse," because solar or lunar eclipses occur only when the Moon traverses this plane. The orbital plane of the Moon around Earth is very slightly inclined (by 5 degrees) relative to the ecliptic plane.

Ecosphere—The soil, water, and air environment in which living beings evolve on Earth.

Eightfold way—Mathematical method in the theory of quarks, which allows for mesons and baryons to be arranged in groups of eight (or 10).

Electromagnetic force—Force transmitted by photons (or particles of light) that acts on charged particles. It is responsible for the property that particles with opposite charges attract each other, while those with like charges repel each other. It binds together atoms and molecules.

Electromagnetic spectrum—The set of all types of radiation, from radio waves (the least energetic) to gamma rays (the most energetic). See Figure 46.

Electron—The least massive stable elementary particle. Electrons carry a negative charge and, together with protons and neutrons, are constituents of atoms. Electrons, as well as neutrinos, belong in the family of leptons.

Electronuclear force—Force resulting from the unification of the electromagnetic, strong nuclear, and weak nuclear forces.

Electroweak force—Force resulting from the unification of the electromagnetic and weak nuclear forces.

Emergent property—Refers to a property of a complex system that cannot be deduced or explained in terms of the properties of its constituents. In other words, the whole is greater than the sum of its parts.

Entropy—Thermodynamical quantity that measures the degree of disorder in a large collection of atoms or molecules.

Equivalence principle—Principle at the basis of Einstein's theory of general relativity; it states that, in space-time, the movement of an object in a gravity field is locally no different from that caused by a uniform acceleration.

Ether—Hypothetical imponderable and elastic fluid, thought in the nineteenth century to be the agent responsible for transmitting light and electricity. The Michelson-Morley experiment, which demonstrated that the speed of light is invariant, invalidated the concept of ether.

Exclusion principle—Discovered by the Austrian physicist Wolfgang Pauli, it states that no two similar particles of a given type (such as electrons or neutrons) can be in the same state—that is to say, with the same positions and velocities.

Fermion—Particle with half-integer spin ($1/2$, $3/2$, . . .), to be distinguished from bosons, which have an integer spin (0, 1, . . .). Quarks and leptons are fermions.

Field—Set of values taken by a physical quantity at different locations of a given space. Waves are characterized by periodic variations of this field.

Fractal object—Object whose spatial dimension is not an integer. It may also refer to an object whose dimension is an integer but which displays patterns that repeat themselves ad infinitum regardless of magnification.

Galactic disk—Central plane in which most of the visible matter of a galaxy is confined. Stars and interstellar gases revolve around the galactic center in the plane of the disk.

Galaxy clusters—Collection of thousands of galaxies bound by gravity.

Galaxy—Large system containing on average 100 billion stars bound together by gravity. It is a fundamental building block of the vast structures in the universe.

General relativity—A theory, developed by Einstein in 1915, that relates accelerated motions to gravity and the geometry of space-time.

Gluon—Particle of spin 1, exchanged between quarks, which transmits the strong nuclear force. Its range is 10^{-13} cm, the size of an atomic nucleus.

Grand unification theories—Physical theories attempting to unify the electromagnetic, strong, and weak nuclear forces into a single force—the electronuclear force.

Gravitational force—Force responsible for the attraction of one material object toward another; it is proportional to the product of the masses of both objects, and inversely proportional to the square of the distance separating them.

Gravitational lens—A celestial body (star, galaxy, quasar, or galaxy cluster) aligned with Earth and a more distant celestial object; its gravitational field deflects the light originating from the distant object, enhancing its brightness and causing its image to split or be distorted (for instance, into circular arcs).

Gravitational wave—Wrinkle in the fabric of space-time, caused, for instance, by the collapse of a star into a neutron star or black hole, or by two black holes merging into one. It propagates at the speed of light (300,000 km/s).

Gravitino—Hypothetical particle of spin $3/2$, predicted by the theory of supergravity.

Graviton—Hypothetical particle of spin 2, which would transmit the force of gravity.

Gravity—Mutual attraction of material bodies or particles.

Great Red Spot—Large swirling mass of gas, with a red brick color, in the southern hemisphere of Jupiter's atmosphere. Its color is due to impurities in ammonia ice crystals. The Great Red Spot has persisted for at least three centuries, and is the result of chaotic phenomena in the Jovian atmosphere.

Hadron—Collective term designating all the particles that are sensitive to the strong nuclear force. Baryons and mesons are examples of hadrons.

Holism—Philosophical doctrine in opposition to reductionism. Whereas reductionism asserts that the whole can be decomposed and analyzed in terms of its constituents considered as fundamental, holism professes, on the contrary, that the whole is fundamental and cannot be reduced to its components, as the whole is sometimes greater than the sum of its components.

Hyperspace—Hypothetical flat space in which the curved space of our universe is embedded.

Idealism—Philosophical doctrine in which any phenomenon external to man is subordinated to thought.

Ideas, world of (or world of Forms)—According to Plato, the world of the senses is changing, ephemeral, and illusory; it is only a pale reflection of the world of Ideas, which is eternal, immutable, and genuine.

Incompleteness theorem—Theorem discovered by the Austrian-American mathematician Kurt Gödel; it states that any arithmetic system contains undecidable propositions that can be neither proved nor disproved by means of the axioms contained within that system.

Inflation period—Very brief period following the big bang (from 10^{-35} second to 10^{-32} second), during which the size of the universe increased exponentially.

Initial conditions—The state of a dynamical system at the start of its evolution.

Interference wave—Superposition of two waves, reinforcing each other when their crests and troughs are in phase, and canceling each other out when they are out of phase.

Interferometer—Instrument based on the phenomenon of wave interference.

Invisible matter (or dark matter)—Matter of unknown nature that emits no radiation at all, but whose presence manifests itself by its gravitational effects on the movements of stars and galaxies. It may constitute as much as 90 to 98 pecent of the total mass in the universe.

Ion—Atom that has lost one or more electrons and, therefore, carries a net positive electrical charge.

Isotope—Alternate form of a chemical element; its atoms contain the same number of protons but a different number of neutrons.

Jovian planet—One of the four giant gaseous planets in the solar system—Jupiter, Saturn, Uranus, and Neptune.

Kaluza-Klein theory—Theory based on extra spatial dimensions folded back onto themselves so tightly that they would escape our notice.

Kelvin, degree—Unit of temperature that is a measure of the degree of agitation of atoms. On the Kelvin (K) temperature scale, the zero point is called "absolute zero" because it characterizes an ideal state in which all atomic movement ceases. Conversion to the Celsius (C) scale is done by subtracting the number 273. For instance, absolute zero corresponds to -273°C; water freezes at 0°C or 273°K, and boils at 100°C or 373°K.

Kuiper belt—A region of the solar system, just beyond Pluto's orbit, containing large amounts of cometary material.

Laser—An acronym for "light amplification by stimulated emission of radiation." It refers to a device that amplifies the intensity of a light beam at a particular frequency (or wavelength).

Lepton—Collective name designating elementary particles with spin $1/2$, on which the strong nuclear force has no effect. Electrons and neutrinos are examples of leptons.

Light-year—The distance covered by light (which travels at 300,000 km/s) in one year; it is equal to 9,460 billion kilometers, or 5,879 billion miles.

Linear system—System in which changes in the initial conditions lead to proportional changes in the final state.

Magnetic field—Force field created by electrical charges in motion, which, in turn, acts on the orbits of electrically charged particles.

Materialism—Philosophical doctrine professing that nothing exists besides matter, and that the mind itself is wholly material.

Meson—A particle with integer spin $(0, 1, \ldots)$, sensitive to the strong nuclear force; it is composed of one quark and one antiquark.

Meteor—Also commonly called "shooting star," it is the fiery trail produced in the sky by a small asteroid entering the atmosphere and burning up because of friction against the air. If the asteroid does not completely disintegrate in flight, it falls to the ground in the form of a charred rock called meteorite.

Meteorite—Fragment of a small asteroid that failed to completely burn up as it passed through Earth's atmosphere.

Molecule—Combination of one or more atoms, bound by the electromagnetic force.

Morphogenesis—Structural development of an embryo into the various parts of the body.

Muon—Subatomic particle with a positive or negative charge and spin $1/2$, and which behaves like an electron but is approximately 200 times heavier. Both muons and electrons belong in the family of leptons. The muon has a lifetime of about one-millionth of a second. It decays into electrons and neutrinos. Muons are produced in accelerators by means of collisions or decay processes of other particles, or by colliding cosmic rays.

Neutrino—Elementary particle without electrical charge and with an extremely small mass. It is subject to the weak nuclear force only and interacts very weakly with ordinary matter.

Neutron star—Celestial object that is extremely compact (its radius is 10 km) and dense (1 million billion, or 10^{15}, g/cm3). Made almost entirely of neutrons, it spins rapidly and emits two light beams, one of which sweeps by Earth during each revolution. This sweep manifests itself in a succession of periodic signals, hence the other name—"pulsar"—given to this object.

Neutron—Electrically neutral particle composed of three quarks; together with protons, it is a constituent of atomic nuclei. A neutron is 1,838 times more massive than an electron, and has almost the same mass as a proton (which has 1,836 times the mass of an electron). A free neutron has a lifetime of about 15 minutes. It decays into a proton, an electron, and an anitineutrino. When inside an atomic nucleus, however, it does not decay, being then as stable as a proton.

No-return radius of a black hole—Radius defining the horizon sphere of a black hole. Once this radius is crossed, neither matter nor light can escape the black hole.

Nonlinear system—System in which changes in initial conditions do not produce proportional changes in the final state.

Nucleon—A constituent of the atomic nucleus. It can be either a proton or a neutron.

Occam's razor—The notion that a simple explanation for a phenomenon is more likely to be true than a more complicated one. The term "razor" refers to "shaving off"—that is to say, eliminating any superfluous hypothesis. Occam's razor is attractive because it satisfies our sense of beauty and elegance.

Oort's comet cloud—Large spherical region centered on the Sun and located at a distance of about 50,000 times the Earth–Sun distance; it holds a large number of comets, occasionally sending one wandering into the solar system.

Parallel universes—Two or more universes existing simultaneously, but completely disconnected from our own and, therefore, not accessible to observation. Quantum mechanics as well as certain theories of the big bang predict the existence of such parallel universes.

Periodic table—List of chemical elements in increasing order of atomic numbers, grouped in columns according to their reactive properties. It was discovered by the Russian chemist Dmitri Mendeleev.

Phase space—Abstract space with as many dimensions as there are parameters characterizing the dynamical state of a system (position and velocity of each member). One point in phase space completely determines the state of a system at a given time. As the system evolves, that point describes a more or less complex figure in phase space.

Photon—Elementary particle associated with light. With no mass or charge, it always travels at the speed of light, or 300,000 km/s. Depending on its energy, the particle can be, in order of decreasing energy, a gamma, X, ultraviolet, visible, infrared, or radio photon (Figure 46).

Piezoelectric crystal—A crystal producing an electric voltage when subjected to a compression or an expansion.

Planck's length—Equal to 1.62×10^{-33} cm, it is the dimension at which space becomes a quantum foam (Figure 50).

Planck's time—Equal to 10^{-43} s, it is the shortest time interval that can exist. If two events are separated by less than Planck's time, it is impossible to tell which of the two occurred first.

Planet—A body orbiting around a star in a solar system. Unlike stars, planets do not possess their own internal energy source, such as nuclear energy. The radiation they emit is due nearly entirely to reflection of the light from the star.

Planetesimal—A solid body, whose size ranges from a few millimeters to a few kilometers, such as existed at the time the solar system was forming 4.5 billion years ago. It results from interstellar dust grains coalescing. Planets are eventually produced by accretion of planetesimals.

Poincaré's plane—Vertical plane in phase space slicing through the trajectory of the point representing the state of a dynamical system. The patterns traced by the intersections of that trajectory with Poincaré's plane make it possible to study the evolution of the system and determine whether or not it is chaotic.

Positron—Antiparticle of the electron, with a positive electrical charge.

Precession (of the Earth)—Slow movement (its period is 26,000 years) of Earth's spin axis, describing a cone; it is caused by the gravitational interaction of the Moon and Sun with Earth's equatorial bulge.

Proton—Particle with a positive charge, composed of three quarks. Together with the neutron, it is a constituent of atomic nuclei. A proton is 1,836 times more massive than an electron.

Pulsar—*See* neutron star.

Quantum foam—Structure of space on a scale comparable to Planck's length (10^{-33} cm), which can only be described in terms of probabilities.

Quantum mechanics—A branch of physics that describes the structure and behavior of atoms and their interactions with light; probabilities play a preeminent role. In this theory, the energy, spin, and other quantities are quantized—that is to say, they can vary only in discrete amounts that are multiples of a unit value. The phenomena predicted by quantum mechanics include quantum fuzziness, wave-particle duality, quantum fluctuations, and virtual particles.

Quantum theory of gravity—Theory (yet to be developed) that would unify the two pillars of modern physics—quantum mechanics and general relativity. Such a theory would enable us to venture beyond Planck's wall, which currently constitutes a barrier to our knowledge. It would make it possible to describe physical phenomena on scales comparable to Planck's length (10^{-33} cm) and Planck's time (10^{-43} s).

Quantum vacuum—Space filled with virtual particles and antiparticles that appear and disappear in extremely short life and death cycles; their existence is related to the energy fuzziness resulting from the uncertainty principle.

Quark—Hypothetical particle with a fractional electrical charge that is either positive or negative, equal to $1/3$ or $2/3$ of the charge of an electron. No quark has ever been seen in a free state. Quarks combine in groups of three, bound together by the strong nuclear force, to form a proton or a neutron. Alternatively, a quark combines with an antiquark to form a meson. Six different kinds of quarks are known—up, down, strange, charm, bottom, and top.

Quasar—Celestial object, with the appearance of a star (the name is a contraction of "quasi-star"), whose light exhibits a large redshift. Most astronomers think that quasars are the most distant and brightest objects in the universe, and that their energy comes from a supermassive black hole with a billion solar masses, devouring the stars of the associated galaxy.

Radioactivity—Process by which certain atomic nuclei decay under the influence of the weak nuclear force, with emission of subatomic particles and gamma rays.

Reductionism—Method for studying a physical system by decomposing it into its most elementary constituents considered as fundamental. Even though it has proved extremely valuable in many branches of science, it has limitations and must be supplemented by a holistic and unifying approach.

Resonance—A phenomenon occurring when an object is subjected to periodic gravitational perturbations from another object. For instance, a resonance is established when two objects orbit around a third with periods that are multiples or fractions of each other.

Second law of thermodynamics—A law stating that the total entropy of a closed system must always increase, or at least remain constant. It can never decrease.

Singularity—Mathematical point without volume and of infinite density. Physics as we know it breaks down at such a point. According to the theory of general relativity, the universe started from a singularity, and any object that collapses into a black hole must also end up as a singularity.

Space-time—Four-dimensional structure resulting from the unification of space and time.

Special relativity—A geometrical theory, developed by Einstein in 1905, dealing with relative motions; it establishes an intimate connection between time and space, which are no longer absolute and universal, but depend on the motion of the observer. Masses also depend on motion. In this theory, the speed of light remains the same (300,000 km/s) for any observer.

Spin—Self-rotation of a subatomic particle.

Spiral galaxy—Galaxy with a spherical collection of stars (called "bulge") in the middle of a disk containing stars, gases, and interstellar dust. Young, massive, luminous stars trace out pretty spiral arms across the disk (see Figure 34).

Standard model—A fruitful theory, accepted by most physicists, that describes quarks, leptons, and their interactions.

Strange attractor—Subset of states, characterized by a fractal structure, toward which a chaotic dynamical system evolves, regardless of its initial conditions.

Strong nuclear force—The most intense of the four fundamental forces in Nature; it binds quarks to form protons and neutrons, and protons and neutrons themselves to form atoms. Its range is 10^{-13} cm, the size of an atomic nucleus. When particles are far apart, the strong force pulls them together, while it pushes them apart when they get too close. The strong force does not affect photons and electrons.

Superconductivity—A state of matter in which electrical currents can flow without any resistance.

Superfluid—A fluid flowing without any resistance. Whereas the atoms or molecules of an ordinary fluid experience random movements that dissipate kinetic energy into heat, those of a superfluid possess ordered and coherent movements, which eliminates any energy dissipation.

Supergravity theory—Theory of Everything, which attempts to unify the theory of supersymmetry with gravity.

Supernova—Explosive death of a massive star (with more than 1.4 times the mass of the Sun) after it has exhausted its nuclear fuel. The envelope of the star is ejected outward, while the core collapses to form either a neutron star (when the star has a mass between 1.4 and 5 solar masses) or a black hole (in the case of a star with a mass greater than 5 solar masses).

Superstring theory—A theory based on the notion that elementary particles of matter are not points but vibrations of infinitesimally small (10^{-33} cm) bits of string.

Supersymmetry theory—A theory professing that fermions and bosons, which appear to us to have different behaviors and properties, are actually closely related (they are said to be "symmetrical"). In this theory, any particle has a superpartner. For instance, the electron's superpartner is the "selectron," that of the lepton is the "slepton," and so on.

Symmetry—Situation in which certain properties of an object, a subatomic particle, or a force remain invariant.

Tachyon—Hypothetical particle traveling faster than light.

Teleology—The concept that things tend toward a final cause. The word comes from the Greek *telos*, meaning "end."

Teleonomy—Character of living matter that causes it to fulfill a design or purpose.

Terrestrial planet—One of the four small planets in the solar system that have a solid crust—Mercury, Venus, Earth, and Mars.

Theory of Everything—Physical theory that attempts to describe the four fundamental forces in Nature (electromagnetic, strong nuclear, weak nuclear, and gravitational) as different manifestations of one and the same force.

Three-body problem—Problem involving the orbits of three massive objects interacting gravitationally with one another. It was while working on this problem that Henri Poincaré discovered chaotic orbits.

Tidal force—Differential force due to the gravitational attraction of one body on another. Here on Earth, the to-and-fro cycle of the tides is caused by differential gravity forces exerted by the Moon and the Sun on the waters of the oceans.

Titius-Bode law—Numerical sequence that gives approximately the distances of the various planets and the asteroid belt to the Sun. It fails to yield the correct answer for Neptune and Pluto.

Turbulence—Disordered movement of the elements of a liquid or gas.

Turing test—Test proposed by the British mathematician Alan Turing for determining whether a machine is or is not endowed with intelligence.

Uncertainty principle—Discovered by the German physicist Werner Heisenberg, it states that the velocity and position of a particle cannot be measured simultaneously with arbitrary precision, no matter how sophisticated our measuring instruments. It is sometimes referred to as quantum fuzziness. The uncertainty principle also applies to the energy and lifetime of a particle. The fuzziness of the energy allows for the existence of virtual particles and antiparticles.

Virtual particle—Particle created as a pair with its matching antiparticle (the total electrical charge must be conserved and remain zero) by borrowing energy from an adjacent region of space. In accordance with the uncertainty principle, the amount of energy borrowed must be returned very quickly, so that the virtual particles disappear in a very short time and cannot be detected by our instruments. Virtual particles can materialize into real particles when energy is being fed in, as was the case in the first moments of the universe.

Vitalism—Doctrine according to which biological systems cannot be reduced to collections of molecules and their interactions, but possess a life principle distinct from both the soul and the organism.

W—Electrically charged particle that transmits the weak nuclear force.

Wave-particle duality—The property of light or matter to behave sometimes as waves, sometimes as particles.

Weak nuclear force—Short-range force (about 10^{-14} cm) responsible for radioactivity. It transforms one particle into another. For instance, it is responsible for the decay of a free neutron into a proton in about 15 minutes.

White dwarf—Compact celestial object (its diameter is about 10,000 km, comparable to that of Earth) resulting from the collapse of a star with less than 1.4 times the mass of the Sun, after it has exhausted its nuclear fuel. By virtue of the exclusion principle, the electrons in the white dwarf cannot be squeezed arbitrarily close together and exert a pressure opposing the effect of gravity, preventing the white dwarf from shrinking to a smaller size.

Wino—Hypothetical supersymmetrical partner of the W boson.

Wormhole—Hypothetical "tunnel" resulting from the topology of space. Such a wormhole would exist in a hyperspace and connect two distinct locations of a single universe or two different universes.

Z—Neutral particle that transmits the weak nuclear force.

« BIBLIOGRAPHY »

The following list is not intended to be exhaustive, but constitutes a useful starting point for those who want to learn more.

John D. Barrow, *Pi in the Sky: Counting, Thinking, and Being* (New York: Oxford University Press, 1992).

P. C. W. Davies, *The Cosmic Blueprint: New Discoveries in Nature's Creative Ability to Order the Universe* (New York: Simon & Schuster, 1988).

P. C. W. Davies, *The Mind of God: The Scientific Basis for a Rational World* (New York: Simon & Schuster, 1992).

Paul Davies and John Gribbin, *The Matter Myth: Dramatic Discoveries that Challenge our Understanding of Physical Reality* (New York: Simon & Schuster, 1992).

James Gleick, *Chaos: Making a New Science* (New York: Viking Books, 1987).

Gordon L. Kane: *The Particle Garden: Our Universe as Understood by Particle Physicists* (Reading, Mass.: Addison-Wesley, 1995).

Benoit B. Mandelbrot, *Fractals: Form, Chance, and Dimension* (San Francisco: W. H. Freeman, 1977).

Jacques Monod, *Chance and Necessity: An Essay on the Natural Philosophy of Modern Biology*, trans. Austryn Wainhouse (New York: Knopf, 1971).

Roger Penrose, *The Emperor's New Mind: Concerning Computers, Minds, and the Laws of Physics* (New York: Oxford University Press, 1989).

Ivars Peterson, *Newton's Clock: Chaos in the Solar System* (New York: W. H. Freeman, 1993).

Henri Poincaré, *Science and Hypothesis* (New York: Dover Publications, 1952).

Ilya Prigogine, *The End of Certainty: Time, Chaos, and the New Laws of Nature* (New York: Free Press, 1997).

Bernard Pullman, *The Atom in the History of Human Thought* (New York: Oxford University Press, 1998).

David Ruelle, *Chance and Chaos* (Princeton, N.J.: Princeton University Press, 1991).

Erwin Schrödinger, *What Is Life?* (New York: Cambridge University Press, 1967).

Ian Stewart, *Does God Play Dice?: The Mathematics of Chaos* (New York: Penguin Books, 1989).

Kip S. Thorne, *Black Holes and Time Warps: Einstein's Outrageous Legacy* (New York: W. W. Norton & Company, 1994).

Trinh Xuan Thuan, *The Secret Melody* (New York: Oxford University Press, 1995).

Steven Weinberg, *The Discovery of Subatomic Particles* (New York: Scientific American Books, 1983).

Anthony Zee, *Fearful Symmetry: The Search for Beauty in Modern Physics* (New York: Macmillan Publsishing Co., 1986).

« PHOTO CREDITS »

« INDEX »